WITHDRAWN

**Computational Methods and Problems
in Aeronautical Fluid Dynamics**

Computational Methods and Problems in Aeronautical Fluid Dynamics

Proceedings of a Conference held at the University of Manchester in September 1974, organised by the Institute of Mathematics and its Applications

Edited by

B. L. HEWITT
British Aircraft Corporation Limited

C. R. ILLINGWORTH
University of Manchester

R. C. LOCK
Royal Aircraft Establishment, Farnborough

K. W. MANGLER
University of Southampton

J. H. McDONNELL
British Aircraft Corporation, Limited

CATHERINE RICHARDS
The Institute of Mathematics and its Applications

F. WALKDEN
University of Salford

1976

ACADEMIC PRESS
London   New York   San Francisco
A Subsidiary of Harcourt Brace Jovanovich, Publishers

ACADEMIC PRESS INC. (LONDON) LTD.
24/28 Oval Road,
London NW1

United States Edition published by
ACADEMIC PRESS INC.
111 Fifth Avenue
New York, New York 10003

Copyright © 1976 by
The Institute of Mathematics and its
Applications

All Rights Reserved

No part of this book may be reproduced in any form by photostat, microfilm, or any other means, without written permission from the publishers

TL
573
C614

Library of Congress Catalog Card Number: 76–1086
ISBN: 0-12-346350-5

Printed in Great Britain by
J. W. Arrowsmith Ltd., Bristol

PREFACE

Although economic incentives to continual improvement of computational methods have been particularly strong in aeronautical fluid dynamics, the rate of improvement actually achieved was (at least until recently) much slower than in engineering generally and, indeed, slower than in some other branches of fluid dynamics itself. For example, almost any pipeline calculation, or other computational problem in one-dimensional fluid dynamics, however complicated in certain respects, brings in only two independent variables (x and t) and accordingly proves far more tractable than the necessarily three-dimensional problems posed by aircraft geometries. Again, problems at low Reynolds number are much easier to treat computationally because they lack those large ratios between the length scales which generally characterise different aspects of a high-Reynolds-number aeronautical fluid flow.

There are two other branches of fluid dynamics whose great economic importance justified the allocation of huge resources and the expenditure of vast efforts, and where impressive results have in fact emerged rather more quickly. On the one hand, the very short lived motions involved in *blast* fluid dynamics proved amenable to methods specially devised to evaluate accurately those high amplitude, short lived transients. On the other hand, the *meteorological* fluid dynamicists succeeded in producing remarkably effective multi-level models of atmospheric processes that can be integrated forward in time to give weather forecasts of real value for at least 3 days ahead.

Admittedly there are important corners of fluid dynamics where the computational difficulties proved even greater than in the aeronautical application. *Ship* fluid dynamics is a case in point: the boundary condition at the free surface generates difficulties (again owing to a variety of length scales) even when it is linearised and far greater difficulties when as often happens it cannot properly be linearised. Most *biological* applications of fluid dynamics are especially difficult because they involve four independent variables (time as well as three space dimensions), together with large motions of highly flexible boundaries. The computational study of the detailed fluid dynamics of *turbulence* is another field where four independent variables are unavoidable, and difficulties arising from a wide variety of length scales are at their most extreme.

Somewhere in the middle of the range of difficulty presented by all these different branches of computational fluid dynamics lies the *aeronautical* division of the subject. A review of its status is now timely, because big steps forward during the last decade have once more raised the question whether computational studies will soon reach the stage when they will make possible a major reduction in the very high cost of aeronautical model-making for wind-tunnel tests. Interested readers will be able to form their own judgement on this important question from the thorough survey of the current state of the art given in the present volume.

Some advantages of aeronautical fluid flows from a computational standpoint include (*i*) an often justifiable elimination of the time variable, restricting the number of independent variables to three; and (*ii*) a frequent restriction of phenomena characterised by a small length scale to quite thin boundary layers and wakes. Complexities in the shapes of external surfaces of aircraft and their component parts have, in the past, gone far to counteract these advantages, but methods of handling such complexities of shape are now far advanced, as indicated in this book. Still greater difficulties, of course, continue to be posed by problems associated with predicting the separation characteristics of three-dimensional boundary layers, and with studying vortex breakdowns and shock-wave/boundary-layer interactions. Broadly speaking, in relation to the two classical aeronautical quests, for low drag and high lift, the continuing difficulties are greater in the latter quest than in the former!

Computational methods, if the phrase be used in its widest sense, have of course had a very long history in aeronautical fluid dynamics. Gifted individuals, although calculating long ago without any programmable computer, made remarkably good use of similarity solutions which could reduce key problems to the numerical solution of ordinary differential equations. At the same period, a wider range of linear problems could be handled by means of linear combinations of singular solutions. Various two-dimensional problems, if elliptic, might be treated by conformal mapping or by the relaxation method; and, if hyperbolic, by the numerical method of characteristics. Theorems on the stability of numerical methods such as that began to become known.

In the meantime, some nonlinear equations governing flows in thin layers were being handled in an approximate integrated form to simplify the computation. Other nonlinear problems could be converted into linear problems by special mappings like the "hodograph transformation." A "special function" flavour permeates this "Ancient History" of our subject, because of the major rôle played in it by volumious mathematical tables of the so-called special functions.

The "Medieval History" (development of the subject from 1950 to 1965, say) brought in many new themes, out of which a few of continuing importance may be mentioned here. In one bold gambit an obvious advantage was deliberately thrown away through the idea that even a steady flow might usefully be computed in practice by re-introducing the time as a variable. For elliptic problems, in particular, this could replace a tricky iteration process founded on the classical idea of relaxation by a straight-forward convergence to an asymptotic large-time limit.

Other work was founded on the idea that the vorticity had great merit as a dependent variable in many problems where the region of space occupied by vorticity was quite limited. This proved particularly useful for certain wing shapes like those of Concorde.

At the same time, expansions in series began to yield more information, partly as a result of "matching" techniques. Later, it became feasible to compute enormous numbers of terms in a series expansion and then sum the series by various techniques far outside its radius of convergence in the usual sense. In the meantime, a variety of mappings continued to play a useful auxiliary rôle in this new computational era.

But this book is concerned above all with the "Modern History" of the subject: its development over the past 10 years and in the immediate future. Readers will find many old themes and many new themes emerging in this Modern History as it is authoritatively surveyed in the chapters which follow. The aircraft industry today is one of the most lively areas of application of computational methods on a big scale, whether in fluid dynamics, structural analysis, power-plant analysis or systems analysis. The conclusion that their effectiveness in the first of these fields is still growing very rapidly indeed can be readily derived from a study of this book.

University of Cambridge   James Lighthill

Contents

Preface	v
SIR JAMES LIGHTHILL	
Practical Requirements in Industry	1
H.P.Y. HITCH	
Introductory Numerical Analysis	15
L. FOX	
Methods for Elliptic Problems in External Aerodynamics	53
R.C. LOCK	
Compressible Subcritical Flow through Axially Symmetric Sharp-lipped Orifices and Nozzles	100
G.M. ALDER	
Subsonic Flows in Turbomachines	117
H. MARSH	
Finite Element and Difference Methods for Cascades	140
M.J. O'CARROLL and L.A. MORGAN	
The Finite Element Method applied to Fluid Mechanics	158
J.H. ARGYRIS and P.C. DUNNE	
Free Vortex Sheets	198
K.W. MANGLER	

Some Problems of Unsteady Flow about
Aircraft 214
G.J. HANCOCK
Transonic Flows 242
M.G. HALL
An Extended Integral Equation Method for
the Unsteady Transonic Flow Past a Two-
dimensional Aerofoil 270
D. NIXON
Relaxation near a Sonic Line 290
A. ROBERTS
A Transonic Hodograph Theory for Aerofoil
Design 327
J.W. BOERSTOEL
Supersonic Flows 354
F. WALKDEN
Applications of Linearised Supersonic Wing
Theory to the Calculation of Some Aircraft
Interference Flows 383
M. PURSHOUSE and R.K. NANGIA
Steady Supersonic Flowfields with
Embedded Subsonic Regions 424
R.W. MACCORMACK, A.W. RIZZI and M. INOUYE
Fluid Dynamics Applications of the Illiac
IV Computer 448
R.W. MACCORMACK and K.G. STEVENS, Jr.
Numerical Computation of Steady Boundary
Layers - A Survey 466
D.B. SPALDING

Numerical Solution of Turbulent Swirling
Flows 492
D.G. LILLEY

PRACTICAL REQUIREMENTS IN INDUSTRY

H. Hitch
(British Aircraft Corporation Limited)

THE AERODYNAMIC JOB

The aerodynamic effort (excluding wind tunnel effort) for a typical design, from early inception to Certificate of Airworthiness release, is of the order of 200 man-years. Of this, about 25 man-years are directly associated with fundamental fluid flow problems in determining the aircraft's detailed shape. These problems are most severe when the greatest aerodynamic efficiency (as measured, say, by load/drag) is needed, particularly in the highly competitive world of the subsonic civil jet transport. The wing design, complete with fuselage and intersection geometry, is the most important aspect, around which most of the work centres. This effort is augmented by some £50 000 of computer usage, with computers of the ICL 1906A or IBM 370/155 class.

There is also a wind tunnel effort which amounts to about £2 million, about half of which is in more or less direct support of the aerodynamic fluid flow estimates.

FLUID DYNAMICS
Effort Per Project

Aerodynamics office: 25 man-years	£ 125 000
Computer support	£ 50 000
Wind tunnel support	£1 000 000
Total	£1 175 000

Mean time between projects: 6 years

The cost of doing the standard job in today's environment is about £1 million and we may anticipate that in the future this effort will recur once every 6 years or so.

THE CURRENT METHODS, POTENTIAL FLOW SOLUTIONS

Almost without exception the current methods are based on potential flow solutions.

POTENTIAL FLOW SCHEMA

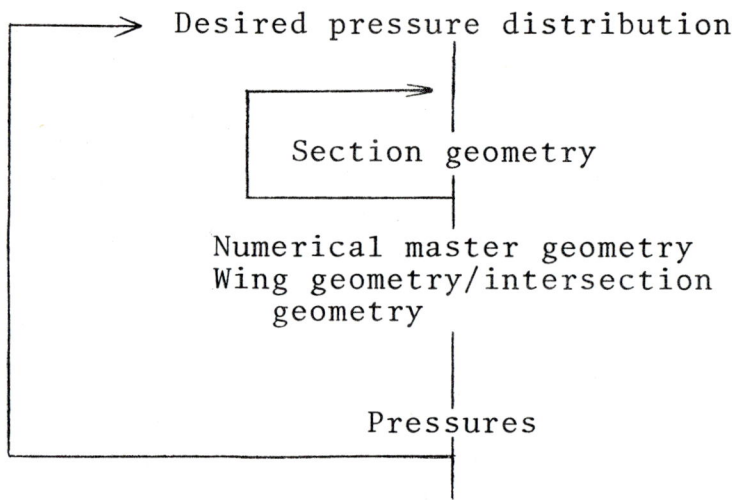

The general procedure is to choose from experience and test information on real flows that have small Reynolds number, in the wind tunnel, the desired pressure distribution at the design Mach number, and then to calculate, using potential flow procedures based on a few, say four, spanwise control stations, the geometry that will achieve this pressure distribution. This is an iterative process in the original design, and its speed has been increased by the introduction of computing and graphics on line [1]. The geometry that results from the process in the first place applies only at the control stations, and geometric mathematical processes, typified by the numerical master geometry system [2], are introduced to make a full three-dimensional consistent reconciling geometry. This is then used in potential flow solutions to give the pressures in detail everywhere on the surface for both the design case and a whole range of off-design cases. Modifications to the geometry of the design are then made in order to reconcile the differences between the desired and calculated pressure distributions.

Such processes based on potential flow solutions have been more or less standard over the past 15 years and have been much refined, the core solutions being (in the UK) based on Dr. Weber's original work at the RAE [3].

In the early days some firms relied heavily on computational processes to cut the total design time and did less wind tunnel testing. The resulting designs were not always right! The "Whitcombe bodies" on the Coronado bear witness to the modifications that may be necessary in this event (Fig.1).

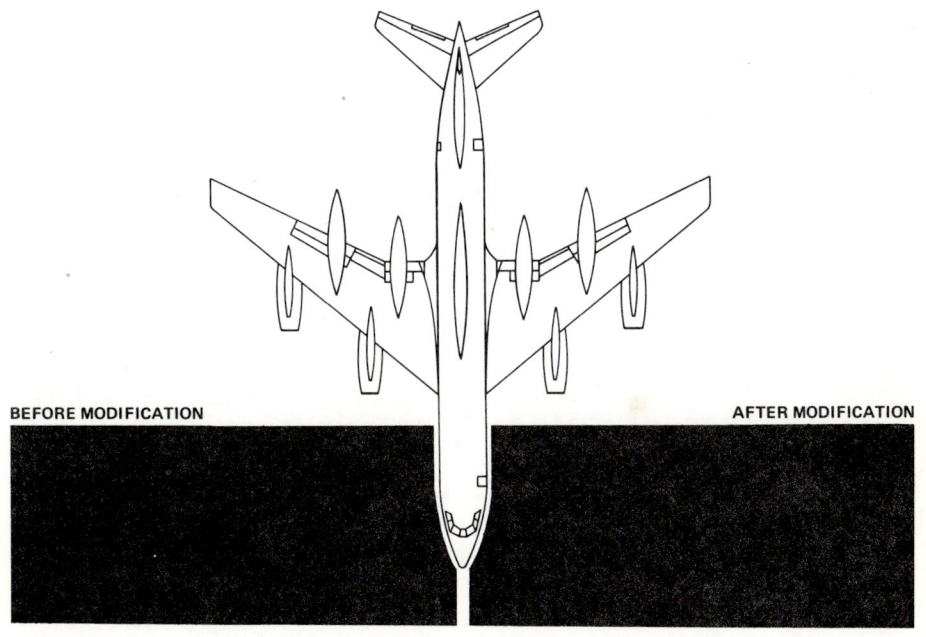

Fig. 1. CONVAIR-CORONADO 990

Less dramatic, but equally important, are the changes in the VC10 shape which, if made after the design was frozen, would have improved the drag (Fig. 2).

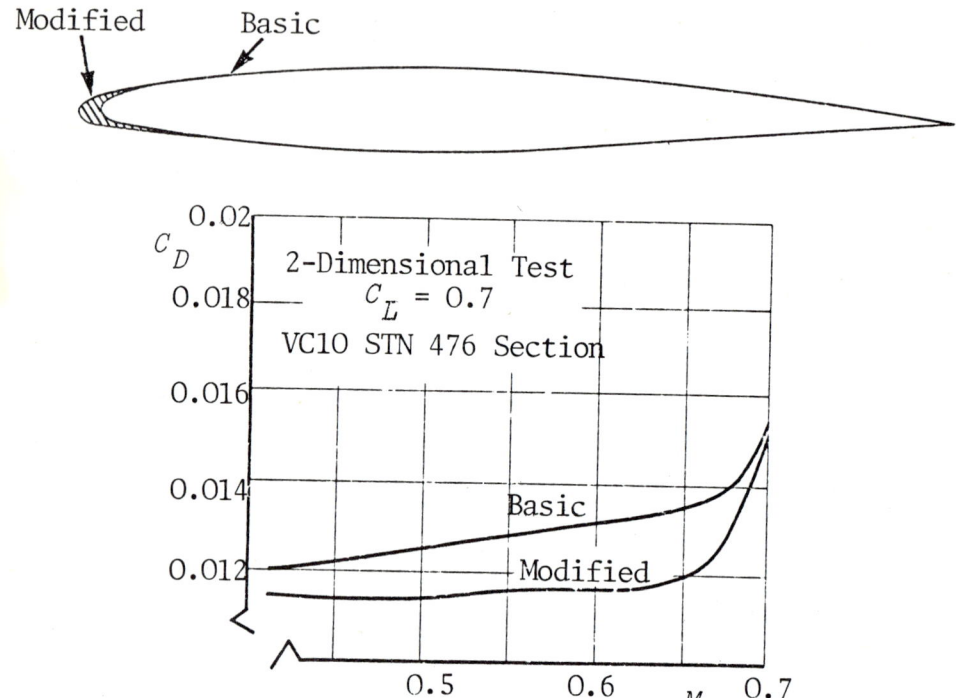

Fig. 2. Drag coefficient C_D versus Mach number M, for a constant lift coefficient of 0.7

The more recent developments in potential flow solutions, for example those of Roberts [4], allow fuselage, fuselage-wing junction, and wing, geometry to be handled in an economic manner on current computers, and produce pressure distributions where detail can be relied upon.(Fig. 3). These solutions are only first order ones.

The need now in connection with potential flow solutions is to devise procedures that are cheap, simple to operate for the purpose of find-

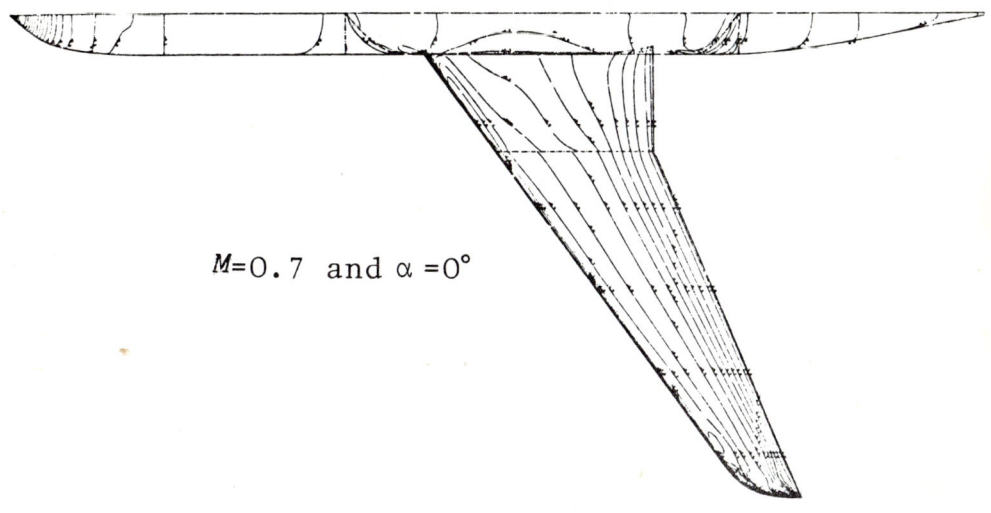

Fig. 3. A practical wing-body combination. The thin lines are pressure contours for a Mach number of 0.7, and zero incidence

ing incremental solutions, and capable of quickly evaluating the changes in shape and pressure produced by changes in pressure and shape, respectively. It is acceptable to do a few full calculations costing say £1000 per shot, but not to use such methods to refine a design and produce exchange ratios when many minor variations are needed.

VISCOUS FLOWS

The actual flow is viscous and potential flow solutions are only approximations to it. If viscous effects are not properly taken into account the shape produced can be considerably wrong. Drag coefficients can be out by 10% or more. The difference in shape shown by Fig. 2 implies a drag change of this order, and this matters!

VISCOUS FLOW

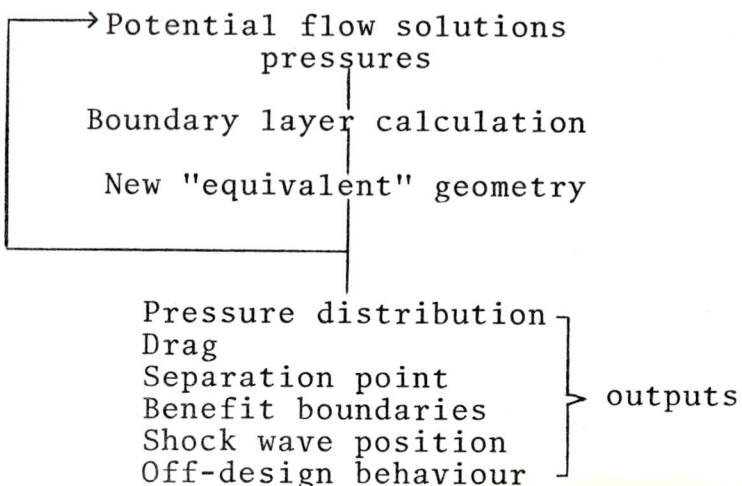

From potential flow solutions and general boundary layer calculations, a boundary layer thickness can be estimated and used to modify the geometry obtained from potential flow solutions. This is now standard practice, and one iteration is usually accepted in order to finalise the design. This procedure yields a pressure distribution in detail, from which new boundary layer conditions are calculated, and then drag, low speed separation, buffet onset and shock wave formation are deduced. But this procedure is a whole pot-pourri of aerodynamic witchcraft and cookery. It contains no real statement about the process about separation, drag, buffet or the sensitivity of the design to these things, and in any case the boundary layer calculations are only two-dimensional ones.

This is barely acceptable in today's design procedures, and, with the advent of aft-loaded sections (supercritical wings) as proposed for the immediate future, it will become highly unacceptable.

SUPERCRITICAL WINGS

The search for improved aerodynamic efficiency (more lift for a given drag without unacceptable side effects) has led to the development of supercritical wings and much effort is being deployed to discover how best to exploit them (Fig. 4).

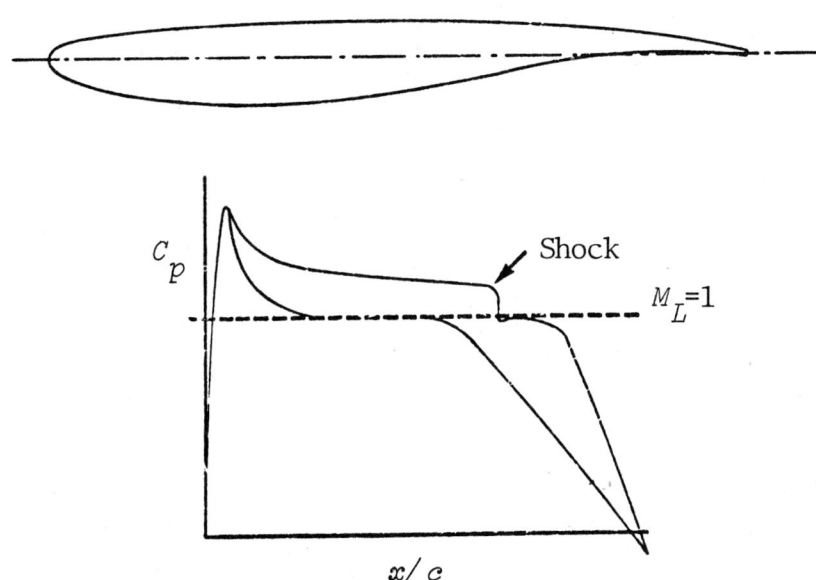

Fig. 4. Supercritical wing section. The graph shows the pressure coefficient C_p on the upper and lower surfaces of the wing as functions of the (non-dimensional) chordwise coordinate. The broken line shows the level of C_p corresponding to a local Mach number of 1

It seems clear that on a typical 150/200 seater subsonic transport some 2 or 3% improvement in standard formula direct operation cost will result. This is as much as any other single item can achieve and is probably the least controversial.

It should be mentioned here that recent advances in the inviscid field include transonic potential flow solutions [5]. Such solutions are basically used in the same kind of iterative procedure as the one involved in conventional potential methods, but have the twin drawbacks of being two-dimensional (at present) and of requiring 15 minutes of present day computing power per run. The effect of this is to increase the cost of a single cycle of the design loop by about 80%.

It becomes vital, therefore, to develop computational procedures which can be used by the design teams and are applicable to such designs. We need computational methods that produce reliable drag estimates for three-dimensional, compressible, viscous flows over real geometries, in which separation and shock conditions are produced en route.

Furthermore, for such procedures to be acceptable in the design process, they must not be excessively costly or time consuming, and should, for preference, be usable on the normal in-house computers. It is to be expected that some 20 runs of such a process per design would be required, so each run should cost less than, say, £2000, or roughly double the current cost of a full potential flow run. The time taken for such a run is a function of the computer and its loading, but any run taking more than 1 hour (or perhaps 2 hours at night) would be most unpopular, and 20 minutes would be a good figure at which to aim.

The scheme essentially represents the nub of the aircraft industry's practical requirements. It is not likely to be achieved in this form for a long time (perhaps 10 years or more) but it is hoped that progressively closer interim procedures will be developed.

To add insult to injury, the innocent phrase "real geometries" should be taken to include devices such as slats, Kruger flaps (Fig. 5), triple slotted flaps (Fig. 6), ordinary control surfaces with realistic gaps and perhaps speed brakes. To show goodwill, the corresponding solutions need only be for non-oscillating geometries - for the time being!

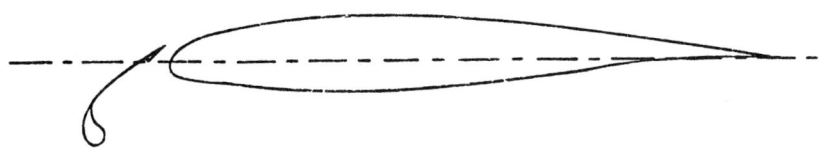

Fig. 5. Krüger slat

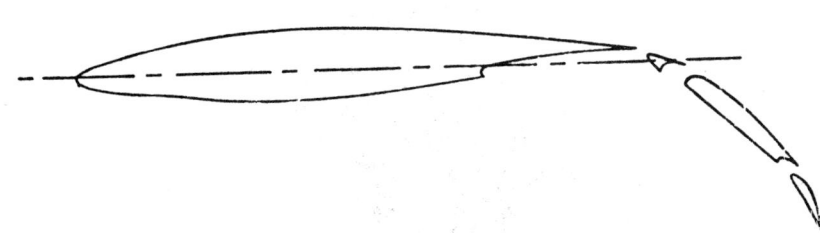

Fig. 6. Triple slotted flap

THE WIND TUNNEL

Previous designs have leaned heavily on the wind tunnel. A typical wind tunnel programme on a new subsonic commercial aircraft design costs about £2 million and involves two or three low speed models, four or five high speed models and various other partial models (for example, intakes and control surfaces). A model costs up to £100 000 to design and build and therefore some 30% of the wind tunnel cost is in model preparation and 70% in testing and analysis: the bigger models each take about one year to produce.

The proportion of wind tunnel effort supporting the main fluid flow design procedures is

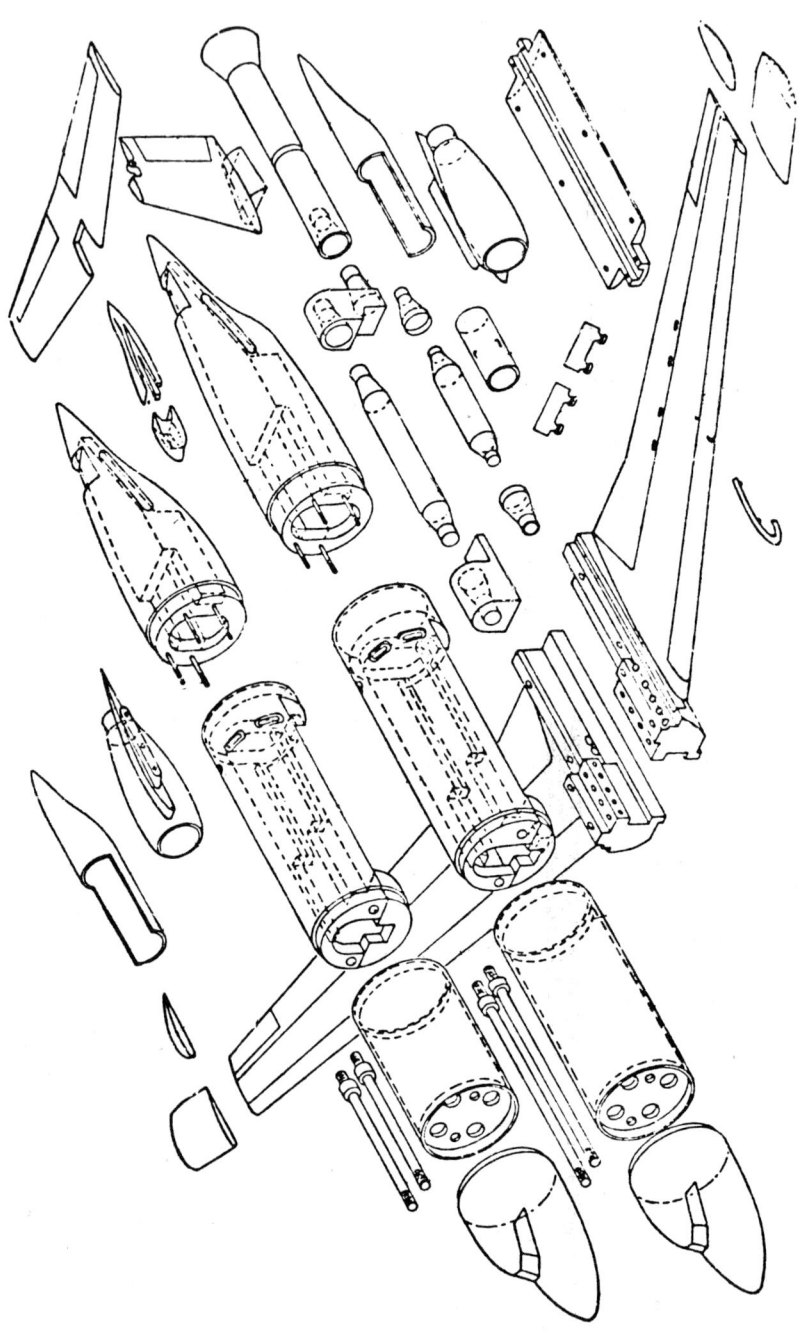

Fig. 7.3-11 Wind tunnel model

about 50% and it is important to be able to modify models and get results back to the test originator quickly (Fig. 7). A recent development is numerical machine tool control, and on-line data collection with computer analysis and graphic displays have considerably improved this development. These benefits have, however, been "cashed" in terms of more elaborate models and more involved testing. For example, pressure plotting has been developed into an everyday operation, with spark erosion drilling typically allowing 240 pressure holes to be arranged in a 14 inch half span model.

These developments of that analog computer, the wind tunnel, make a strong reply to the challenge from the processes using digital computers to replace it.

But with a few notable exceptions, the wind tunnel has a major deficiency in having only a relatively low Reynolds number. A typical test Reynolds number is 6 million for a high speed tunnel; flight Reynolds numbers are in the range of 20 to 40 million, and this difference is significant especially for supercritical wing designs. It can be said that the existing European tunnels cannot prove the new designs as well as they proved those in the past. This points to the need for the LAWS 5m tunnel costing some £40 million, proposed for early 1980. It would have a Reynolds number of about 40 million and would enable satisfactory wind tunnel tests to be carried out.

But what do we do in the meantime? Presumably we have to develop digital computational procedures and be satisfied with checks at low Reynolds numbers in tunnels in the UK and a few checks in tunnels outside Europe at more appropriate Reynolds numbers.

COMPUTATIONAL DEVELOPMENTS

Industrial design teams have come to enjoy the use of digital computers in a multi-access

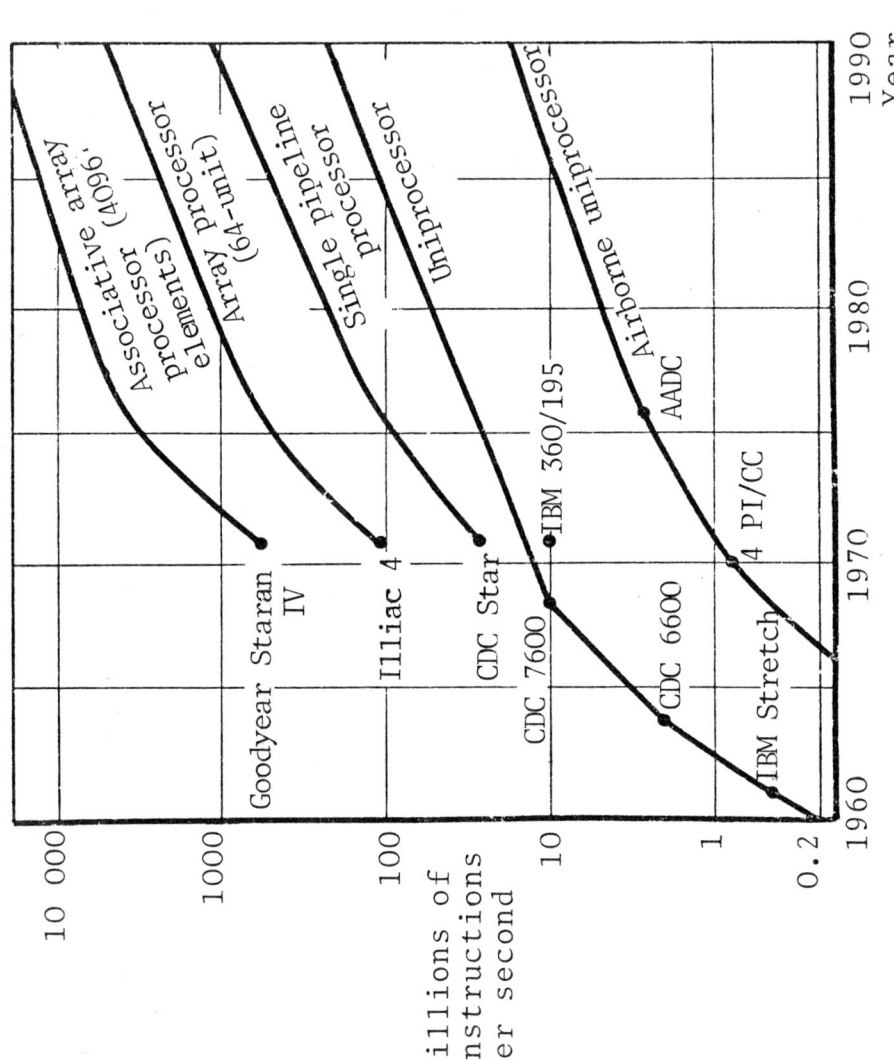

Fig. 8. Processor performance forecast [6]

Practical Requirements in Industry

mode and have learnt to take advantage of the power they afford. Typically, computer usage adds 30% to the hourly manpower rate on fluid flow work, and a major job such as a full potential flow solution on a complete wing/fuselage geometry performed on an ICL 1906A or an IBM 155 computer involves a cost of about £1000.

The computational processes which it seems desirable to develop would demand much more computer effort than existing ones. At a guess the required effort would be 100 times as large, and in this case current computers would be required to spend about 100 hours on each problem. The solution of such problems is therefore not feasible with current equipment; computers of some 100 times the speed will be needed to bring the problem down to the acceptable duration of one hour or less.

Unfortunately, the next round of computer development is likely to be some way from this target and an improvement of 10 times the power (perhaps for only twice the cost) is more likely. Fortunately the array processing machine such as Univac's Illiac IV and CDC's Star have a 100 times factor, or more, which at least shows the target to be possible [6] (Fig. 9). However the economics are an open question.

THE WIND TUNNEL VERSUS THE COMPUTER

It is fashionable to debate whether wind tunnels should give way to digital computation. At the moment, neither looks good enough (in the UK at least) to cope with the developments needed in fluid dynamics: the wind tunnel because its Reynolds number is too low, and the computer because there is no satisfactory computational method. However on the assumption that both will develop so as to be able to cope, early in 1980 say, it is most probable that the design will be done on a computer and then tested in a tunnel. The iterative activity is likely to be transferred from tunnel to computer, and the tunnel used to produce quantities of results for off-design cases.

CONCLUDING REMARKS

Effort in fluid flow computational procedures should move away from potential flow solutions to viscous flow solutions for real geometries (including control surfaces) involving transonic flows with shocks and separation intrinsically within them.

REFERENCES

1. Hughes, P.F. and Davies, R.S., "The Optimisation of Wing Design," Royal Aeronautical Society Symposium on Optimisation in Aircraft Design (1972).
2. Sabin, M.A., "An existing System in the Aircraft Industry: The British Aircraft Corporation Numerical Master Geometry System," *Proc. Roy. Soc. Lond. A.*, **321**, 197-205 (1971).
3. "Method for Predicting the Pressure Distribution on Swept Wings with Subsonic Attached Flow," Royal Aeronautical Society TDM 6312 (cf reference 1).
4. Roberts, A. and Rundle, K., "Computation of Incompressible Flow About Bodies and Thick Wings Using the Spline Mode System," British Aircraft Corporation (CAD) Aero MA 19 (1972).
5. Krupp, J.A., "The Numerical Calculation of Plane Steady Transonic Flows Past Thin Lifting Aerofoils," Boeing Science Res. Lab. Report D180-12958-1 (1971).
6. Turn, R., "Computers in the 1980's - Trends in Hardware Technology," IFIP Conference paper (1974).

INTRODUCTORY NUMERICAL ANALYSIS

L. Fox

(Oxford University Computing Laboratory)

INTRODUCTION

Problems in fluid dynamics can rarely be solved in closed mathematical form, that is, we can hardly ever find exact mathematical expressions for the velocity, pressure or whatever else we seek. Hence the need for numerical *methods*, which hopefully can produce results to any desired accuracy if we are prepared to spend enough human and computing machine time. Hence also the need for numerical *analysis*, since we find that some methods, at first sight very attractive, fail completely to produce anything worthwhile.

Fluid dynamics is a fertile field for the numerical analyst. There is a very large number of different problems and, as Richtmyer and Morton [1] observe, there are almost as many methods as there are problems. The numerical analyst who would become expert in fluid dynamical calculations will find that virtually every bit of his knowledge is relevant and required. He needs to be expert in numerical linear algebra, especially the solution of large sets of algebraic equations with particular types of matrices of rather sparse form. These equations, moreover, will often be linearised forms of a nonlinear system, so that a knowledge is also needed of iterative methods, and the convergence thereof, for solving very large numbers of nonlinear equations. He will find himself solving ordinary differential equations and every form of partial differential equation, evaluating integrals in one and more dimensions, dealing with singularities, trying to determine boundaries or interfaces not known in advance and even worrying about the formulation of the relevant problem in respect of boundary or initial conditions, choosing the required solution from a possible non-unique situation, and so on.

On a more mundane but still important level, some of the problems are so huge that a skilled machine programmer can make all the difference between success and abject failure. For example, such a problem might involve a system of four non-linear partial differential equations, with three space and one time independent variable, and with boundaries in space and time virtually at infinity. Only recently, in fact, have we had in this country computing machines of sufficient power to cope with such problems and so far, I suppose, we have not produced much in the way of good and accurate program packages for even just a few of the important problems.

I shall not be able to comment on all numerical aspects of fluid dynamics, but I hope to mention a few of the problems and some of the methods used to cope with them. All my material exists in published literature, but two or three fairly recent items, published in the numerical analysis literature, might not be well known to those who do not normally read such publications.

INCOMPRESSIBLE FLOW

Let us first look at two-dimensional incompressible flow with vorticity, for which the Eulerian equations are given by

$$\left. \begin{array}{l} \dfrac{\partial u}{\partial t} + u \dfrac{\partial u}{\partial x} + v \dfrac{\partial u}{\partial y} = - \dfrac{1}{\rho} \dfrac{\partial p}{\partial x} + \nu \nabla^2 u \\[1em] \dfrac{\partial v}{\partial t} + u \dfrac{\partial v}{\partial x} + v \dfrac{\partial v}{\partial y} = - \dfrac{1}{\rho} \dfrac{\partial p}{\partial y} + \nu \nabla^2 v \end{array} \right\}, \quad (1)$$

in which the velocity components u and v satisfy the continuity equation

$$\frac{\partial u}{\partial x} + \frac{\partial v}{\partial y} = 0, \qquad (2)$$

and the vorticity ζ is given by

$$\zeta = \frac{\partial u}{\partial y} - \frac{\partial v}{\partial x}. \qquad (3)$$

The introduction of the stream function ψ, for which

$$u = \frac{\partial \psi}{\partial y}, \quad v = -\frac{\partial \psi}{\partial x}, \qquad (4)$$

and the use of the so called conservation form of the equations gives rise to the pair of partial differential equations

$$\left. \begin{array}{l} \dfrac{\partial \zeta}{\partial t} = -\dfrac{\partial}{\partial x}(u\zeta) - \dfrac{\partial}{\partial y}(v\zeta) + \nu \nabla^2 \zeta \\ \nabla^2 \psi = \zeta \end{array} \right\}, \qquad (5)$$

in which ν is the kinematic viscosity, inversely proportional to the Reynolds number.

Now how might we solve these equations, the first of which is essentially parabolic in time, and the second elliptic in space? Well, given enough initial conditions at $t = 0$, we can compute the right hand side of the first of (5), and this gives an approximation to ζ at the end of the first time step $t = \Delta t$. Then, with some appropriate boundary conditions involving ψ at time Δt, we can compute from the second of (5) an approximation to ψ everywhere at time Δt, and then return to the first of (5) to advance another step in time.

Two numerical problems arise immediately. First, is our time-stepping process stable? Normally we use finite difference equations, with finite intervals Δt, Δx and Δy, and stability generally requires that as these intervals are taken smaller and smaller the resulting finite difference solutions get closer and closer to the true solutions of the differential equations. Some surprising things can happen, and we find that great care must be exercised in the choice of the finite difference equations.

Second, what methods are most convenient for the numerical solution of the second of (5), an elliptic boundary value problem? There are various finite difference methods which have a long history, and in more recent times the finite element method has attracted much attention. Both methods give rise to the solution of large sparse sets of algebraic equations, for which iterative methods are commonly needed but for which direct methods are often more satisfactory when they can be applied. In the next two sections I consider separately the two problems associated with equations (5).

The initial-value problem

For simplicity, and to avoid sacrificing a view of the wood by concentrating too much on the trees, let us examine the simpler one-dimensional equation

$$\frac{\partial \zeta}{\partial t} = -\frac{\partial}{\partial x}(u\zeta) + \nu \frac{\partial^2 \zeta}{\partial x^2}, \qquad (6)$$

and assume for further simplicity that u is a known constant. The simplest thing we can do, consistent with reasonable accuracy, is to use forward differences in time and central differences in space, producing the finite difference equation

$$\frac{\zeta_m^{n+1} - \zeta_m^n}{\Delta t} = -\frac{(u\zeta_{m+1}^n - u\zeta_{m-1}^n)}{2\Delta x} + \nu \frac{(\zeta_{m+1}^n - 2\zeta_m^n + \zeta_{m-1}^n)}{(\Delta x)^2}, \qquad (7)$$

where ζ_m^n is the value of ζ at the mesh point $m\Delta x$, $n\Delta t$. This permits the direct evaluation of mesh values on the line $(n + 1)\Delta t$ from the known values on $n\Delta t$, and is therefore known as an explicit method.

For stability we require that there shall be no amplification of rounding errors or any other types of error, and perhaps not even of the solution which is specified at $t = 0$. The most common and probably the easiest form of analysis is that of von Neumann. We consider the Fourier decomposition

$$\zeta_m^n = \sum_j A_j e^{i\beta_j m\Delta x}, \qquad (8)$$

and consider just one harmonic $Ae^{i\beta m\Delta x}$. At the next time step this will be $\lambda A e^{i\beta m\Delta x}$, and if no harmonic is to be amplified we need $|\lambda| \leq 1$. Substitution in (7) gives

$$\frac{\lambda-1}{\Delta t} = -\frac{iu\sin\beta m\Delta x}{\Delta x} + \frac{2\nu(\cos\beta m\Delta x - 1)}{(\Delta x)^2} \qquad (9)$$

and a little tedious computation then gives the conditions

$$\frac{u\Delta x}{\nu} \leq 2, \quad \frac{\nu\Delta t}{(\Delta x)^2} \leq \tfrac{1}{2} \qquad (10)$$

for a guarantee of stability. They give a limit on the permissible size of the space increment, and an even more severe restriction on the time increment. Notice, in particular, that if $\nu = 0$ then this method is completely unstable!

All sorts of strange things happen with finite difference equations. We might, for example, replace the left hand side of (7) by the central time-difference expression

$$\frac{\partial \zeta}{\partial t} \simeq \frac{\zeta_m^{n+1} - \zeta_m^{n-1}}{2\Delta t}, \qquad (11)$$

which is considerably more accurate in terms of the size of the local truncation error. And yet a similar analysis shows that with $\nu \neq 0$ this explicit method is unstable for all Δx and Δt, verifying the known fact that there is an unending war between local accuracy and stability. Perhaps even more surprising, this method is stable, when $\nu = 0$, if $u\Delta t/\Delta x \leq 1$, which is not such a crippling restriction on Δt.

There is a very large literature on finite difference methods for initial-value problems, in which their accuracy, stability and computational convenience are discussed. Some is written by people who would call themselves mathematicians or numerical analysts, but much has also been done by scientists who have used their intuition and physical "know-how" to good effect. We hear, for example, about "upwind differencing," which involves the approximations

$$\frac{\partial}{\partial x}(u\zeta) = \begin{cases} \dfrac{u\zeta_m^n - u\zeta_{m-1}^n}{\Delta x} & u > 0 \\ \dfrac{u\zeta_{m+1}^n - u\zeta_m^n}{\Delta x} & u < 0 \end{cases} \qquad (12)$$

which have better stability properties, are connected with the characteristics of the corresponding hyperbolic equation and conserve important things if the sign of u is constant. Even more useful is another upwind-difference approximation

$$\frac{\partial}{\partial x}(u\zeta) = \frac{u_R \zeta_R - u_L \zeta_L}{\Delta x}$$

$$u_R = \tfrac{1}{2}(u_m + u_{m+1}), \quad u_L = \tfrac{1}{2}(u_{m-1} + u_m)$$

$$\zeta_R = \zeta_m \text{ for } u_R > 0, \quad \zeta_R = \zeta_{m+1} \text{ for } u_R < 0$$

$$\zeta_L = \zeta_{m-1} \text{ for } u_L > 0, \quad \zeta_L = \zeta_m \text{ for } u_L < 0$$

(13)

which implies a non-constant velocity, and is both conservative and transportive, that is, a physical perturbation in viscosity is propagated only in the direction of the velocity.

Of the various other methods I will mention only two. The first is a class of implicit methods, in which values on the new line are computed from a set of simultaneous algebraic equations. Of these the most famous is the Crank-Nicolson method, which for our simple problem uses the approximation

$$\frac{\zeta_m^{n+1} - \zeta_m^n}{\Delta t} = \tfrac{1}{2}\left(\frac{\partial \zeta_m^{n+1}}{\partial t} + \frac{\partial \zeta_m^n}{\partial t}\right) \qquad (14)$$

in which the two terms on the right hand side are first obtained from the given differential equation in terms of first and second space derivatives and then approximated by central differences. The resulting formula is extremely accurate and unconditionally stable, although more arithmetic is involved in the computation at each step.

In fact the new mesh values are obtained by solving linear equations with a triple-diagonal matrix, for which there are good and fast algorithms. But in the case of two space dimensions the

Crank-Nicolson formula gives rise to the solution of something like Laplace's equation at each step, and this is a more formidable undertaking. It can be avoided by using the so called Alternating Direction Implicit (ADI) method, for which typical equations are given by

$$\left.\begin{array}{l} \dfrac{\zeta^{n+\frac{1}{2}} - \zeta^n}{\frac{1}{2}\Delta t} \\[6pt] = -u\,\dfrac{\partial \zeta^{n+\frac{1}{2}}}{\partial x} - v\,\dfrac{\partial \zeta^n}{\partial y} + \nu\,\dfrac{\partial^2 \zeta^{n+\frac{1}{2}}}{\partial x^2} + \nu\,\dfrac{\partial^2 \zeta^n}{\partial y^2} \\[10pt] \dfrac{\zeta^{n+1} - \zeta^{n+\frac{1}{2}}}{\frac{1}{2}\Delta t} \\[6pt] = -u\,\dfrac{\partial \zeta^{n+\frac{1}{2}}}{\partial x} - v\,\dfrac{\partial \zeta^{n+1}}{\partial y} + \nu\,\dfrac{\partial^2 \zeta^{n+\frac{1}{2}}}{\partial x^2} + \nu\,\dfrac{\partial^2 \zeta^{n+1}}{\partial y^2} \end{array}\right\} \quad (15)$$

the right hand sides being replaced by central difference expressions. The first equation is implicit for values at half time step on a line in the x direction, and the second for values at full time step on a line in the y direction, and the computation on each of these lines involves the solution of a triple-diagonal system. The method is stable, accurate and economic, and is a typical example of the activities of numerical analysts of the more practical kind. Like some other methods it can be extended without prohibitive computation to the case of three space dimensions.

The boundary value problem

Turning now to the second of (5), we ask what are the available and convenient methods for the boundary value problem? If we use finite difference methods, approximating to the Laplacian by the famous five-point formula, then the problem is reduced to the solution of linear equations which, without too much ambiguity, I shall write in the standard form

$$A\underset{\sim}{z} = \underset{\sim}{b} \, . \qquad (16)$$

In our current problem $\underset{\sim}{z}$ is the required vector of mesh values of ψ, $\underset{\sim}{b}$ is a vector of known quantities involving the mesh values of ζ and the boundary conditions on ψ, and A is a probably large but especially sparse matrix with the typical partitioned form

$$A = \begin{bmatrix} B & I & & & & \\ I & B & I & & & \\ & I & B & I & & \\ & & \cdot & \cdot & \cdot & \\ & & & I & B & I \\ & & & & I & B \end{bmatrix}, \qquad (17)$$

in which B is probably a triple-diagonal matrix.

For example, for the mesh shown in Fig. 1, with ψ known on the boundary, the natural ordering of the points produces a matrix with a corresponding partitioned structure shown in (18).

$$\begin{bmatrix}
x & x & & & x & & & & & & & & \\
x & x & x & & & x & & & & & & & \\
 & x & x & x & & & x & & & & & & \\
 & & x & x & & & & x & & & & & \\
x & & & & x & x & & & x & & & & \\
 & x & & & x & x & x & & & x & & & \\
 & & x & & & x & x & x & & & x & & \\
 & & & x & & & x & x & & & & x & \\
 & & & & x & & & & x & x & & & x \\
 & & & & & x & & & x & x & x & & \\
 & & & & & & x & & & x & x & x & \\
 & & & & & & & x & & & x & x & x \\
 & & & & & & & & x & & & x & x
\end{bmatrix} \qquad (18)$$

9	10	11	12
5	6	7	8
1	2	3	4

Fig. 1

This type of problem was solved by Southwell and his colleagues by the so called "relaxation method." The dodges and stratagems of the experts in this technique are not easily applied with modern computers, but in the last 20 years there has been an extensive literature on more systematic iterative methods allied to relaxation. Most such methods use a matrix "splitting"

$$A = M - N, \qquad (19)$$

and obtain successive iterates from the equation

$$M \underset{\sim}{z}^{(n+1)} = N \underset{\sim}{z}^{(n)} + \underset{\sim}{b}. \qquad (20)$$

The choice of M and N is decided by the necessity for quick and accurate solution of the linear equations (20), so diagonal, triangular or at worst triple-diagonal forms are preferred for M, and by the need for rapid convergence of the iterative process. This depends on the modulus $|\lambda_{max}|$ of the largest eigenvalue of the matrix $M^{-1}N$. If $|\lambda_{max}| < 1$ the method converges, and the rate of convergence is best for smallest $|\lambda_{max}|$. If this is near unity, which is the case in most problems of the type we are considering, the convergence is very slow and we have to be very careful about the termination of the iteration. It is not safe to stop when the next "correction" is negligible. For example, if the largest eigenvalue λ is real, then for sufficiently large r we can show that

$$z^{(r+2)} - z^{(r+1)} \sim \lambda (z^{(r+1)} - z^{(r)}), \quad (21)$$

and from this it is easy to see that the real error in our current $z^{(r)}$ is not $z^{(r+1)} - z^{(r)}$ but something like $(1-\lambda)^{-1}$ times this amount. If $\lambda = 0.99$ the true correction is 100 times the current estimate!

In attempts to obtain satisfactory rates of convergence, various methods have been used and analysed. Perhaps the most famous is the method of successive over-relaxation (SOR) which introduces a parameter w into the iteration equation and chooses a best value w_{opt} to secure fastest convergence. In many cases w_{opt} can be determined quite accurately with a little analysis or experimental computation, and if λ_{max} is reduced from 0.99 to 0.90, say, the gain is very worthwhile.

The ADI method already mentioned for the parabolic equation can also be adapted in an iterative scheme for the elliptic problem, and one or more parameters can be introduced to accelerate the rate of convergence. The theoretical justification of this method is valid only for rectangular boundaries, clearly unlikely in our particular problem, but in practice the method appears to have more general success.

There are many variants of these methods, and some new developments which are too new to be well tested. One of these, for example, is the so called "method of chaotic relaxation," where with suitable hardware and software the computer can perform iterations simultaneously in different parts of the field. There may well come significant improvements in the standard techniques as our computers become more efficient in this sense.

In recent times numerical analysts have become interested in the finite element method, used much earlier by engineers and reputed, at least for some problems, to be superior to the finite difference method. The latter basically approximates the differential equation at selected points of the region, whereas the finite element method seeks a solution as a linear combination

$$u = \sum_{i=1}^{N} a_i \phi_i \qquad (22)$$

of basis functions $\phi_i(x,y)$, where N is a finite approximation to infinity. There are many possible choices of the $\phi_i(x,y)$, and many corresponding ways to determine the a_i. The finite element method generally seeks a function of type (22) which minimises some quadratic form derived from the differential equation. For the second of (5), for example, the required solution minimises the integral

$$I(\psi) = \tfrac{1}{2} \iint \{(\tfrac{\partial \psi}{\partial y})^2 + (\tfrac{\partial \psi}{\partial y})^2 - 2\psi\zeta\} dx dy \qquad (23)$$

over the closed region. Substituting (22) in (23) and equating to zero the derivatives with respect to the parameters a_i gives rise to a set of linear algebraic equations for the required a_i. In the finite element method the ϕ_i are chosen in a special way, for example as linear functions over small finite elements, quite often triangles, into which the region is subdivided. The resulting linear equations are sparse, being similar in their structure to those of the finite difference method, but the triangular elements can have any suitable shapes and can therefore be used with convenience in irregular regions. Since this method will be described fully in a later paper in this volume I content myself here with this brief introduction, with the remark that either iterative or direct methods can be used for the solution of the linear equations.

For various reasons, a direct solution of the finite difference or finite element equations would be far preferable to an iterative solution especially where, as here, we have the same matrix on the left hand side and a sequence of different right hand sides. A Gauss elimination method, for example, which is equivalent to the matrix decomposition

$$A = LU, \qquad (24)$$

where L is a lower triangular matrix and U upper triangular, permits the solution of the equations (16) in terms of forward and backward substitution from the triangular sets of equations

$$L\underset{\sim}{w} = \underset{\sim}{b}, \quad U\underset{\sim}{b} = \underset{\sim}{w}. \qquad (25)$$

Here the computation in (24) is the longest part of the work, and need be done only once for a sequence of different right hand sides $\underset{\sim}{w}$.

The trouble with elimination is the time it takes, associated with the fact that some of the many zero positions in the matrix get "filled up" with non-zero numbers and so there is a corresponding need for a large amount of machine storage. Recently, however, there have been some important developments in direct methods for solving special classes of sparse systems, and since they may not be well known I give here a brief introduction and summary of this work.

The first relevant method is probably that of Hockney [2], who considers Poisson's equation

$$\nabla^2 \phi = g, \text{ subject to } \phi = 0 \text{ on the rectangular boundary } x=0,1, \ y=0,m. \qquad (26)$$

He observes that ϕ can be expanded as a Fourier series in either direction so that, for example,

$$\phi(x,y) = \sum_k \bar{\phi}_k(y) \sin(\pi k x/l), \qquad (27)$$

with a similar expansion for $g(x,y)$. Substitution in (26) produces a set of ordinary differential equations of the form

$$\frac{d^2 \bar{\phi}_k(y)}{dy^2} - \left(\frac{\pi k}{l}\right)^2 \bar{\phi}_k(y) = \bar{g}_k(y). \qquad (28)$$

Here k goes from 1 to ∞, but if a corresponding analysis is performed on the corresponding finite difference equations the number of values of k is equal to the number of mesh points in the y direction, and (28) becomes a triple-diagonal system of linear equations. For suitable choices of the mesh length, moreover, the symmetry in the sine functions permits very fast solution of (28) (Fourier analysis) and the final computation of (27) (Fourier synthesis).

Hockney and others have developed this method to a considerable extent, and a comprehensive account is given by Buzbee, Golub and Nielson [3], which I summarise briefly here. Consider the partitioned form (17) and suppose that the vectors

$$\underset{\sim}{z}^T = [\underset{\sim}{z}_1^T, \underset{\sim}{z}_2^T, \ldots, \underset{\sim}{z}_q^T], \quad \underset{\sim}{b}^T = [\underset{\sim}{b}_1^T, \underset{\sim}{b}_2^T, \ldots, \underset{\sim}{b}_q^T]$$

are partitioned similarly. B and I are square ($p \times p$) and symmetric matrices and A is square of size $p \times q$.

Now there is an orthogonal matrix Q which "diagonalises" B, that is,

$$Q^T B Q = \Lambda, \quad Q^T Q = I, \qquad (29)$$

where Λ is a diagonal matrix whose elements are the eigenvalues of B, the columns of Q being the eigenvectors of B. Expansion of (16) in the form

$$\left.\begin{array}{rl} B\underset{\sim}{z}_1 + \underset{\sim}{z}_2 & = \underset{\sim}{b}_1 \\ \underset{\sim}{z}_1 + B\underset{\sim}{z}_2 + \underset{\sim}{z}_3 & = \underset{\sim}{b}_2 \\ \cdots\cdots\cdots\cdots\cdots\cdots\cdots & \\ \underset{\sim}{z}_{q-1} + B\underset{\sim}{z}_q & = \underset{\sim}{b}_q \end{array}\right\} \qquad (30)$$

and use of (29) gives

$$\left.\begin{array}{rl} \Lambda\, \underset{\sim}{\bar{z}}_1 + \underset{\sim}{\bar{z}}_2 & = \underset{\sim}{\bar{b}}_1 \\ \underset{\sim}{\bar{z}}_1 + \Lambda\, \underset{\sim}{\bar{z}}_2 + \underset{\sim}{\bar{z}}_3 & = \underset{\sim}{\bar{b}}_2 \\ \cdots\cdots\cdots\cdots\cdots\cdots\cdots & \\ \underset{\sim}{\bar{z}}_{q-1} + \Lambda\, \underset{\sim}{\bar{z}}_q & = \underset{\sim}{\bar{b}}_q \end{array}\right\}, \qquad (31)$$

where $\underset{\sim}{\bar{z}}_i = Q^T \underset{\sim}{z}_i$ and similarly for $\underset{\sim}{\bar{b}}_i$. Then, if $\underset{\sim}{\bar{z}}_i$ has components $\bar{z}_{1,i}, \bar{z}_{2,i}, \ldots, \bar{z}_{p,i}$, and similarly for $\underset{\sim}{\bar{b}}_i$, we can write, for $i = 1, 2, \ldots, p$, the scalar equations

$$\left.\begin{array}{rl} \lambda_i\, \bar{z}_{i,1} + \bar{z}_{i,2} & = \bar{b}_{i,1} \\ \bar{z}_{i,1} + \lambda_i\, \bar{z}_{i,2} + \bar{z}_{i,3} & = \bar{b}_{i,2} \\ \bar{z}_{i,2} + \lambda_i\, \bar{z}_{i,3} + \bar{z}_{i,4} & = \bar{b}_{i,3} \\ \cdots\cdots\cdots\cdots\cdots\cdots\cdots & \\ \bar{z}_{i,q-1} + \lambda_i\, \bar{z}_{i,q} & = \bar{b}_{i,q} \end{array}\right\}, $$
(32)

and, if we use the further notation

$$\Gamma_i = \begin{bmatrix} \lambda_i & 1 & & \\ 1 & \lambda_i & 1 & \\ & \cdot & \cdot & \cdot \\ & & 1 & \lambda_i \end{bmatrix}, \quad \underset{\sim}{\hat{z}}_i = \begin{bmatrix} \bar{z}_{i,1} \\ \bar{z}_{i,2} \\ \cdot \\ \bar{z}_{i,q} \end{bmatrix}, \quad \underset{\sim}{\hat{b}}_i = \begin{bmatrix} \bar{b}_{i,1} \\ \bar{b}_{i,2} \\ \cdot \\ \bar{b}_{i,q} \end{bmatrix}, \quad (33)$$

We can write (32) in the condensed form

$$\Gamma_i \hat{\underline{z}}_i = \hat{\underline{b}}_i, \quad i = 1, 2, \ldots, p. \tag{34}$$

The algorithm therefore consists of: (i) finding the eigenvalues and eigenvectors of B; (ii) computing $\bar{\underline{b}}_i = Q^T \underline{b}_i$, $i = 1, 2, \ldots, q$; (iii) solving (34) for $i = 1, 2, \ldots, p$; and (iv) computing $\underline{z}_i = Q\bar{\underline{z}}_i$, $i = 1, 2, \ldots, q$. Steps (ii) and (iv) can be performed rapidly by using the fast Fourier transform technique of Cooley and Tukey [4] in certain cases, and Q is often known from analytical considerations. The solution of the special triple-diagonal systems (34) is also quick and easy.

Further simplifications are obtained by Hockney's technique of "odd-even" reduction, when $q = m-1$ and $m = 2^{k+1}$. A group of three of the equations in (30) is

$$\left. \begin{array}{l} \underline{z}_{j-2} + B\underline{z}_{j-1} + \underline{z}_j = \underline{b}_{j-1} \\ \underline{z}_{j-1} + B\underline{z}_j + \underline{z}_{j+1} = \underline{b}_j \\ \underline{z}_j + B\underline{z}_{j+1} + \underline{z}_{j+2} = \underline{b}_{j+1} \end{array} \right\}, \tag{35}$$

and on multiplying the successive equations by I, $-B$ and I and adding, we find that

$$\underline{z}_{j-2} + (2I - B^2)\underline{z}_j + \underline{z}_{j+2} = \underline{b}_{j-1} - B\underline{b}_j + \underline{b}_{j+1}, \tag{36}$$

from which the "odd-order blocks," usually odd-ordered lines of mesh values, have been eliminated. Equally important, this new system of equations can be solved by precisely the method just discussed. The eigenvectors of $2I - B^2$ are those of B, and the eigenvalues are $2 - \lambda_i^2$.

The process can be repeated again and again,

stopping perhaps with equations for just one block. The arithmetic of all this, however, needs considerable care in order to avoid numerical instabilities, and a method due to Buneman [5], although mathematically equivalent to the corresponding unstable methods, has arranged its arithmetic and numerical steps in a stable way.

This method can be used for many more difficult problems. For example, it can be adapted to treat Neumann or periodic boundary conditions. Also, the matrix I in (17) can be replaced by a more general symmetric T, and although the situation is easier when $AT = TA$ (so that T and A have the same eigenvectors) this condition is not essential. It can be applied to the more general self adjoint equation

$$\frac{\partial}{\partial x}\{f(x)\frac{\partial u}{\partial x}\} + \frac{\partial}{\partial y}\{g(y)\frac{\partial u}{\partial y}\} + u(x,y) = q(x,y), \quad (37)$$

and it can be extended to problems in three independent variables. Moreover, although by its nature and connection with Fourier series methods it is at its best for rectangular boundaries, it can be used with little extra work for shapes such as that of Fig. 2, and recent work by Hockney [6] and Buzbee, Dorr, George and Golub [7] suggests that it can be adapted for more irregular regions and with computational times an order of magnitude smaller than for iterative methods.

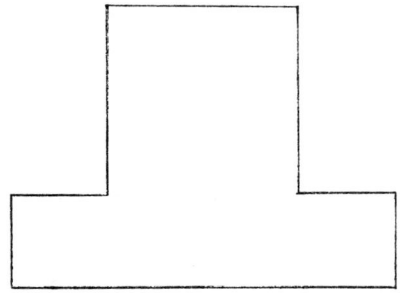

Fig. 2

There have been some even more recent and important developments in the application of Gauss-type elimination methods which in their normal form, as I mentioned earlier, suffer from "fill-up" which produces a corresponding need for a large amount of storage and a consequential large number of numerical operations. By the word "normal" here I mean that the original equations are ordered naturally row by row. But George [8] has introduced some remarkable orderings which reduce these numbers substantially. Consider, for example, the situation of Fig. 3, in which a typical equation, including equations at boundary points, is a linear combination of the values at a point and its nearest neighbours in the horizontal, vertical and both diagonal directions, as in the nine-point formula for Laplace's equation.

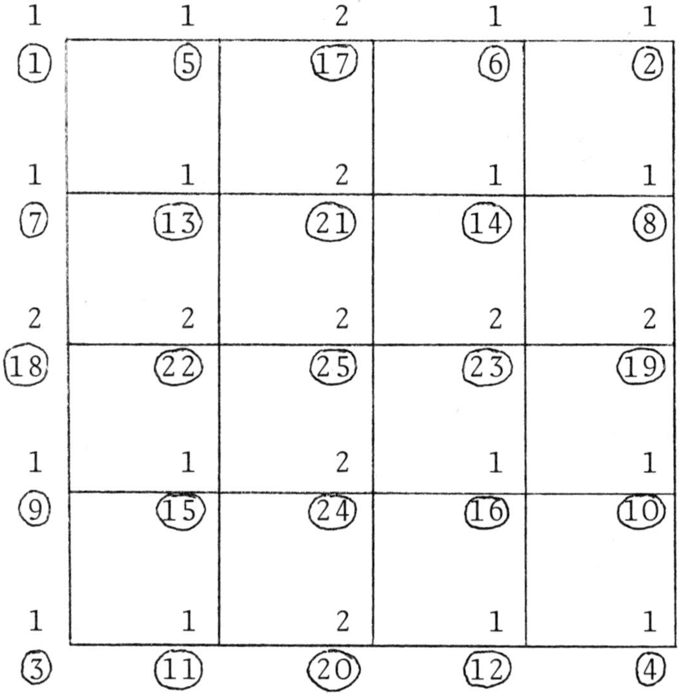

```
 1      1      2      1      1
(1)    (5)   (17)    (6)    (2)

 1      1      2      1      1
(7)   (13)   (21)   (14)    (8)

 2      2      2      2      2
(18)  (22)   (25)   (23)   (19)

 1      1      2      1      1
(9)   (15)   (24)   (16)   (10)

 1      1      2      1      1
(3)   (11)   (20)   (12)    (4)
```

Fig. 3

The points are divided into subsets according to the integers 1,2, ... at the mesh points, the determination of which is given in [8]. Points in subset 1 are then numbered sequentially, followed by those in subset 2, and so on in general. Within each subset the ordering starts with the nodes with the fewest connections to other nodes. A possible ordering (for subsets 1 and 2) is then that shown by the numbers ⓝ in Fig. 3. George shows that this ordering is nearly optimal, and offers substantial improvements over the natural ordering. For example, with 25 equations as in the mesh of Fig. 3, the upper triangular matrix after elimination has a number of non-zero elements above the diagonal which is about 80% of that obtained with the natural ordering, and the corresponding ratio of the amounts of arithmetic involved is about 75%. With 1089 equations, with 32 squares in each direction, these ratios are about 50 and 38%, respectively. This is a substantial gain, which improves steadily with the number of equations, and makes elimination competitive with iteration, especially when, as in some of our current problems, we deal with the same matrix but with many different right hand sides in the linear equations.

George [9], [10] has also worked on schemes using block elimination, with nothing more prohibitive than the solution of triangular systems. These methods are not optimal but are probably easier to code and have fewer red-tape operations. Their relative evaluation has not yet been established.

The steady-state solution

In my introduction I mentioned the possibilities of different formulations of the same problem, and I can exemplify this by looking at methods for determining the steady state solution of (5). Some writers obtain this by using the time-stepping process just discussed, stopping when there is no further change and the steady state has been reached. But others, and this would be my choice, put $\partial \zeta/\partial t = 0$ in (5) and solve the resulting pair of simultaneous partial differential equations, in the independent space variables x and y, given by

$$\left. \begin{array}{l} \nu \nabla^2 \zeta = \dfrac{\partial}{\partial x}(\zeta \dfrac{\partial \psi}{\partial y}) - \dfrac{\partial}{\partial y}(\zeta \dfrac{\partial \psi}{\partial x}) = \dfrac{\partial \zeta}{\partial x}\dfrac{\partial \psi}{\partial y} - \dfrac{\partial \zeta}{\partial y}\dfrac{\partial \psi}{\partial x} \\ \nabla^2 \psi = \zeta \end{array} \right\}, \quad (38)$$

and associated with some boundary conditions. The first of these is nonlinear, implying that some form of iterative method is essential.

Typical of the possible methods is the following, in which the boundary is rectangular and ψ and its normal derivative $\partial \psi/\partial \nu$ are specified at all boundary points.

(i) Take a rectangular mesh and estimate ζ at all mesh points.
(ii) Solve the second of (38), by any suitable method, to find a corresponding estimate for ψ.
(iii) Compute $\partial \psi/\partial x$ and $\partial \psi/\partial y$ at all mesh points, and solve the first of (38), which is now linear in ζ, to get a new approximation to ζ.
(iv) Use this in (i) and repeat the cycle until convergence is hopefully attained.

There is a passing problem relating to boundary conditions for the first of (38), but the specification of the normal derivative of ψ at the boundary, not so far used, permits the computation of an approximate value for ζ on the boundary. For example, at the straight line boundary of Fig. 4, the introduction of the "fictitious" ψ value at point 4 and the approximations

$$\left. \begin{array}{l} h^2 \nabla^2 \psi 0 = \psi_1 + \psi_2 + \psi_3 + \psi_4 - 4\psi_0 \\ 2h \partial \psi 0/\partial y = \psi_2 - \psi_4 \end{array} \right\} \quad (39)$$

give rise to the approximation of ζ_0 in terms of the specified values of ψ_1, ψ_3, $\partial \psi_0/\partial y$ and the value of ψ_2 computed at the previous stage.

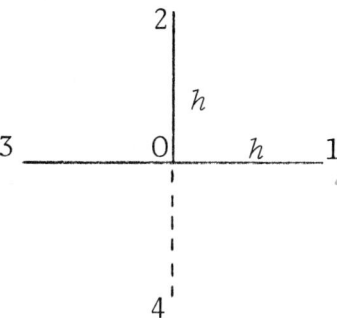

Fig. 4

There is also no problem involved in the computation of ψ from the second of (38), either by SOR type iteration or one of the fast direct methods mentioned earlier. The first of (38) is a little more difficult, and Greenspan [11] recommends the use of either forward or backward-difference approximations for the ψ derivatives, chosen so that the resulting finite difference equations have diagonal dominance, being effectively equivalent to the use of "upwind" differences. This guarantees the convergence of an SOR type iteration, although there is no theory for the optimum relaxation parameter and this has to be determined more or less by trial and error. The fast direct methods, of course, are not easily applicable here.

So what is sometimes called the "inner iteration" of the relevant computer program is guaranteed, but there is no guarantee that the "outer iteration" will converge. Nor is there as yet any analysis of how to choose the first guess, on which everything depends, in order to guarantee this convergence, and Greenspan uses certain *ad hoc* stratagems which succeed in most cases.

Finally, one can formulate this problem in yet another way, by eliminating the vorticity from the original equations and producing for the stream function the single fourth order equation

$$\nu \nabla^4 \psi + \frac{\partial}{\partial x}\left(\frac{\partial \psi}{\partial y}\nabla^2\psi\right) - \frac{\partial}{\partial y}\left(\frac{\partial \psi}{\partial x}\nabla^2\psi\right) = 0, \quad (40)$$

with ψ and $\partial\psi/\partial\nu$ specified at all boundary points. Most authors avoid this method because the corresponding finite difference equations are nonlinear and because there is a tradition that high order equations are more difficult to solve accurately than a set of simultaneous lower order equations. But this might be the best way of solving this problem, with large and fast computers and with better modern techniques for solving nonlinear systems.

This might also depend on the nature of the boundary and of the boundary conditions, and the main point I am making here is that one should bear in mind the fact that several different formulations of a problem are the rule rather than the exception, and that there is a case for analysis and numerical experiment with each one before one can decide on the "best" method.

COMPRESSIBLE FLOW

In the last section we had to deal with parabolic equations and elliptic equations. With compressible flow we find ourselves concerned with hyperbolic equations. A simple form of the Eulerian equations in one space dimension is

$$\frac{\partial \underset{\sim}{U}}{\partial t} + A \frac{\partial \underset{\sim}{U}}{\partial x} = Q, \quad (41)$$

where $\underset{\sim}{U}$ is the vector with components ρ (density), u (velocity), and p (pressure), and

$$A = \begin{bmatrix} u & \rho & 0 \\ 0 & u & 1/\rho \\ 0 & \gamma p & u \end{bmatrix}, \quad (42)$$

where γ is the ratio of specific heats at constant pressure and constant volume. The problem is hyperbolic because the matrix A has real eigenvalues u, $u-c$, $u+c$, where $c = \sqrt{(\gamma p/\rho)}$.

With hyperbolic equations we have real characteristics, whose slopes at any point are the reciprocals of these eigenvalues, and in one space dimension it is quite practicable and attractive to compute and integrate along the characteristics in a step-by-step manner. In more dimensions, however, this is much more difficult, and again we generally prefer to use finite difference equations on a rectilinear mesh.

Just as in our discussion of parabolic equations, so here we seek finite difference equations which are convenient in computation, stable and reasonably accurate locally. Moreover, there is some advantage in using conservative forms of the differential and finite difference equations. There are many possibilities, and I mention what appears to be one of the best of these methods. A conservation form of the equations is

$$\frac{\partial \underline{W}}{\partial t} + \frac{\partial}{\partial x} \underline{F}(\underline{W}) = \underline{0} \tag{43}$$

where

$$\underline{W} = \begin{bmatrix} \rho \\ m \\ e \end{bmatrix} \quad \underline{F}(\underline{W}) = \begin{bmatrix} m \\ (\gamma-1)e - \tfrac{1}{2}(\gamma-3)\frac{m^2}{\rho} \\ \frac{\gamma e m}{\rho} - \tfrac{1}{2}(\gamma-1)\frac{m^3}{\rho^2} \end{bmatrix}, \tag{44}$$

$m = \rho u$, $e = p/(\gamma-1) + \tfrac{1}{2}\rho u^2$

The original Lax-Wendroff method [12] starts with the Taylor series

$$\frac{W_{\sim m}^{n+1} - W_{\sim m}^n}{\Delta t} = \frac{\partial W_{\sim m}^n}{\partial t} + \tfrac{1}{2} \Delta t \frac{\partial^2 W_{\sim m}^n}{\partial t^2} + \ldots, \qquad (45)$$

replacing $\frac{\partial W}{\partial t}$ on the right by $\frac{-\partial}{\partial x} F(W)$ and $\frac{\partial^2 W}{\partial t^2}$ by $\frac{\partial}{\partial x}(\frac{A \partial F}{\partial x})$, where the matrix A is the Jacobian of $F(W)$, that is, $A_{ij} = \partial F_i / \partial W_j$. The x-derivatives are then approximated by differences, in a somewhat careful way, so as to produce for W_m^{n+1} the explicit scheme

$$W_{\sim m}^{n+1} = W_{\sim m}^n - \tfrac{1}{2} \frac{\Delta t}{\Delta x} (F_{\sim m+1}^n - F_{\sim m-1}^n)$$

$$+ \tfrac{1}{2} \left(\frac{\Delta t}{\Delta x}\right)^2 \left[A_{m+\frac{1}{2}} (F_{\sim m+1}^n - F_{\sim m}^n) - A_{m-\frac{1}{2}} (F_{\sim m}^n - F_{\sim m-1}^n) \right]$$

where

$$A_{m+\frac{1}{2}}^n = A(\tfrac{1}{2} W_{\sim m+1}^n + \tfrac{1}{2} W_{\sim m}^n). \qquad (46)$$

But a better two-step Lax-Wendroff scheme, which avoids the computation of the matrix A, uses the equations

$$\left.\begin{array}{l} W_{\sim m+\frac{1}{2}}^{n+\frac{1}{2}} = \tfrac{1}{2} (W_{\sim m+1}^n + W_{\sim m}^n) - \tfrac{1}{2}\frac{\Delta t}{\Delta x} (F_{\sim m+1}^n - F_{\sim m}^n) \\ W_{\sim m}^{n+1} = W_{\sim m}^n - \frac{\Delta t}{\Delta x} (F_{\sim m+\frac{1}{2}}^{n+\frac{1}{2}} - F_{\sim m+\frac{1}{2}}^{n+\frac{1}{2}}) \end{array}\right\}, \qquad (47)$$

which is locally quite accurate and conserves all the relevant quantities.

Stability is tested by the von Neumann

method, previously mentioned for parabolic equations, and applied to the corresponding equations with constant coefficients, that is, with $F(W) = BW$. The amplification factor found earlier for an equation with one dependent variable becomes an eigenvalue of a matrix of order three for three dependent variables. We easily find that the relation between such an eigenvalue μ_i and an eigenvalue λ_i of the matrix B is given by

$$\mu_i = 1 - i \lambda_i \frac{\Delta t}{\Delta x} \sin\alpha - (\lambda_i \frac{\Delta t}{\Delta x})^2 (1 - \cos\alpha), \qquad (48)$$

where α is any angle between 0 and 2π. Moreover it is obvious that the eigenvalues of B are those of A, because the equations and therefore the characteristics are the same in both formulations, and we easily deduce the stability requirement

$$(|u| + \sqrt{(\gamma p/\rho)}) \frac{\Delta t}{\Delta x} < 1. \qquad (49)$$

Notice that this is not nearly so restrictive for the time interval as it is in the parabolic case, when Δx appears as $(\Delta x)^2$, and perhaps for this reason implicit methods are less important in compressible flow than in incompressible flow problems. Moreover the Lax-Wendroff method is fairly easily extensible to problems in more space dimensions, in which implicit methods give rise to substantial amounts of computation.

Shocks

If the flow is smooth the method just outlined is satisfactory and convenient. But disturbances may exist or may develop, the most difficult of these being a shock line or curve across which the density, pressure and velocity become discontinuous, and the differential equations must be supplemented by the so called "jump conditions" at points on these curves of discontinuity. I have neither the space nor the knowledge to describe the various methods used to deal with this problem. The general idea appears to be to introduce into the differential or difference equations extra terms, equivalent, say, to viscosity, which to some extent smooth the shock and allow the relevant quantities to change very rapidly but not discontinuously, while still preserving the conservation laws on which the jump conditions are based. For example, if A is a constant matrix, the scheme (46) may be written in the form

$$\underset{\sim}{W}_m^{n+1} - \underset{\sim}{W}_m^n + \tfrac{1}{2} \frac{\Delta t}{\Delta x} A(\underset{\sim}{W}_{m+1}^n - \underset{\sim}{W}_{m-1}^n)$$
$$- \tfrac{1}{2} \left(\frac{\Delta t}{\Delta x}\right)^2 A^2 (\underset{\sim}{W}_{m+1}^n - 2\underset{\sim}{W}_m^n + \underset{\sim}{W}_{m-1}^n) = \underset{\sim}{0}, \qquad (50)$$

and the idea is to add a term to the differential equation so that in this simple case it becomes

$$\frac{\partial \underset{\sim}{W}}{\partial t} + A \frac{\partial \underset{\sim}{W}}{\partial x} = Q\underset{\sim}{W}, \qquad (51)$$

with Q chosen so that $\underset{\sim}{W}$ satisfies (50) with the addition of a term $O(\Delta t)^4$ or preferably $O(\Delta t)^5$. Lax and Wendroff [12] show how to determine a suitable Q in the more general case.

Relevant to all this work is also the idea of a "weak solution." If $W(x,t)$ is a solution of (43) with initial value $W^{(0)}(x)$, and if $\phi(x,t)$ is a suitably smooth "test function," then multiplication of (43) by $\phi(x,t)$ and integration over all x and all $t > 0$, followed by integration by parts, gives the equation

$$\iint \left(\frac{\partial \phi}{\partial t} \underset{\sim}{W} + \frac{\partial \phi}{\partial x} \underset{\sim}{F}\right) dx\, dt + \int \phi(x,0) \underset{\sim}{W}^{(0)}(x) dx = \underset{\sim}{0}. \qquad (52)$$

Any $\underset{\sim}{W}$ that satisfies this equation for all test functions, and with initial value $\underset{\sim}{W}^{(0)}$, is called a weak solution, and it can be shown that this solution has the required jump conditions at the shock. Moreover the solution of the appropriate finite difference equations converges to a weak solution, as Δx and $\Delta t \to 0$ provided that the finite difference equations are stable (which will involve some restriction on the ratio $\Delta x/\Delta t$). One can then solve these equations, with suitably small Δx and Δt, knowing that the result is a good approximation to the required solution, with the shock virtually "smeared" over a few finite difference intervals.

Analogous methods have been found to be important and convenient in various other scientific contexts. For example in the so called Stefan problem, in which we seek the interface between the solid and liquid region at which latent heat is involved, there are step changes in the temperature gradient that make computation difficult, but this is eased by making slight changes in the differential equation and corresponding difference equations, whose solution then converges to a weak solution of the given problem. Various papers describing this technique can be found in reference [13].

OTHER PROBLEMS

In this section I shall discuss a few items illustrating some other introductory remarks. They include (*i*) a problem in which the solution is not unique, (*ii*) a treatment of a class of free-boundary problems and (*iii*) a discussion of singularities.

Non-uniqueness of solution

One famous example of this kind is provided by the Faulkner-Skan boundary-layer equation, namely

$$\left. \begin{array}{c} y''' + yy'' + \beta\{(y')^2 - 1)\} = 0 \\ \text{subject to the boundary conditions} \\ y(0) = y'(0) = 0, \quad y'(\infty) = 1. \end{array} \right\} \quad (53)$$

This is a third order equation, and we might expect the three given conditions to be sufficient to provide a unique solution. But nonlinearities have peculiar effects, and Hartree [14] showed that for positive values of β the solution is not unique. There is one solution for which

$$y' = 1 + O(e^{-x^2}) \text{ as } x \to \infty, \tag{54}$$

and another for which

$$y' = 1 + O(x^{-n}) \text{ as } x \to \infty. \tag{55}$$

It is of interest to note that some iterative methods of solution [15] converge to that of type (54), even though the first approximation is close to that of type (55). This is good because (54) is the solution required, but of course such success cannot always be guaranteed in advance.

In passing, I mention the so called "shooting method" for solving boundary value problems. In relation to (53) we might try to solve the third order equation, preferably written as three first order equations, with the initial conditions

$$y(0) = 0, \; y'(0) = 0, \; y''(0) = p, \tag{56}$$

and try to find the value of p for which $y'(\infty) = 1$ is satisfied. Although attractive in theory, this method can be difficult and inaccurate when the differential equation has some rapidly increasing solutions, although I have not tested it for this particular example. But when this is not a problem, the convergence of the shooting method can be made quite rapid if we use a Newton type operation, and I mention this because the idea is useful in many contexts.

Suppose that we integrate the differential equation (53) with the initial conditions (56). We seek a value of p for which

$$f(p) = y'(p,\infty) - 1 = 0, \tag{57}$$

the notation implying that y is a function of the parameter p as well as the variable x. To a first approximation, Newton's correction to the required p is given by

$$\delta p = \frac{-f(p)}{\frac{\partial}{\partial p} f(p)}. \tag{58}$$

Choosing "infinity" to be some suitably large value of x we have the numerator in (58) from the first integration, and the denominator is obtained by integrating the derived form, with respect to p, of the original equation and conditions. If $z = \partial y/\partial p$, this derived system is just

$$\left. \begin{array}{l} z''' + yz'' + 2\beta y'z' + y''z = 0, \\ z(0) = z'(0) = 0, \ z''(0) = 1, \end{array} \right\} \tag{59}$$

which is a linear system in which y, y' and y'' have already been computed for a selected p. The correction is then

$$\delta p = -\frac{y'(\infty) - 1}{z'(\infty)}, \tag{60}$$

and we can then start the iteration again with the better starting approximation $p+\delta p$.

A free-boundary problem

I consider next a technique that appears to be quite useful for the determination of a free boundary in some problems in which the boundary is not known in advance but is determined in the course of the computation. Fox and Sankar [16] consider the problem of axisymmetric incompressible flow past a circular disc placed coaxially inside a circular cylinder. The differential equation for the stream function is

$$\frac{\partial^2 \psi}{\partial r^2} - \frac{1}{r}\frac{\partial \psi}{\partial r} + \frac{\partial^2 \psi}{\partial z^2} = 0. \tag{61}$$

The boundary conditions with reference to Fig. 5 are

$$\psi = 1 \text{ on } AB, \quad \psi = 0 \text{ on } FED, \quad \frac{\partial \psi}{\partial z} = 0 \text{ on } AF \text{ and } BC, \qquad (62)$$

and

$$\psi = 0, \quad \frac{1}{r}\left(\frac{\partial \psi}{\partial \nu}\right) = q \qquad (63)$$

on the free boundary DC.

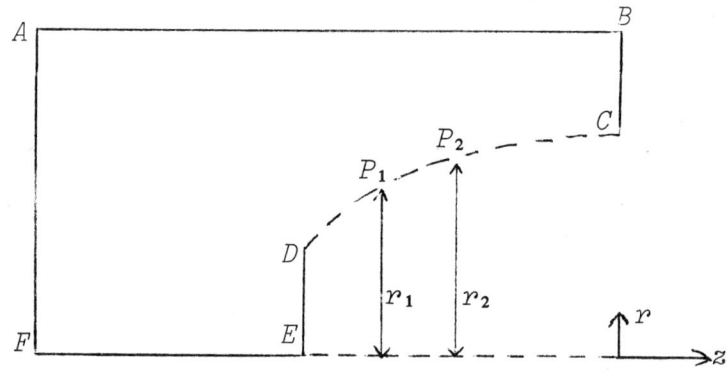

Fig. 5

We have to determine the position of the free boundary DC and the constant q, assuming a knowledge of all the dimensions in Fig. 5. A common procedure is to guess a shape for DC, solve (61), (62) and, say, the first of (63), then use the second of (63) to adjust the first guess at DC, and proceed with the obvious iterative sequence. There is no guarantee that this iteration will converge. It will depend on the first guess for DC, and even more on the way in which successive adjustments to the boundary are made. In fact if we use finite differences, and adjust separately the boundary mesh points P_1, P_2, ... in Fig. 5 so that the second of (63) is temporarily satisfied, we encounter the incidence and rapid development of unwanted oscillations in subsequent adjustments.

What we seek, of course, are the values of $r_1, r_2, \ldots$ such that

$$\frac{1}{r_1}\left(\frac{\partial \psi_1}{\partial \nu}\right) - q = 0, \quad \frac{1}{r_2}\left(\frac{\partial \psi_2}{\partial \nu}\right) - q = 0, \quad \ldots, \quad (64)$$

which form a set of nonlinear simultaneous equations, "simultaneous" because $\partial\psi/\partial\nu$, for example, is a function of all the $r_1, r_2, \ldots$ and not just of r_1. It is this dependence which is not allowed for explicitly in the simple and unsuccessful method of adjustment.

Now in solving a set of nonlinear equations, written for simplicity in the form

$$f_i(r_1, r_2, \ldots r_n) = 0, \quad i = 1, 2, \ldots, n, \quad (65)$$

we would like to use some form of Newton's method. It is easy enough to calculate the f_i, but Newton's method requires the computation of all the derivatives $\partial f_i/\partial r_j$, and in this problem this is rather prohibitive. In [16] we therefore thought of using the generalised Regula-Falsi method, well known for a single equation but ignored for simultaneous equations, and this turned out to be very successful.

For some estimated constant q_e we select $n+1$ different starting curves DC, the ith curve of which we represent by the "point"

$$Q_i = (r_1^{(i)}, r_2^{(i)}, \ldots r_n^{(i)}). \quad (66)$$

For each curve we solve the boundary-value problem, and then compute the $f_j(Q_i)$ given by (64). A "better" curve, obtained from the Regula-Falsi equations, then has the coordinates

$$r_j = \frac{\left| r_j^{(i)}, \; f_1(Q_i), \; f_2(Q_i), \; \ldots, \; f_n(Q_i) \right|}{\left| 1, \; f_1(Q_i), \; f_2(Q_i), \; \ldots, \; f_n(Q_i) \right|}, \quad j=1,2,\ldots,n, \quad (67)$$

where the numerator and denominator are determinants of which the ith row is shown in each case, and i goes from 1 to $n+1$.

This is a "better" curve for this particular q_e, and from it we can compute $\frac{1}{r}\frac{\partial \psi}{\partial \nu}$ at every boundary mesh point and hence determine the average q_a of these quantities. The "goodness" of q_e is determined by the maximum deviation from this average, and we obtain a "best" q_e by repeating the whole process with several q_e and choosing the one for which maximum deviation is smallest.

The whole operation appears at first sight formidable, and certainly well outside the scope of desk computation. But the modern computing machine has such good hardware and software that a few minutes of computation can produce a satisfactory result. The largest part of the work, as I hinted at in my introduction, is the writing of a good computer program for solving this particular boundary value problem with a more or less arbitrary position for the portion DC of the boundary.

Many writers avoid the curved nature of the free boundary by interchanging the dependent and independent variables. In this problem, for example, Brennen [17] introduced a potential function ϕ, and with $f = r^2$ reduced the problem to the solution of the differential equation

$$\frac{\partial^2}{\partial \phi^2} \log f + \frac{\partial^2 f}{\partial \psi^2} = 0. \qquad (68)$$

The boundaries in this case are shown in Fig. 6, and the boundary conditions are

$$\left.\begin{array}{l} f = \psi \text{ on } AF,\ f = 1 \text{ on } AB,\ f = 0 \text{ on } FE,\ \frac{\partial f}{\partial \phi} = 0 \text{ on } BC \\ \frac{\partial f}{\partial \psi} = 0 \text{ on } ED,\ \frac{1}{f}\left(\frac{\partial f}{\partial \phi}\right)^2 + \left(\frac{\partial f}{\partial \psi}\right)^2 = \text{constant on } DC \end{array}\right\}. \quad (69)$$

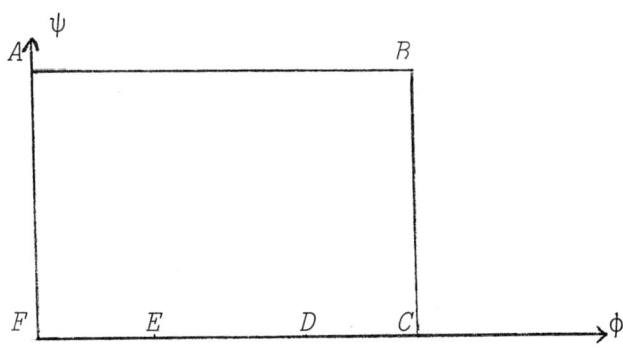

Fig. 6

By specifying the constant velocity the problem is reduced to the determination of the boundary BC. The boundaries are rectangular, but the differential equation and one of the boundary conditions is nonlinear and there is still a singularity at D. There is now complication of two iterative processes, an "inner" iteration for the solution of the nonlinear boundary value problem and an "outer" iteration for finding the position of BC. Brennen got good results but his methods and computations were by no means trivial, and for this particular example the original formulation together with the Regula-Falsi method might well be superior. Alternatively the Regula-Falsi method could be applied to Brennen's method to produce faster iteration to the required position of BC.

A treatment of singularities

The previous problem has a singularity at the point D, where the fluid moves away from the disc. One can show that the free boundary DC is tangential to the disc DE, but the curvature at D is infinite and the fluid changes direction very rapidly at the separation point. The finite difference method will converge to the correct solution as the intervals are reduced, but, without special treatment near the singularity, one is likely to need a very small interval to produce a good result. In this problem I have not been able to find enough about the nature of the singularity to decide what this special treatment should be. For some easier problems, however, Fox and Sankar [18] have extended earlier work by Motz [19] and Woods [20] to produce a technique which avoids the need for small intervals.

Consider the equation, written in polar coordinates and given by

$$\nabla^2 u = g(r,\theta)u, \quad g(r,\theta) = \sum_{n=0}^{\infty} g_n(\theta) r^n, \qquad (70)$$

with boundary conditions on two straight lines meeting at the singular point (the coordinate origin) given by

$$u = F(r) \text{ on } \theta = 0, \quad u = G(r) \text{ on } \theta = \omega, \qquad (71)$$

or by various combinations of u and $\frac{1}{r}\frac{\partial u}{\partial \theta}$. We can show that in the neighbourhood of the origin the solution has the form

$$u = \phi_0(r,\theta) + \sum_{j=0}^{\infty} c_j \phi_j(r,\theta), \qquad (72)$$

where the c_j are constants and each $\phi_j(r,\theta)$ is a combination of terms

$$r^{\alpha_j + n}\{(\log r)A_{\alpha_j,n}(\theta) + B_{\alpha_j,n}(\theta)\}, \qquad (73)$$

where the α_j and the functions $A_{\alpha_j,n}(\theta), B_{\alpha_j,n}(\theta)$ can be obtained quite simply by solving ordinary differential equations.

In the numerical work we take just a few of the more important terms in (72), and let this be the solution in a region I (Fig. 7) near the singular point. We use finite difference methods in the remaining part of the field, and with no particular difficulty we can write down a set of linear equations whose solutions are the mesh values of ϕ in region II and the constants c_0, c_1, ... in region I.

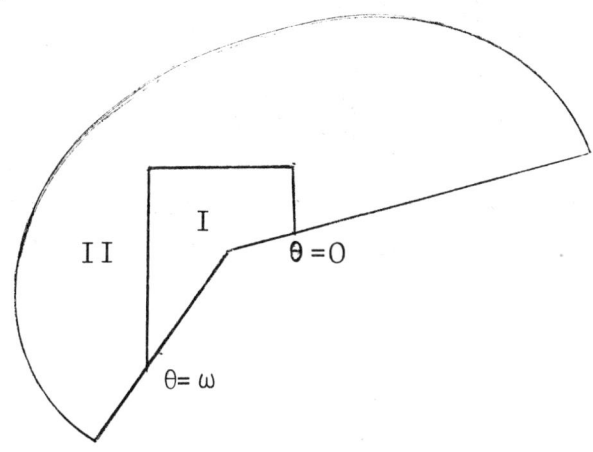

Fig. 7

Examples are given in Fox and Sankar [18] and in Fox [21], and the method is obviously practicable. We do not have an accurate or elaborate error analysis, but this is a fairly common situation in numerical analysis, especially for problems with singularities!

CONCLUSION

I have tried to give some indication of the numerical problems that arise in fluid dynamics, of the areas of numerical analysis involved, and of some of the numerical methods currently being applied. Of course, I have had to leave out many important things. For example I have said very little about boundary conditions, or about the inherent instability of our problems, the extent to which small changes in the data can make large changes in the results. These topics are important, and there is some discussion of them in the comprehensive work by Roache [22] called "Computational fluid dynamics."

With regard to our various methods I suppose

that many of them are still of an *ad hoc* nature, there being little real analysis to support our trial and error and intuitive judgement. I suppose further that there are some problems for which our numerical methods can be dangerous without the associated mathematical analysis, and it is certainly possible to convince oneself that one has obtained a respectable solution when such a solution does not exist. I have heard scientists and applied mathematicians criticise the work of numerical analysts on just these grounds, and the answer is surely to engage not in mutual insults but in closer cooperation. In problems of this kind the numerical man and the mathematical man and the scientific man can all improve their performance by consultation and cooperation. Such cooperation, unfortunately, does seem to be rather rare, and I hope that meetings like this, and the very existence of the IMA, will help to remedy this situation.

REFERENCES

1. Richtmyer, R.D. and Morton, K.W., "Difference methods for initial-value problems," Interscience Publishers, New York (1967).

2. Hockney, R.W., "A fast direct solution of Poisson's equation using Fourier analysis," Stanford Electronics Laboratories Tech. Rep. 0255-1(1964).

3. Buzbee, B.L., Golub, G.H. and Nielson, C.W., "On direct methods for solving Poisson's equation," *Siam J. Numer. Anal.*, **7**, 627-656 (1970).

4. Cooley, J.W. and Tukey, J.W., "An algorithm for machine calculation of complex Fourier series," *Math. Comp.*, **19**, 297-301(1965).

5. Buneman, O., "A compact non-iterative Poisson solver," SUIPR-294, Institute for Plasma Research, Stanford University, California (1969).

6. Hockney, R.W., "The potential calculation and

some applications," *in* Adler, B., Fernbach, S. and Rotenberg, M., *Editors*, "Methods of Computational Physics," Academic Press, New York and London (1970).

7. Buzbee, B.L., Dorr, F.W., George, J.A. and Golub, G.H, "The direct solution of the discrete Poisson equation on irregular regions," *Siam J.Numer.Anal.*, **8**, 722-736 (1971).

8. George, J.A., "Nested dissection of a regular finite element mesh," *Siam J.Numer.Anal.*, **10**, 345-363 (1973).

9. George, J.A., "An efficient band-oriented scheme for solving n by n grid problems," Proceedings of the Fall Joint Computer Conference, 1317-1320 (1972).

10. George, J.A., "On block elimination for sparse linear systems," *Siam J.Numer.Anal.*, **11**, 585-603 (1974).

11. Greenspan, D., "Numerical studies of viscous incompressible flow for arbitrary Reynolds number," University of Wisconsin, Comp.Sc.Tech. Rep.11, (1968).

12. Lax, P.D. and Wendroff, B., "Systems of conservation laws," *Comm. Pure Appl.Math.*, **13**, 217-237 (1960).

13. Ockendon, J.R. and Hodgkins, W.R., "Moving boundary problems in heat flow and diffusion," Clarendon Press, Oxford (1975).

14. Hartree, D.R., "On an equation occurring in Faulkner and Skan's approximate treatment of the equations of the boundary layer," *Proc. Camb.Phil.Soc.*, **33**, 223-239 (1937).

15. Fox, L., "The solution by relaxation methods of ordinary differential equations," *Proc.Camb. Phil.Soc.*, **45**, 50-68 (1949).

16. Fox, L and Sankar, R., "The Regula-Falsi method for free boundary problems," *JIMA*, **12**, 49-54, (1973).

17. Brennen, C., "A numerical solution of axisym-

metric cavity flows," *J. Fluid Mech.*, **37**, 671-688 (1969).

18. Fox, L. and Sankar, R., "Boundary singularities in linear elliptic differential equations," *JIMA*, **5**, 340-350 (1969).

19. Motz, H., "The treatment of singularities of partial differential equations by relaxation methods," *Q. Appl. Math.*, **4**, 371-377 (1946).

20. Woods, L.C., "The relaxation treatment of singular points in Poisson's equation," *Q. J. Mech.*, **6**, 163-185 (1953).

21. Fox, L., "Some experiments with singularities in linear elliptic partial differential equations," *Proc.Roy.Soc. A*, **323**, 179-190 (1971).

22. Roache, P.J., "Computational fluid dynamics," Hermosa Publishers, Albuquerque, New Mexico (1972).

METHODS FOR ELLIPTIC PROBLEMS IN EXTERNAL AERODYNAMICS*

R.C. Lock

(Aerodynamics Department, Royal Aircraft Establishment)

INTRODUCTION

After the historical, practical and numerical introductions in the earlier papers in this volume, we come to the first of those in which details are given of what can actually be done to compute some of the information about the aerodynamics of an aircraft needed by the designer. Although the scope of this paper is restricted to cases in which the flight speed of the aircraft is sufficiently low for the flow over it to remain everywhere subsonic (the perhaps more glamorous topics of transonic and supersonic flow are left to later papers) this speed range obviously still remains important in practice. But even with the restriction to purely subsonic flows, implying that the equations of motion are entirely elliptic and therefore supposedly straightforward from a numerical point of view, the practical problems involved are still extremely difficult, mainly because of the complicated nature of the shapes (and hence the boundary conditions) involved. Fig. 1, showing a Lockheed 1011 coming in to land with flaps and undercarriage down, emphasises this point; Fig. 2 (taken from [1]) shows about the most advanced shape that can currently be tackled theoretically - already involving appreciable simplification compared with the real thing. Because

---

* © IMA; Controller HMSO London, 1975.

Fig. 1.

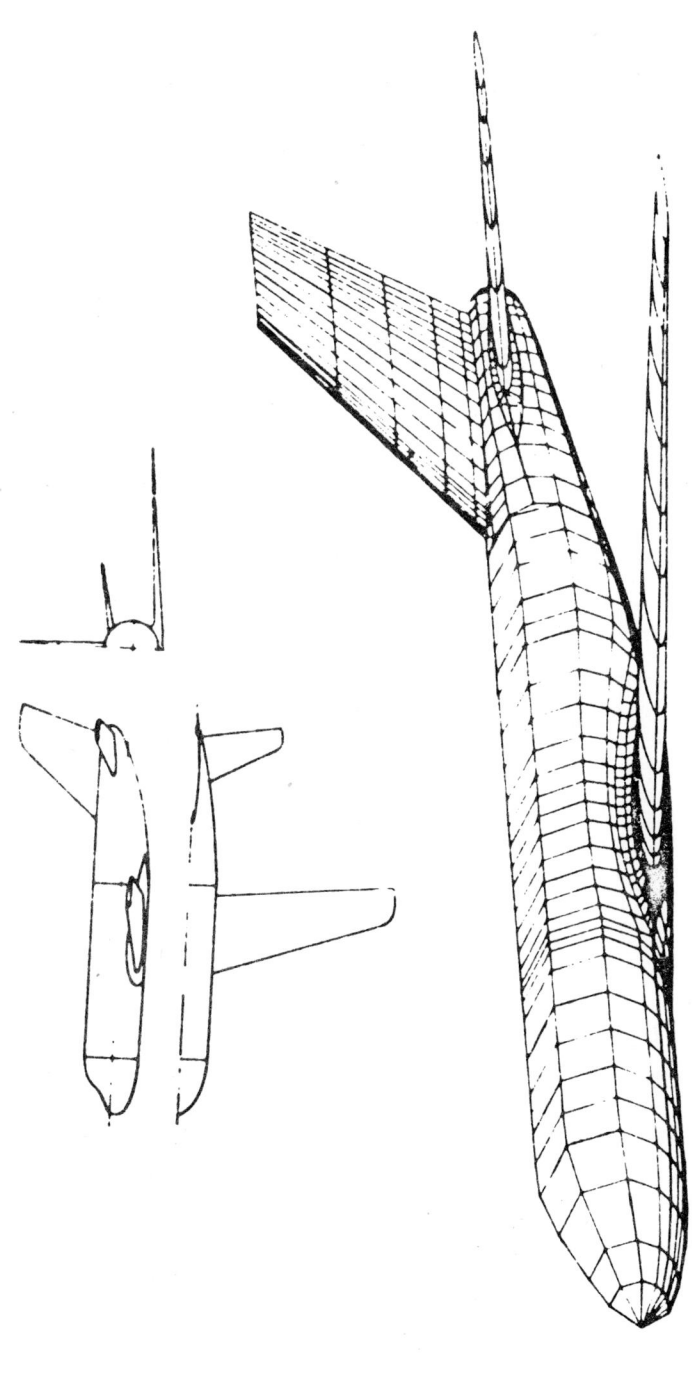

Fig. 2. General arrangement and panel plot of wing-body-tail configuration

of the complexity of the boundary conditions, the vast majority of methods are still restricted essentially to the case of incompressible flow, compressibility being roughly allowed for either by replacing Laplace's equation by the still linearised Prandtl-Glauert equation, or by incorporating empirical compressibility corrections. Solutions to the full nonlinear equations of motion have so far only been obtained for two-dimensional problems and for a few highly simplified three-dimensional shapes.

As our problems are so difficult for inviscid flow, it is not surprising that little progress has yet been made in the calculation of the viscous flows which are of course our real interest. It is true that, provided the boundary layer remains attached over all or most of the body surface, the classical displacement concept can be used, in principle, to combine inviscid flow theory and boundary layer theory into methods for solving the complete interactive problem. But, in fact, this principle has only been put into practice for the simpler two-dimensional problem, in which recent developments have indeed led to methods which are adequate for most practical purposes, even if some fundamental problems, mainly connected with the precise details of the flow near the trailing edge of an aerofoil, remain to be resolved. However, such methods are outside the scope of this paper.

METHODS FOR INCOMPRESSIBLE FLOW

The greater part of this paper is therefore devoted to a review of some of the more important methods for the inviscid incompressible flow problem. In this case the most efficient methods of solution are obtained by representing the flow by a distribution of singularities - sources, doublets or vortices - placed either on or inside the body and also (for lifting systems) on the vortex wake behind it. In this way the equation of motion (Laplace or Prandtl-Glauert) is automatically satisfied, as are the boundary conditions at infinity, and the whole problem is reduced to one of satisfying the boundary conditions on the body and wake. As a result, the dimensions of the problem are effectively reduced mathematically by one; and this is essentially the reason for the improved numerical efficiency of such a procedure compared with that of a so called "field solution" in which the equation of motion is solved explicitly (for example by a finite difference method) with the appropriate boundary conditions.

Methods of this type can conveniently be divided into two classes, according to whether the singularities are placed either (A) inside the body, for example in the chord plane of a wing or along the centre line of a fuselage; or (B) on the "wetted surface" of the body.

In either case the representation of the vortex wake requires a distribution of doublets (or vorticity) on its surface, the shape of which is initially unknown. To calculate this properly leads to a difficult "free surface" problem, of the type considered by Professor Mangler[*]; and for this reason it is standard practice to assume that the shape of the wake is simplified in a convenient way, although this means that the correct physical boundary conditions on it are then violated. Fortunately, there is evidence that the effect of such a simplification on the lift or pressure distribution on a wing is usually small.

Wing theory using internal singularity distributions (Type A)

This way of representing the flow about a thin wing, by placing a source distribution and a doublet or "load" distribution on the mean chordal plane of the wing, is of course the basis of classical linearised theory. It can also be used to obtain higher order approximations, as demonstrated recently by Weber [2] and Sells [3], [4] at RAE.

---

[*] Pp.198-213 of this volume

Their method is an iterative one and in principle is extremely simple. One starts with a suitable first approximation to the source and load distributions, obtained by linearised theory with some empirical improvements. Next, the velocity perturbations produced by these singularity distributions are calculated accurately (using the numerical methods of Sells [5] and Ledger [6]) at a suitable number of control points on the surface of the wing, and the resulting error in the exact boundary condition at these points is determined. From a knowledge of this error over the wing surface, a correction is found to the corresponding boundary condition on the mean plane, leading to a better approximation to the strengths of the singularity distributions. The process may then be repeated, if required, until the error in the surface boundary condition becomes acceptably small everywhere; normally however further repetitions are unnecessary. The difficulty that occurs at the leading edge of the wing, arising from the singular nature of the boundary condition there, is circumvented by means of the device suggested by Lighthill [7], of terminating the singularity distributions at a short distance from the edge, proportional to the local radius of curvature.

For the special case of an uncambered wing [3], the analysis involved in this method becomes particularly simple, and is summarised below (Fig. 3). The velocity vector at any point on the wing surface is defined as

$$\underset{\sim}{U} = \underset{\sim}{U}_\infty + \underset{\sim}{u} = (\cos\alpha, 0, \sin\alpha) + (u,v,w) \quad (1)$$

(with $|U_\infty|$ taken for convenience to be unity) and must satisfy the boundary condition $\underset{\sim}{U}.\underset{\sim}{n} = 0$ on the wing surface $z = Z(x,y) = \pm z_t(x,y)$, where $\underset{\sim}{n}$ is the normal to the surface. This boundary condition may be approximated to second order accuracy by the equation

$$\frac{\partial Z}{\partial x} \cdot \{1 + u(x,y,Z)\} + \frac{\partial Z}{\partial y} \cdot v(x,y,Z) = \alpha + w(x,y,Z) \quad (2)$$

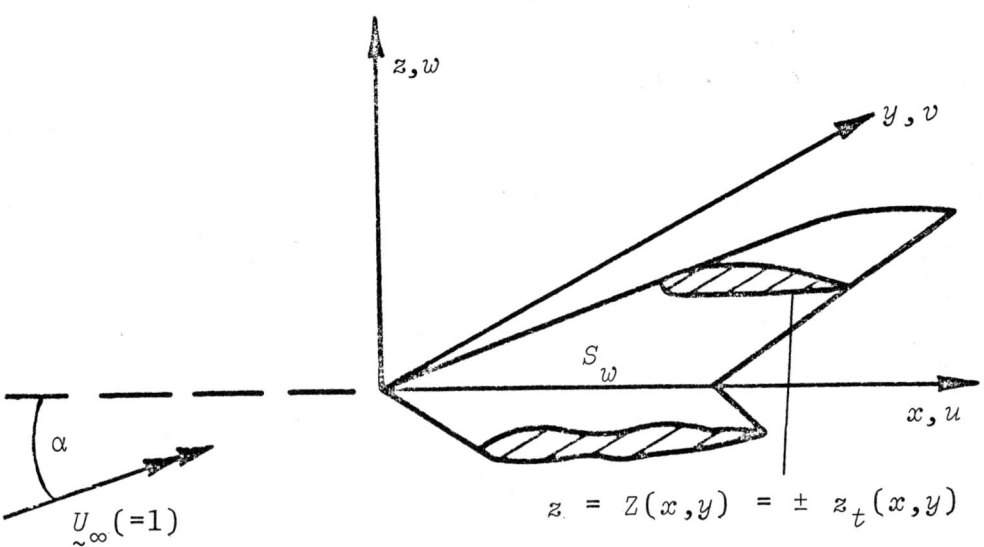

Fig. 3.

We write
$$\underset{\sim}{u}_t = (u_t, v_t, w_t) \tag{3}$$
and
$$\underset{\sim}{u}_l = (u_l, v_l, w_l) \tag{4}$$
for the velocity perturbation vectors produced by the source distribution (of strength $\sigma(x,y)$ per unit area) and load distribution (of strength $l(x,y)$ per unit area), respectively; so that

$$\underset{\sim}{u} = (u_t \pm u_l,\ v_t \pm v_l,\ w_t \pm w_l) \tag{5}$$

where the upper and lower signs correspond to the upper and lower surfaces of the wing. Then, by adding and subtracting the boundary conditions for the upper and lower surfaces, we find that the corresponding boundary conditions, which determine σ and l, become uncoupled, and are, respectively,

$$R_t \equiv (1 + u_t)\frac{\partial z_t}{\partial x} + v_t\frac{\partial z_t}{\partial y} - w_t = 0 \tag{6}$$

and
$$R_l \equiv u_l\frac{\partial z_t}{\partial x} + v_l\frac{\partial z_t}{\partial y} - w_l - \alpha = 0 \tag{7}$$

both to be satisfied on $z = z_t(x,y)$.

First, to find the source distribution satisfying (6) we start with a suitable first approximation $\sigma^{(1)}$ to $\sigma(x,y)$, obtained either from linearised theory

$$\left[\sigma^{(1)} = 2\frac{\partial z_t}{\partial x}\right]$$

or from a modification based on the "RAE standard method" [8] (see [3]). By using Ledger's method [6], we then compute the components of the perturbation velocity $\underset{\sim}{u}_t$, produced by this source distribution, at a number of control points on the surface $z = z_t$; these components will not satisfy equation (6) exactly but leave a residual error, denoted by $R_t^{(1)}(x,y,z)$. In order to obtain a better approximation to $\sigma(x,z)$, the approximation

$$R_t^{(1)}(x,y,0) \simeq R_t^{(1)}(x,y,z_t)$$

is then made, leading to a correction to σ

$$\Delta\sigma = 2R_t^{(1)}(x,y) \qquad (8)$$

and to the second approximation

$$\sigma^{(2)} = \sigma^{(1)} + \Delta\sigma . \qquad (9)$$

This process can then be repeated until R_t becomes acceptably small at all control points.

The procedure followed to determine the load distribution, $l(x,y)$, in order to satisfy the second boundary condition (7), is basically similar; but here a problem in lifting surface theory has to be solved at each stage, to determine l from a knowledge of w_l on $z = 0$. For this purpose a vortex lattice method was chosen - essentially similar to that of Hedman [9] but with improvements suggested by G. Carr-Hill (RAE, unpublished work) because of its speed and flexibility.

When the final approximations to the source and load distributions, together with the velocity perturbations $\underset{\sim}{u}_t$ and $\underset{\sim}{u}_l$ they produce on $z = z_t$, have been computed, the required velocity vectors on the upper and lower surfaces of the wing can be obtained from equations (1) and (5).

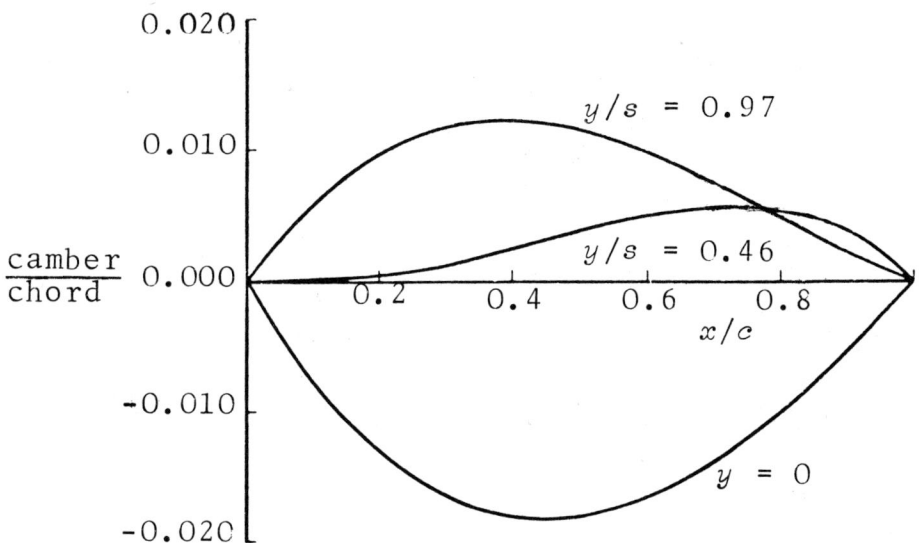

(b) camber distribution

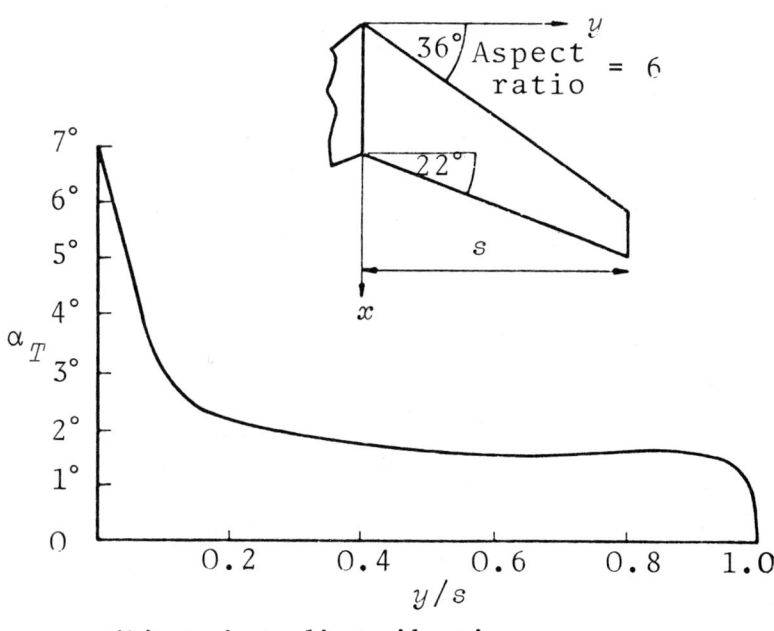

(b) twist distribution

Fig. 4. RAE Wing "B"

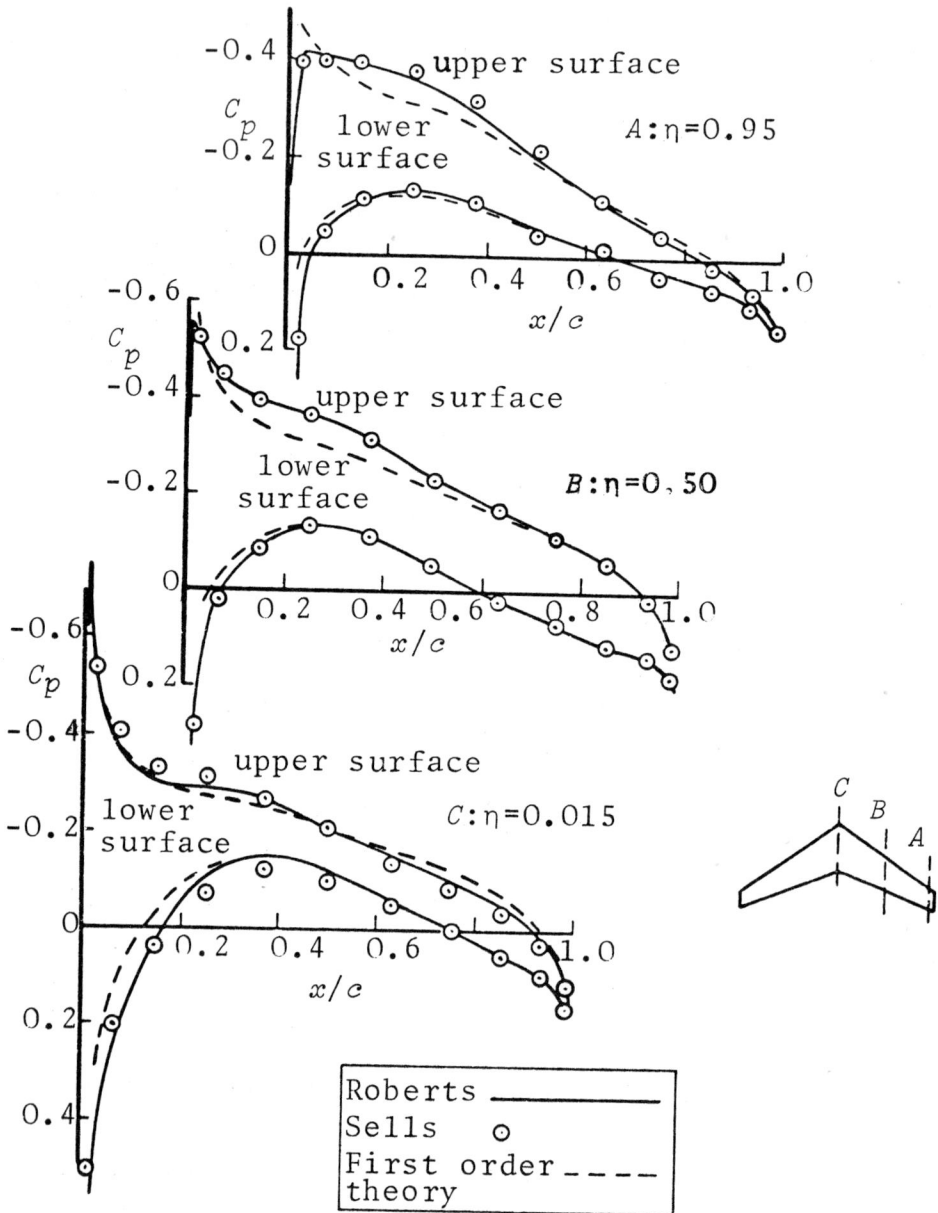

Fig. 5. Pressure distribution on RAE Wing "B": $\alpha=0$, $M=0$

The method has been extended to the general case of cambered wings by Sells [4]. The main difference here is that, to avoid doubling the computing time by making separate calculations of velocity on both upper and lower surfaces, the main calculations are still made on $z = z_t$, and the velocity components on the actual wing surface $z = z_s \pm z_t$, are approximated by means of Taylor series expansions from $z = \pm z_t$. As for many practical wings the camber (z_s) is small compared with the thickness ordinate (z_t), the loss of accuracy caused by this additional approximation is usually small.

An example of the use of this method in its general form is shown in Figs. 3 and 4. These refer to a research wing designed recently at RAE, known as "Wing B." It has a simple thickness distribution - (RAE 101 section, 9% thick), but the camber and twist distributions are complicated, as shown in Fig. 4. Fig. 5 compares the chordwise pressure distributions at the three spanwise positions A, B and C as calculated by Sells' method (circles) with results by the method of Roberts and Rundle [10] (full line). This method is described later but for present purposes can be regarded as "exact." The agreement between these results is good, with a slight deterioration near the wing root, probably caused by the rapid variation of twist in this region. The dashed lines show the results of plain linearised theory and the improvement achieved by the new second order method is obvious. Computing times for Sells' method are relatively low, about one third of those for Roberts' method and perhaps 2 or 3 times those for the linearised theory calculations.

Methods involving surface singularity distributions (Type B)

The pioneer work in this area was done by A.M.O. Smith and J.L. Hess of the Douglas company, and since (in its latest form) it still remains one of the best methods available, I shall describe its basic principles. [Numerous papers on the method have been written: for a comprehensive review of the original (non-lifting) method see [11]; [12] is the latest report on the extended (lifting) method.]

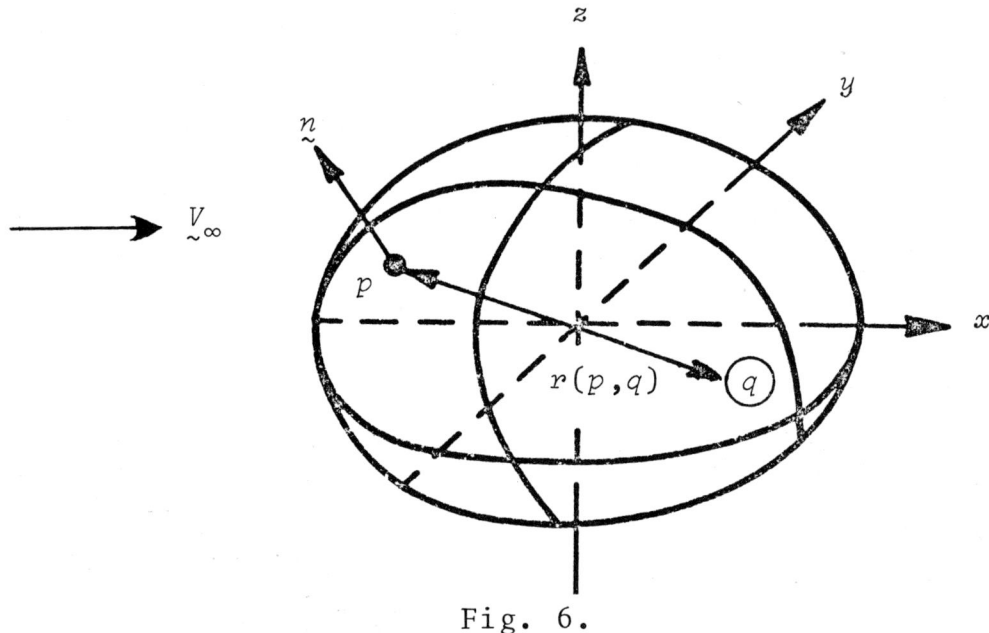

Fig. 6.

We start from the expression for the velocity potential ϕ at a point p arising from a distribution of sources, of strength σ per unit area, over the surface S of the body about which the flow is to be calculated:

$$\phi(p) = -\frac{1}{4\pi} \int\int_S \frac{\sigma(q)\,dS}{r(p,q)} \tag{10}$$

Here q is a variable point on S and $r(p,q)$ is the distance from p to q. Taking the gradient of ϕ, adding the free stream velocity $\underset{\sim}{V}_\infty$ and resolving normal to S, we find that the basic boundary condition - that the component of total velocity normal to S at the point p vanishes - leads to the integral equation

$$\tfrac{1}{2}\sigma(p) - \frac{1}{4\pi} \int\int \frac{\partial}{\partial n}\left(\frac{1}{r(p,q)}\right) \sigma(q)\,dS = -\underset{\sim}{n}(p) \cdot \underset{\sim}{V}_\infty \tag{11}$$

where $\partial/\partial n$ denotes differentiation along the direction of the unit outward normal, $n(p)$, to S at p. To solve this integral equation numerically the shape of the body must first be approximated in some convenient way. Smith and Hess chose the simplest possible way, illustrated in Fig. 7.

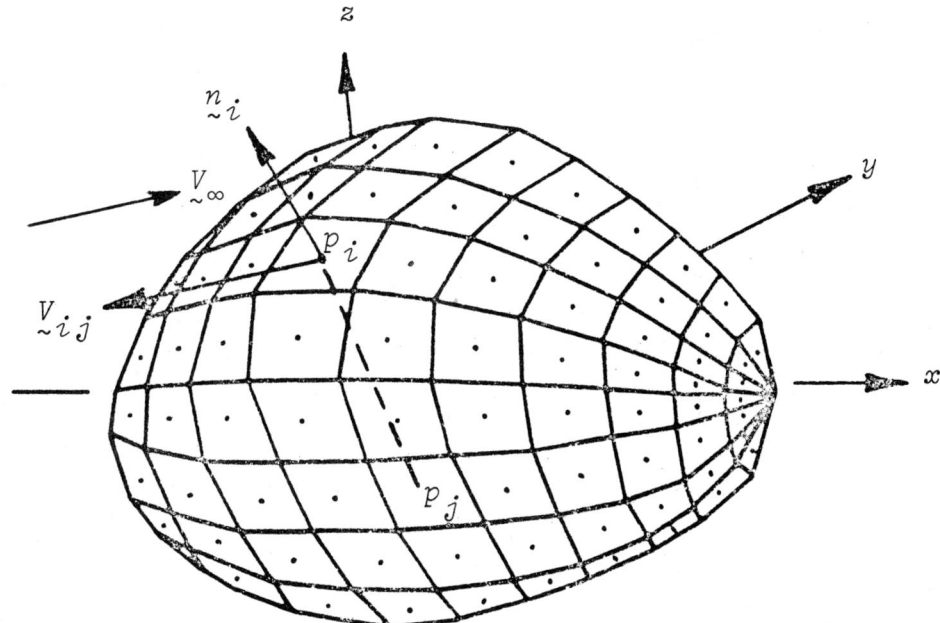

Fig. 7. Approximation of body surface by elements

The surface shape is represented by a number of plane quadrilateral panels, or "facets," whose corners are derived from the set of points used to define the body. These basic elements are numbered in a convenient way, say from 1 to N. It is assumed that on the jth panel a source distribution of constant strength σ_j is placed; the integral equation representing the boundary condition is then satisfied approximately at one control point p on each panel, normally taken to be its centroid. Let $\underset{\sim}{V}_{ij}$ be the velocity vector induced at p_i, the control point on the ith panel, by a unit source distribution on the jth panel. Then the basic integral equation is approximated by the finite sum

$$\tfrac{1}{2}\sigma_i + \sum_{j(\ne i)} \sigma_j\, \underset{\sim}{n}_i \cdot \underset{\sim}{V}_{ij} = -\underset{\sim}{n}_i\, \underset{\sim}{V}_\infty \quad (i=1\ldots N) \quad (12)$$

This may be expressed in matrix form

$$A\sigma = B, \quad (13)$$

where the matrix of *influence coefficients* A is given by

$$A_{ij} = \begin{cases} \underset{\sim}{n}_i \cdot \underset{\sim}{V}_{ij} & (i \ne j) \\ \tfrac{1}{2} & (i = j) \end{cases} \quad (14)$$

the right hand side is the column vector B given by

$$B_i = -\underset{\sim}{n}_i\, \underset{\sim}{V}_\infty \quad (15)$$

and σ is the unknown column vector of source strengths which is to be determined.

Because of the simple way in which the problem is discretised, exact analytic expression for the velocity vectors $\underset{\sim}{V}_{ij}$ and influence coefficients A_{ij} can be determined. However, these expressions are complicated and hence time consuming to compute; and experience has shown that they need only be used at points close to the centroid of each element. For more distant points, a multipole expansion, up to quadripole terms, is used.

The remainder of the problem consists in solving the set of linear equations for the unknown source strengths σ_i, after which the total velocity $\underset{\sim}{V}_i$ at the control point on the ith panel can be calculated from

$$\underset{\sim}{V}_i = \underset{\sim}{V}_\infty + \sum_{j=1}^{N} \sigma_j \underset{\sim}{V}_{ij} \quad (16)$$

The $N \times N$ matrix of influence coefficients A_{ij} has no non-zero elements, but is usually diagonally dominant. For values of N not too large direct Gaussian elimination can be used; Smith and Hess recommend doing this for $N < 500$. For larger values of N standard iterative relaxation methods (for example point Gauss-Seidel) are usually successful for non-lifting problems. Direct matrix inversion has of course the merit that solutions for any number of additional "onset" flows (new right hand sides to the equation) can be obtained with little extra computing time.

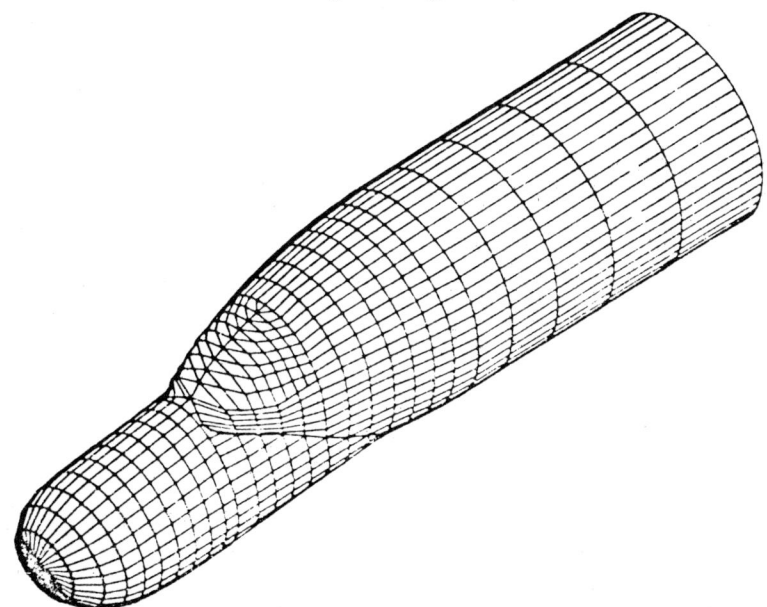

Fig. 8. Elements used to approximate a C-135 fuselage with a large radome

Figs. 8 and 9 illustrate the versatility of the method. Fig. 8 shows the complicated fuselage shape about which the flow is to be calculated, and the distribution of panels used to represent it. Fig. 9 gives a comparison between calculated and measured pressure distributions along various lines on the fuselage. The agreement between theory and experiment is remarkably good, and it would be difficult to conceive a simpler method capable of dealing adequately with such a problem.

The way in which lift can be introduced into the method is best considered through the corresponding, but simpler, two-dimensional problem. First, the matrix A of influence coefficients for the source panels is calculated and inverted, and

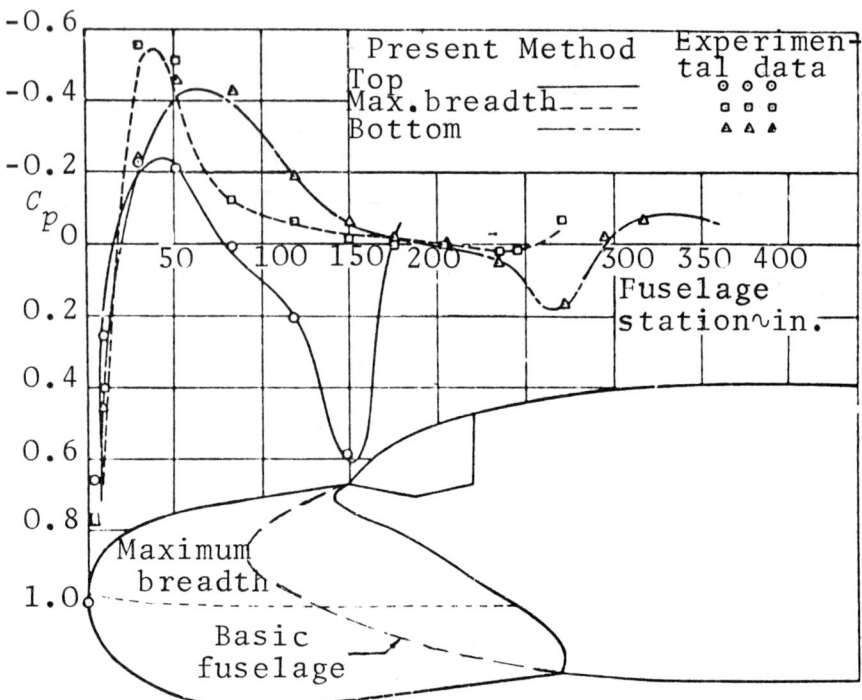

Fig. 9. Comparison of calculated and experimental pressure distributions on a C-135 fuselage with a large radome at zero angle of attack

the column vector of source strengths, $\underset{\sim}{\sigma}_0$ say, is determined from $\sigma_0 = A^{-1} B_0 ,$ (17) with the right hand side B_0 formed with the "onset" flow given by the uniform free stream. This solution does not (in general) satisfy the Kutta condition at the trailing edge. So a second "onset flow" is considered, caused by a distribution of vorticity of unit strength placed on the surface of the aerofoil, to be more precise, placed on the linear panels approximately the shape of the aerofoil*, leading to a second source "vector" $\underset{\sim}{\sigma}_1$, say. The final solution is then given by the linear condition $\underset{\sim}{\sigma} = \underset{\sim}{\sigma}_0 + \lambda \underset{\sim}{\sigma}_1$, the parameter λ being determined so that the Kutta condition is satisfied to some suitable approximation - for

---

*More precisely, placed on the linear panels approximating the shape of the aerofoil.

example by requiring that the velocities at the two control points on the panels immediately adjacent to the trailing edge, on the upper and lower surfaces, shall be equal.

The method can easily be extended to the multi-aerofoil problem by introducing further additional circulatory flows, **one** for each separate aerofoil, and satisfying the Kutta condition at each trailing edge. A typical result is shown in Fig. 10, for an aerofoil with a leading edge slot and slotted flap. Considering that this is a problem for which some representation of the effects of the boundary layers and wakes is almost essential, the agreement with experiment is as good as could be expected.

The extension of the method to deal with three-dimensional lifting problems has been reported recently by Hess [12]. The scheme adopted is a logical extension of the means that I have described of introducing lift into the two-dimensional problem (Fig. 11). The wing surface is subdivided into chordwise strips each approximated in the usual way by quadrilateral (trapezoidal) elements, on each of which is placed a constant source density as before. First, the solution σ_0 is found for the source "vector" with a uniform "onset flow" (the free stream) over the wing and fuselage; this solution will normally violate the Kutta condition along the trailing edge of the wing. Next, L of additional "onset flows" are considered in succession; in each of these a distribution of "bound" vorticity, of constant (unit) strength in the chordwise direction, is introduced round one particular lifting strip of the wing, with zero vorticity on all the others.

Suppose that the source distributions for the onset flows defined by these lifting strips are σ_k ($k = 1,...L$). Then the complete solution for the

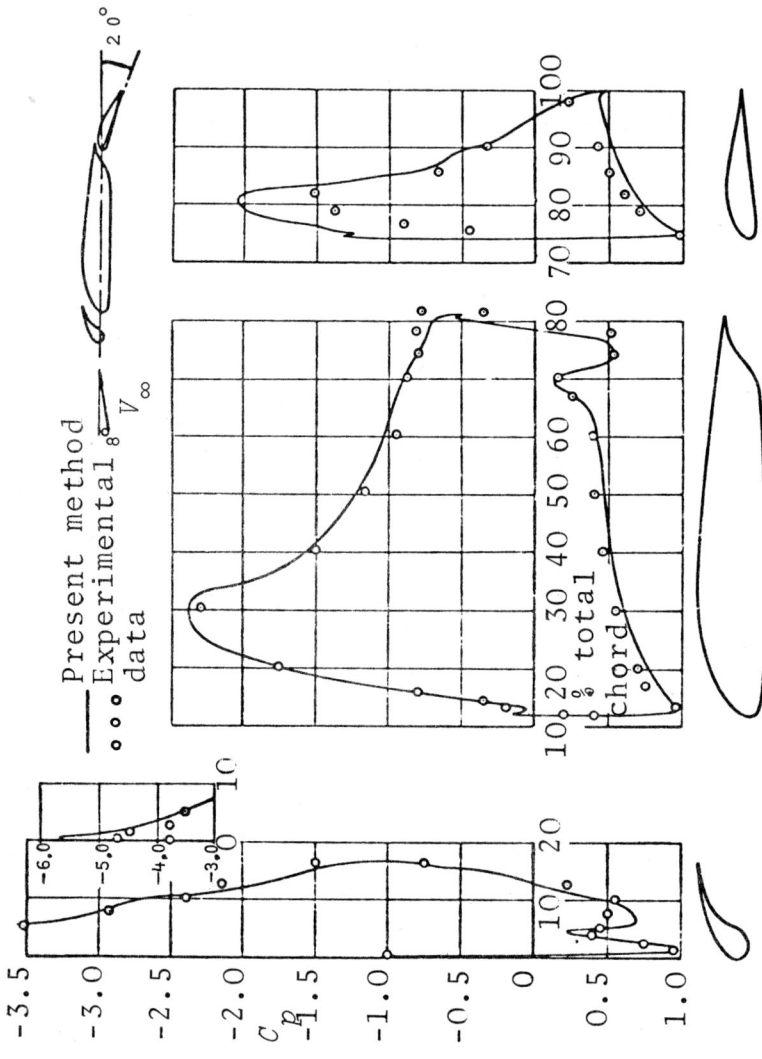

Fig. 10 Comparison of calculated and experimental pressure distributions on an NACA 23012 aerofoil with fixed slot and with flap deflected 20° at 8° angle of attack

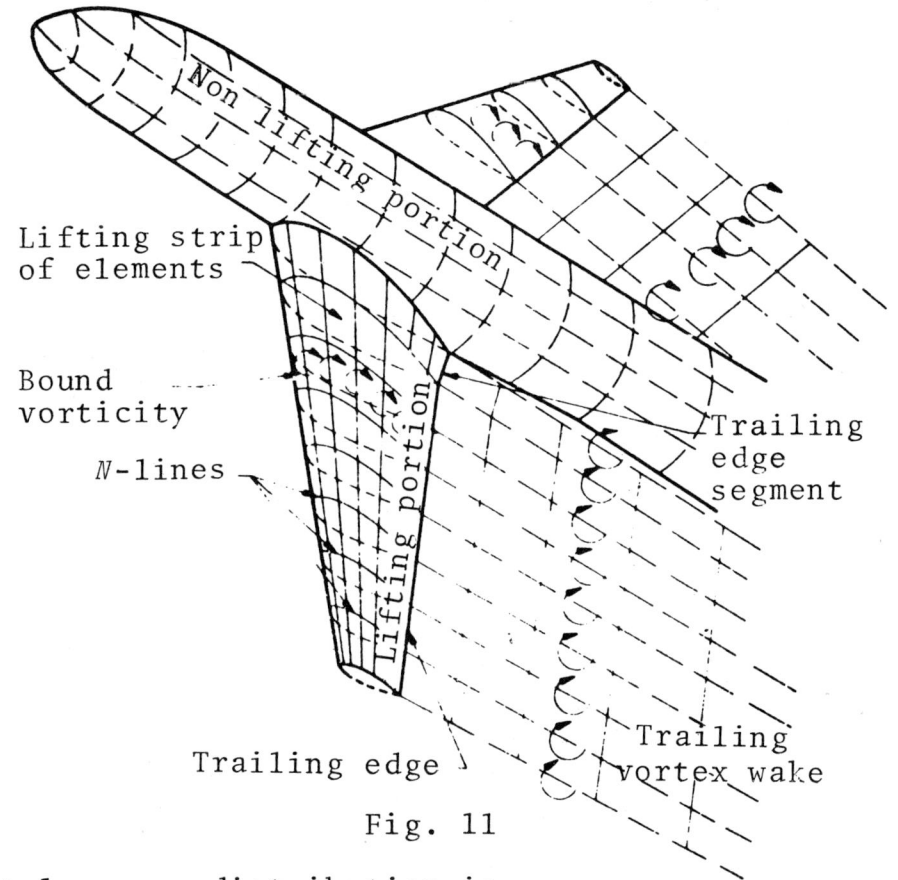

Fig. 11

total source distribution is

$$\underset{\sim}{\sigma} = \underset{\sim}{\sigma}_0 + \sum_{k=1}^{L} \lambda_k \underset{\sim}{\sigma}_k \qquad (18)$$

the parameters λ_j (the strengths of the vortex strips) are determined by satisfying the Kutta condition at L points along (and just upstream of) the trailing edge, one on each strip. The complete flow is that produced by adding the separate flowfields of the free stream, of the individual vortex strips and their associated trailing vortices, and of the source panels on the wing and fuselage.

Some results obtained with this method are shown in Figs. 12, 13 and 14. Fig. 12 refers to

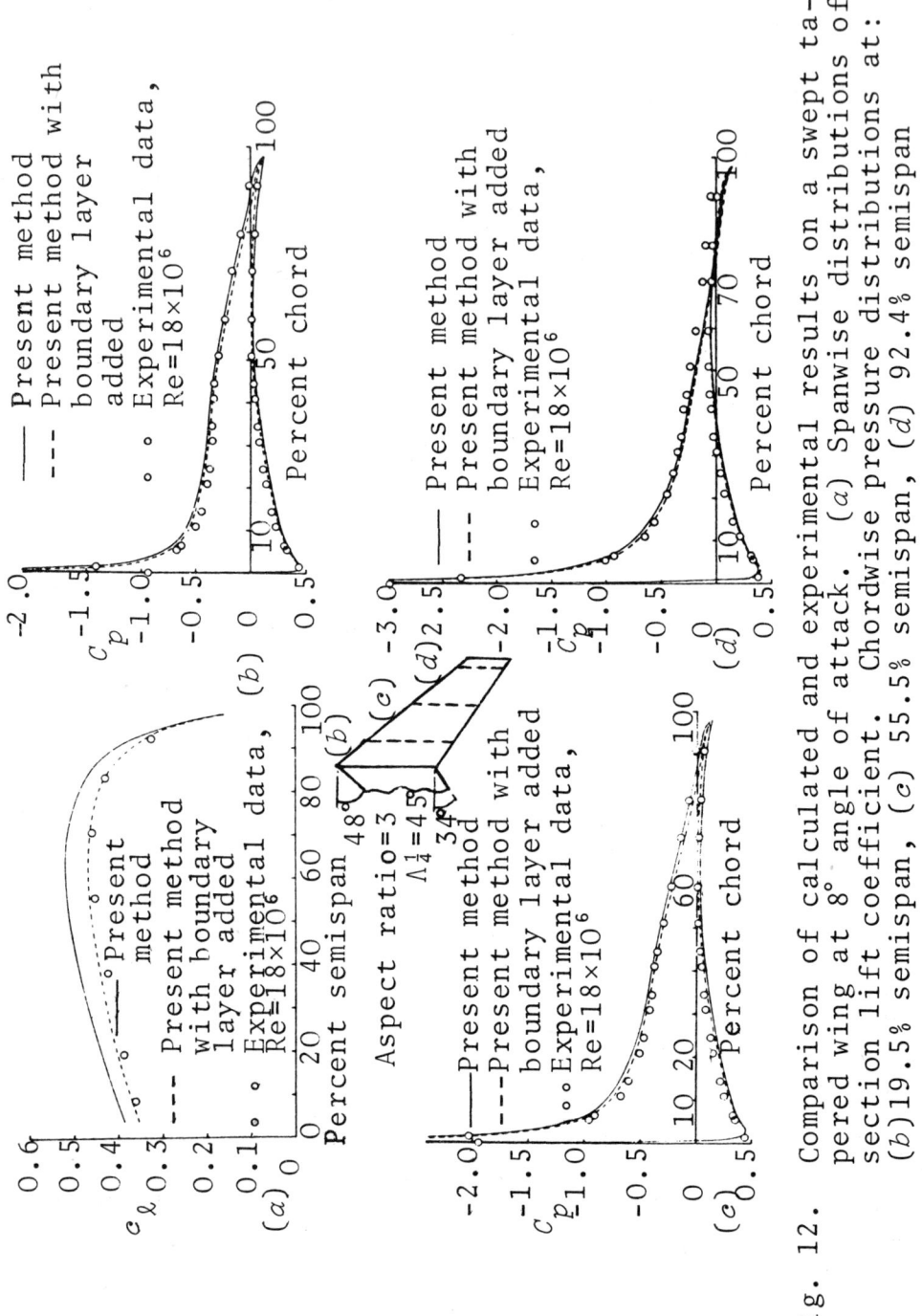

Fig. 12. Comparison of calculated and experimental results on a swept tapered wing at 8° angle of attack. (a) Spanwise distributions of section lift coefficient. Chordwise pressure distributions at: (b) 19.5% semispan, (c) 55.5% semispan, (d) 92.4% semispan

Elliptic Problems

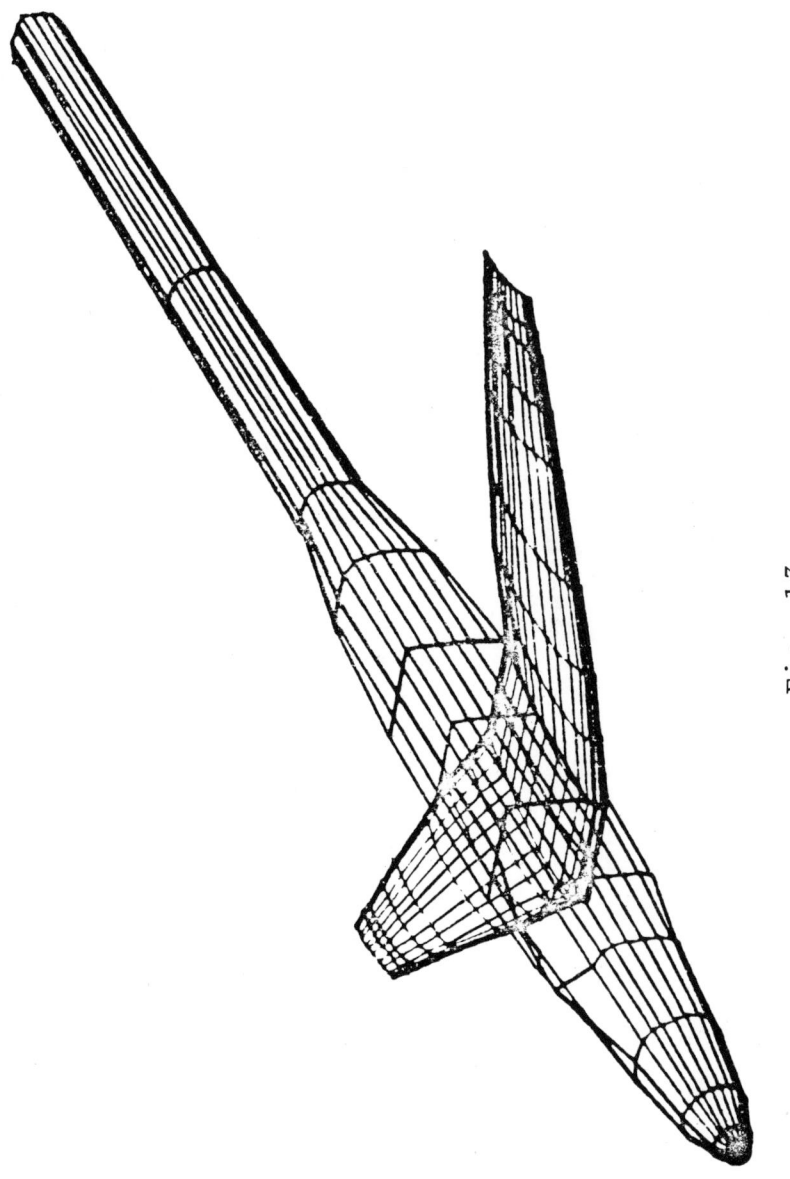

Fig. 13

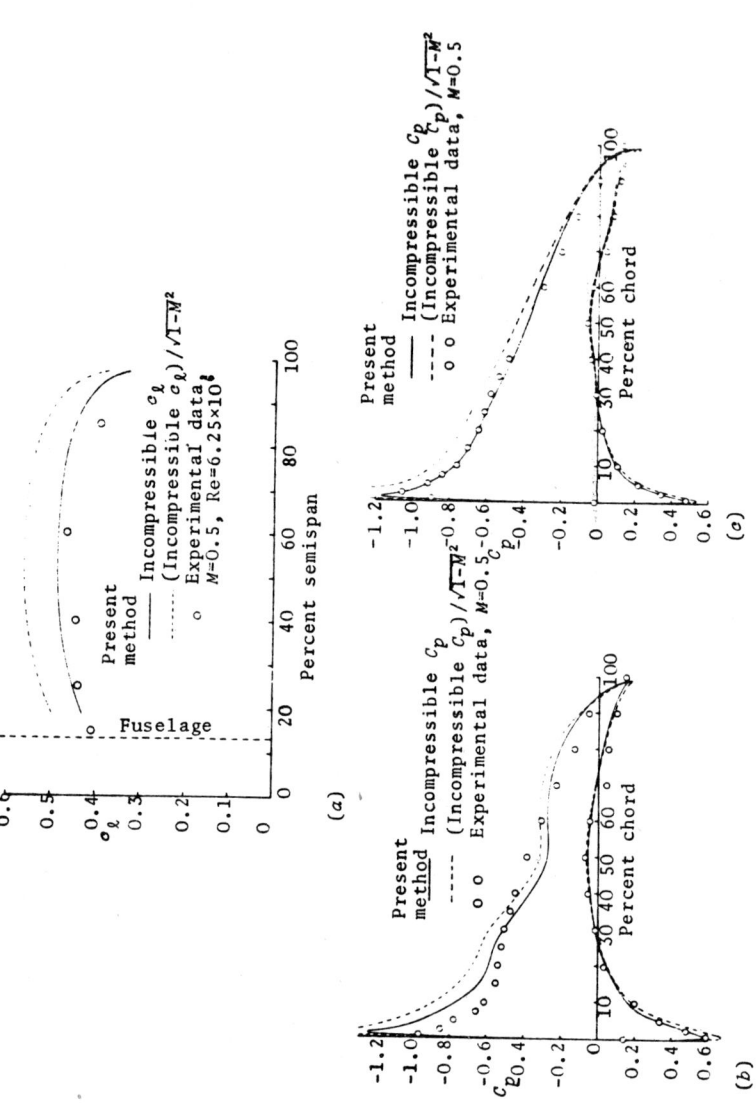

Fig. 14. Comparison of calculated and experimental results on a conventional wing mounted as a low wing on a fuselage at 6.9° angle of attack and a freestream Mach number of 0.5. (a) Spanwise distributions of section lift coefficient. Chordwise pressure distributions at: (b) 15% semispan, (c) 25% semispan

an isolated wing; here a simple correction for the displacement effect of the boundary layer has been applied, and as a result the agreement between theory and experiment is good, both with regard to the spanwise loading (Fig. 12(a)) and the chordwise pressure distribution at various stations along the span (Figs. 12(b), (c) and (d)). Results for a wing-fuselage combination, shown in Fig. 13 are given in Fig. 14. Here a comparison with experiment for the basic (incompressible) solution, given by the full lines, is not meaningful, since effects of compressibility are not negligible (the free stream Mach number is 0.5). The dashed lines, which have been derived by applying a simple (two-dimensional) "Prandtl-Glauert" correction to the solution for incompressible flow, show about the right difference from the measurements, bearing in mind the probable size of viscous effects.

Before the method of Hess (which I have just described) was completed, considerable progress in dealing with the lifting problem had been made elsewhere, initially by Rubbert and Saaris of the Boeing company [13]. Later, a similar technique, but with important improvements with regard to the numerical treatment, was developed by Labrujere and others at NLR, Amsterdam [14]. The way in which lift is introduced into these methods is shown in Fig. 15. Source panels on the surface of the body are used as before, while the circulation is represented by a system of quadrilateral vortex panels placed *inside* the wing, usually on the mean "camber" surface; behind the trailing edge the wake is modelled as a number of discrete vortex lines, one for each chordwise strip on the wing. The relative variation in strength of the internal "bound" vortices in the chordwise direction is prescribed in advance; just one unknown parameter is associated with each strip, corresponding to the total circulation round it, or to the local lift, and these parameters are determined so that the combined flow, produced by the free stream, source panels and vortex system, satisfies a

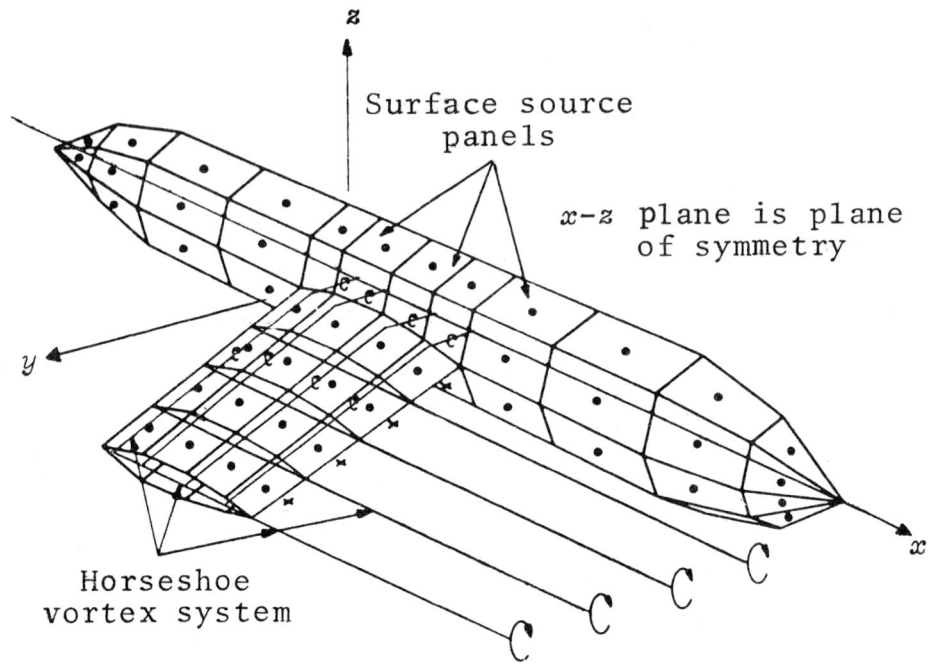

Fig. 15. Schematic view of selected system of singularity distributions. + Wake boundary condition points. • Surface boundary condition points

"Kutta" condition at a number of control points situated just downstream of the trailing edge, to ensure that the flow leaves the trailing edge in a smooth and regular manner. After some numerical experiments, the following form for the Kutta condition was selected: that the direction of the net velocity vector at each control point in the wake [chosen to lie about 0.003.(chord length) downstream of the trailing edge, half-way between adjacent trailing vortices] shall lie in the surface defined by the bisectors of the local trailing edge angles. Although this condition appears to violate the criterion of Mangler and Smith [15] which says that the trailing vortex sheet must emerge, at the trailing edge itself, **parallel** either to the upper or lower surface of the wing (depending on the sign of the trailing vorticity **shed** there)-there are some empirical grounds for supposing it to be justifiable in practice [14].

An important feature of the NLR method is an improved numerical technique for solving the large number of simultaneous equations for the source and vortex strengths. From a study of certain dominant properties of the matrices, an iterative scheme of block relaxation has been devised, which leads to a dramatic reduction in computing time as compared with the direct matrix inversion schemes used in the Boeing method. For example, when the total number of singularities, source and vortex panels, is 1000 - a typical value for a wing-fuselage combination - the computing time is reduced by a factor of about 5.

Figs. 16 and 17 show results for an example of the complicated shape to which this method has been applied.

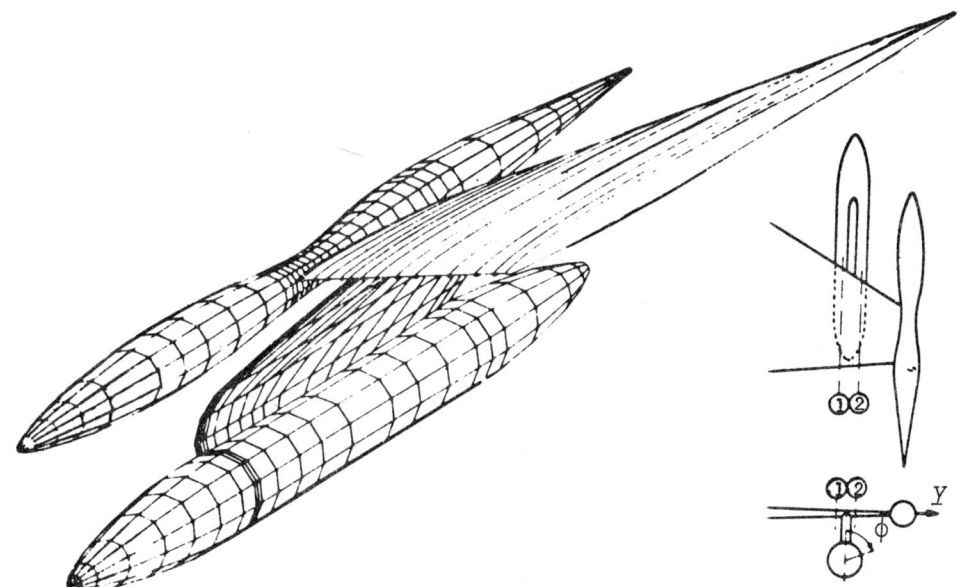

Fig. 16. General arrangement and panel plot of wing-tip tank-pylon-store configuration with sections ① and ② indicated

Fig. 16 shows the shape of the wing, with pylon, store and wing-tip tank added; Fig. 17 shows the effect of adding these objects on the spanwise lift distribution of the wing and on the pressure distribution at two stations on the wing close to its junction with the pylon. It is clear that good general agreement between theory and experiment has been obtained.

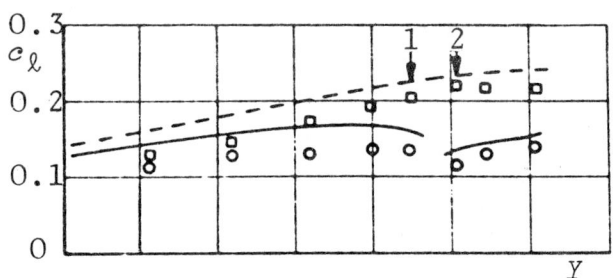

Spanwise lift distribution.

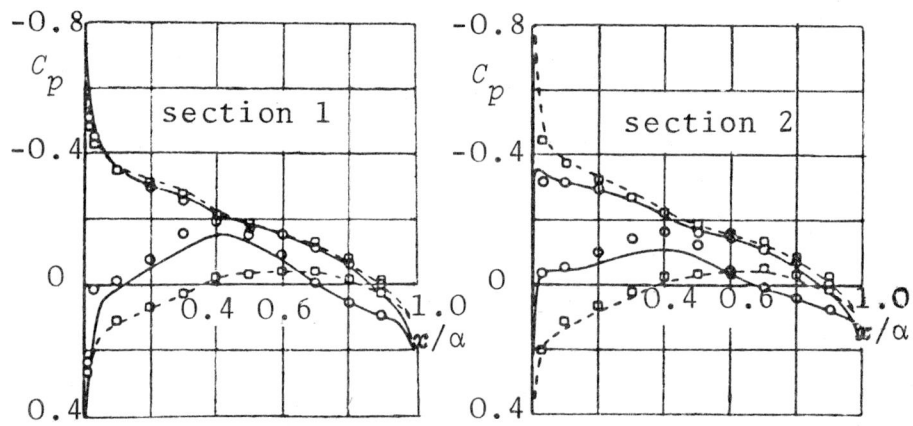

Chordwise pressure distributions.

Fig. 17. Mach number = 0.3 $\alpha = 3°$
——— with pylon/store
----- without pylon/store $\Big\}$ calculation

○   with pylon/store
□   without pylon/store $\Big\}$ experiment
$Re = 6.8 \times 10^6$

Another example is the combination of wing, fuselage, tailplane and vertical fin (Fig. 2). Some comparisons with experiment for this configuration are shown in Fig. 18.

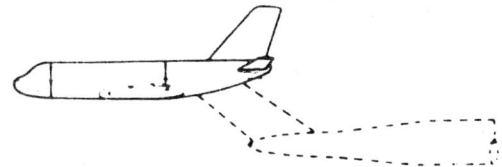

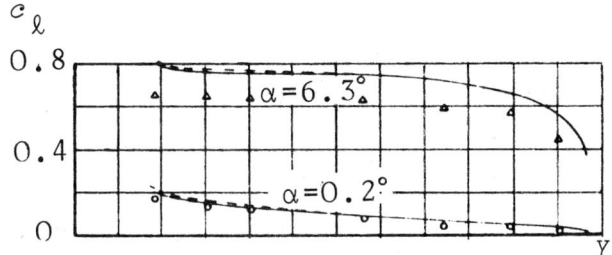

Spanwise lift distributions on wing

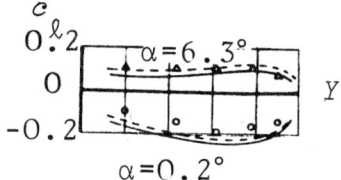

Spanwise lift distributions on stabiliser

Fig. 18. Mach number = 0.4
——— without sting ⎫
- - - - - with sting ⎬ calculation
○ ◻ △ experiment, $Re = 1.6 \times 10^6$

The upper part shows the spanwise variation of local lift coefficient; the discrepancy between theory and experiment at the higher angle of incidence would be expected, because of the neglect of viscous effects in the theory (note the relatively small Reynolds number, 1.6×10^6, of the experiment). The lower part gives corresponding results for the tailplane ("stabiliser"). The lift on the tailplane is a quantity which would be expected to be affected by the precise position of the trailing vortex sheet behind the wing; so an attempt was made to calculate this properly, by the method of [16]. However, in this case it was found that the change produced in the predicted lift on the tailplane was negligible. Thus there remains a discrepancy between theory and experiment which can only partly be accounted for by viscous effects. Some of this discrepancy was traced to the effect of the sting used to support the model, as shown by the dashed lines in Fig. 18.

To conclude this section on "panel" methods, I shall describe a further method, which has some superficial points of resemblance to its predecessors I have just mentioned, but has also some fundamental differences. This is the method developed at the British Aircraft Corporation, Weybridge, by Roberts and Rundle [10],[17]. The common feature of the earlier methods has been the extremely simple, but relatively crude, way in which the body shapes are approximated and the singularity distributions are represented. In Roberts' method the representations are different. First, the body shape is approximated, not by a number of plane "facets" but by a surface interpolation scheme based on bicubic splines. Similarly, the flow perturbations required to satisfy the boundary conditions are produced by *continuous* distributions of sources and doublets, again built up from a number of basic "spline modes," as shown in Fig. 19. The surface is divided into panels by a system of grid lines; the points at which these intersect are chosen as so called "collocation" points, at which the surface boundary condition is to be satisfied. With each point is associated a local source or doublet distribution specified by the basic spline mode whose shape is shown here; this extends over 16 neighbouring panels.

For a non-lifting problem, the procedure is similar to that used in the "facet" methods: the influence coefficients are first calculated, giving the velocity component normal to the sur-

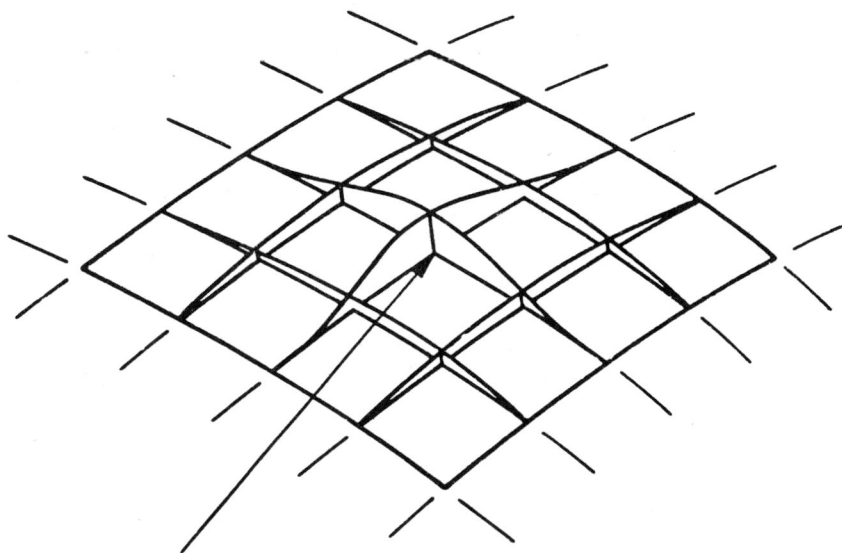

Position of associated collocation point

Fig. 19. A basic spline mode. Each spline mode extends over 16 panels. The associated collocation point, at which the boundary condition is to be satisfied, is at the centre of this set of 16 panels

face induced by each source mode at all collocation points (including its own); next the boundary condition of zero normal velocity for the combined flow (including the free stream) is satisfied at each collocation point; this leads to a set of linear equations for the strengths of the source modes. These equations are solved by direct matrix inversion, and the resultant velocity at any point - on or off the surface - can then be calculated.

Lift is introduced into the problem by doublet sheets, also built up from spline modes (Fig. 20). These cover the wake and extend forward inside the wing along its camber surface. The shape of the wake need not be planar, although with the method in its present form this must be prescribed

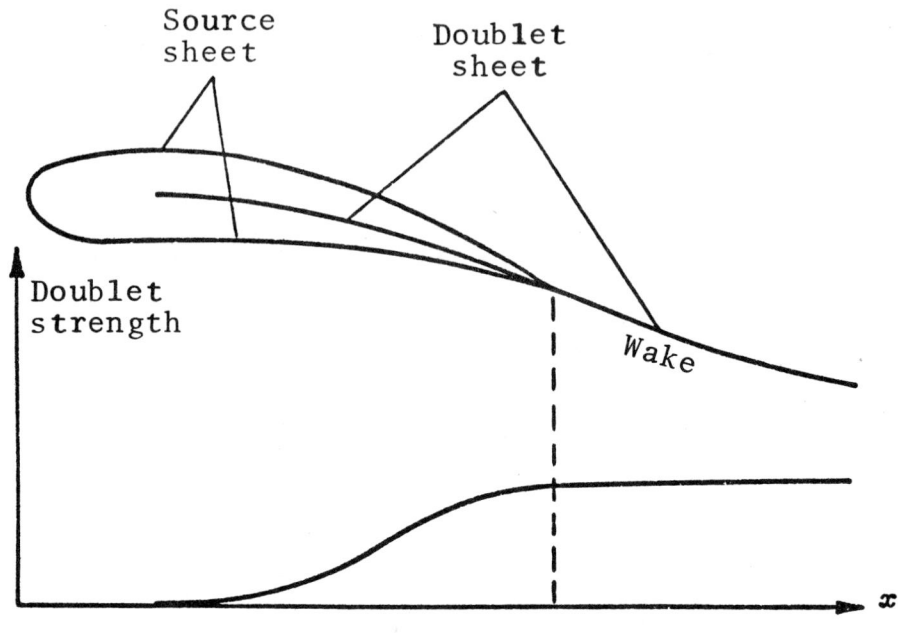

Fig. 20.

in advance. In the wake, the strength of the doublet sheet is constant along streamwise sections, corresponding to the vortex lines on the wake; but the correct boundary conditions on the wake will not normally be satisfied. Ahead of the trailing edge, inside the wing, the streamwise variation of doublet strength is as shown; the *shape* of this curve is prescribed in advance, but there is an arbitrary scaling factor, a function of spanwise position, determined in the course of the calculations. The way in which the Kutta condition is satisfied at the trailing edge is an important feature of the method and is different from the explicit conditions used in the other panel methods. With the particular continuous form chosen here for the variation of doublet strength through the trailing edge, it can be shown that the velocity at the trailing edge is finite if and only if the source strength varies in a particular way as the trailing edge is approached. The source modes adjacent to the trailing edge are thus restricted in accordance with this relation;

and this provides an automatic implicit form of the Kutta condition.

All the features of the method described so far represent clear improvements over the corresponding parts of the alternative "facet" methods. However, there is a serious disadvantage when it comes to calculating the velocity fields and influence functions arising from the individual source or doublet modes. As these modes are defined as bicubic spline functions placed on surfaces whose shapes are approximated in a similar way, there is no possibility of obtaining analytic expressions for the required quantities. Instead, they have to be expressed as surface integrals over the panels, and these are evaluated numerically by double Gaussian quadrature, successively subdividing the panels in both directions until adequate accuracy is obtained. As a result of this complexity, the time taken to evaluate the influence coefficients, and subsequently to calculate the surface velocity or pressure distribution, is greater than that taken by the corresponding calculations in the "facet" methods, and will dominate the over-all computing time - relative to that taken by the matrix inversion - unless the total number of modes is very large, of the order of 1000.

In extending the method to deal with wing-fuselage combinations, several alternative mathematical models were tried [10]. In the most promising all the source modes are placed on the wetted surface, as in the "facet" methods; but in the present method the correct singular behaviour of the source strength at the junction of the wing with the fuselage, according to the theory of Craggs and Mangler [18], is incorporated explicitly (at least approximately), so improved accuracy should be achieved near such a junction. There is still some uncertainty as to the best way of extending the doublet sheet across the fuselage, but fortunately the calculated pressure distributions seem to be insensitive to the particular

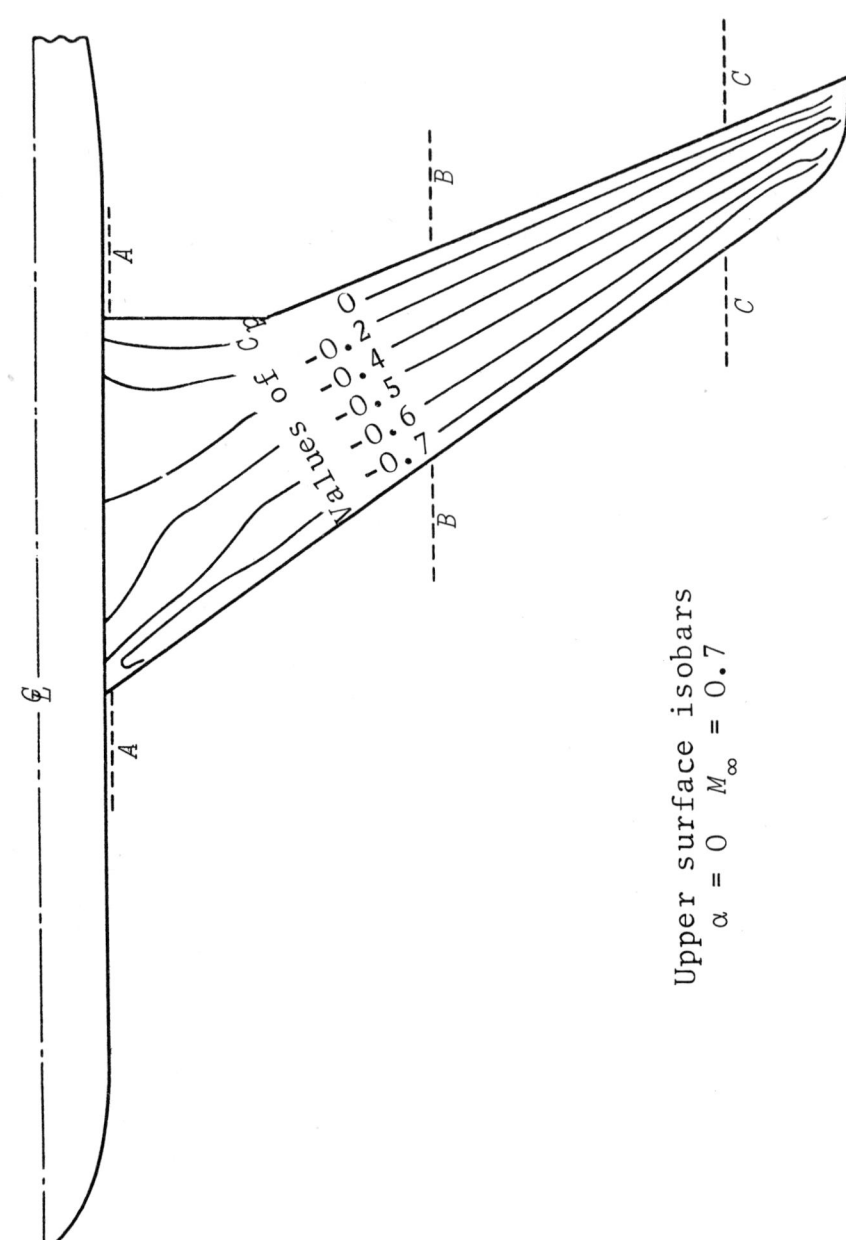

Fig. 21. A practical wing-body combination

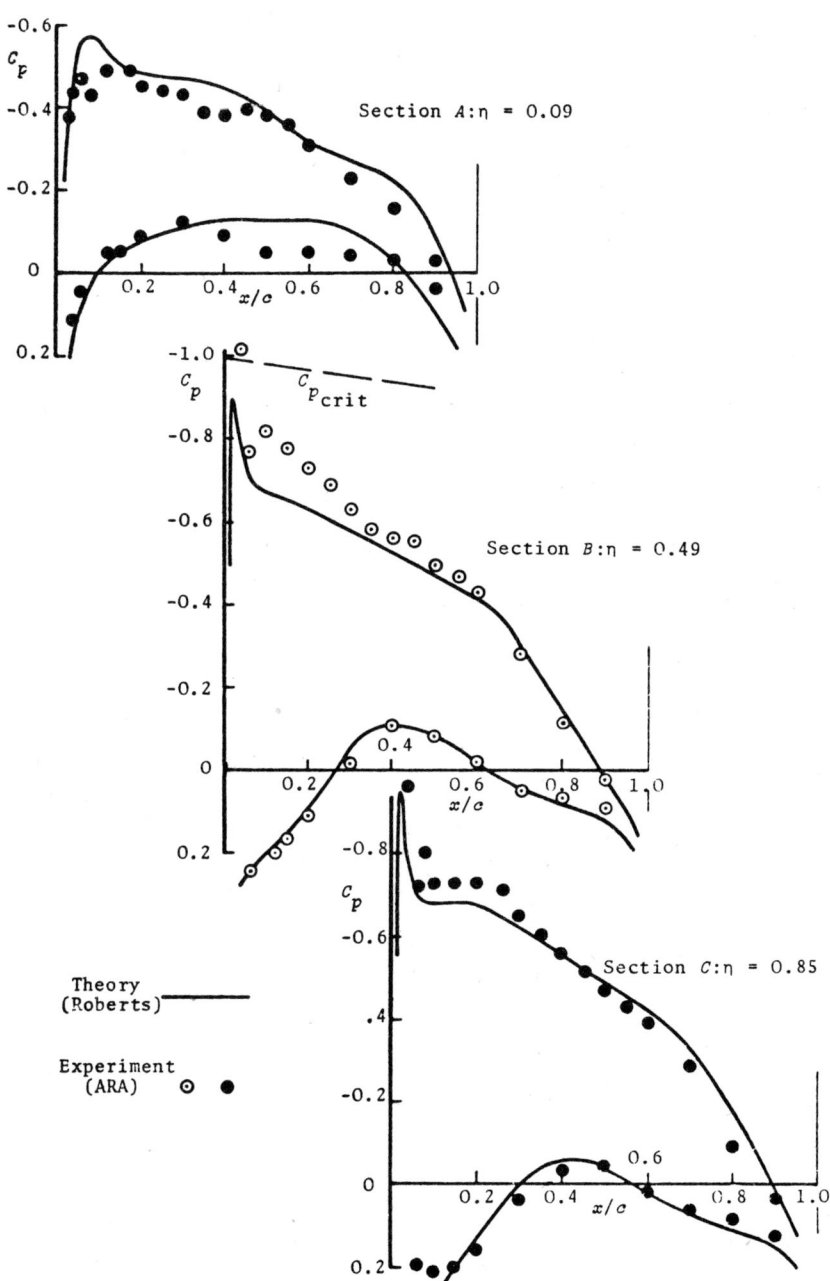

Fig. 22. Results for configuration of Fig. 21

variant used. As an example of the application of this method to a practical aircraft shape, results are shown in Figs. 21 and 22 for a wind tunnel model, tested at the Aircraft Research Association. Fig. 21 shows the shape of the model, together with a calculated isobar pattern on the upper surface of the wing, at a Mach number of 0.7 and zero incidence. Fig. 22 compares calculated and measured chordwise pressure distributions, at the spanwise positions marked on Fig. 21. Such a comparison is confused by the neglect in the theory both of viscous effects and of the nonlinear effects of compressibility; one can only say that a reasonable estimate of the pressure is predicted.

To summarise this section on "panel" methods, I shall consider the relative computing speeds of the methods mentioned. These are compared in Fig. 23. For a given number of panels, the Douglas and Boeing "facet" methods, both of which normally use direct matrix inversion (for lifting problems), are almost identical in speed: the dramatic reduction in computing time achieved by the NLR iterative method is obvious. For a given number of singularity modes, the BAC (Roberts) method is the slowest, particularly when the number is relatively small (less than 1000) so that the calculation of the influence coefficients dominates the computing time. However, this is not a valid comparison as the accuracy with a given number of modes will be much greater than with a "facet" method. It is difficult to make a reliable estimate on this point since not enough systematic investigations of accuracy or convergence have been made with any of the methods; but from a limited amount of evidence it seems probable that, for a given degree of accuracy, a facet method will need about twice the number of panels as the spline method for a two-dimensional problem, and four times the number for a three-dimensional problem. If this is generally true, then the spline method can be considered to be definitely superior to the facet methods involving direct matrix inversion, and possibly competitive with the iterative NLR

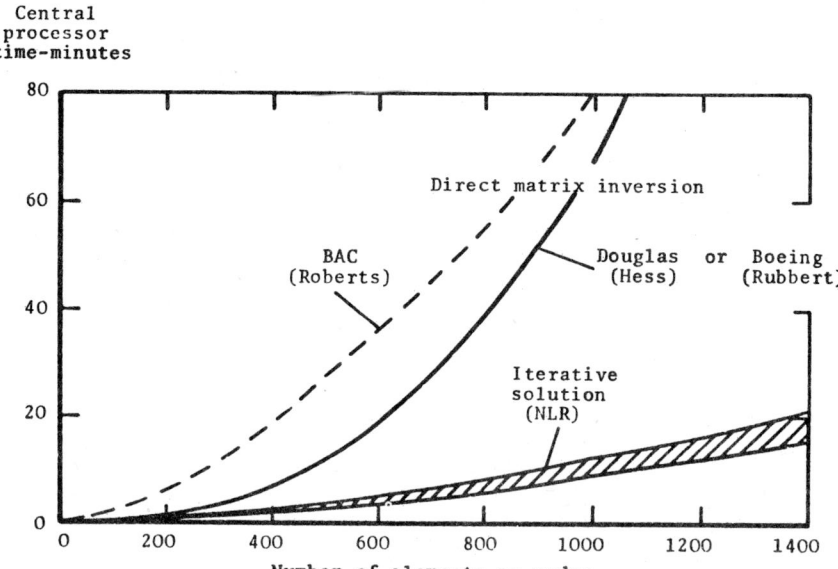

Fig. 23. Equivalent computing times on CDC 6600

method; particularly considering that all methods in which direct matrix inversion is used need little extra computing time to obtain solutions to additional problems involving the same matrix of influence coefficients - such as occur when calculating the effects of a boundary layer or of a small change in the shape of the body.

COMPRESSIBLE POTENTIAL FLOWS

The numerical methods described so far in this paper have been restricted to incompressible or low speed flows, for which the partial differential equation for the velocity potential is linear; although the speed range over which these methods are of practical value can be substantially extended by the use of empirical compressibility corrections, as for example in the NLR method [14]. I feel there is little to be gained, for subcritical flows, in attempting to solve any intermediate approximation to the equations of motion, such as those involving one or more "second order" terms; in this section I consider only methods for solving the full equations of potential flow, with boundary conditions applied properly on the surface of the body. For the three-dimensional problem there are as yet no general practical methods available; but for the simpler two-dimensional problem the position is much more satisfactory. The pioneer method for this problem was that of Sells [19], and since this has certainly not yet been superceded I shall describe it briefly.

The method starts with a conformal mapping (which is normally done numerically) of the region exterior to the aerofoil in the physical (z) plane on to the *interior* of the unit circle in the working (σ) plane (where $z = x + iy$ and $\sigma = re^{i\theta}$); in this way the unbounded region of the physical flow is transformed into a finite closed region suitable for numerical work. A uniform computing grid in (r,θ) is used.

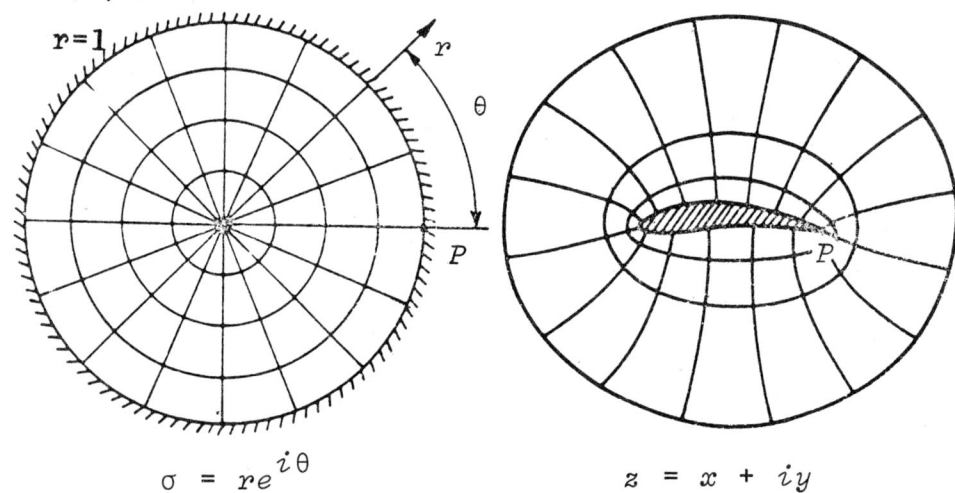

(a) Regular grid in the working (σ) plane

(b) Grid in the physical (z) plane

Fig. 24

Fig. 24 shows (a) this grid in the computing plane and (b) the corresponding contours in the physical plane. Clearly the grid has a relatively fine mesh size near the aerofoil, and particularly near the leading and trailing edges, where the variables are changing most rapidly; this is an important feature of the method.

One independent variable is taken to be the stream function, ψ, so that the equation of continuity is automatically satisfied; the velocity components (u_r, u_θ) in the physical plane, in the directions normal to the curves r = constant, θ = constant, respectively, are given by

$$u_r = \frac{1}{B}\cdot\frac{1}{\rho r}\cdot\frac{\partial \psi}{\partial \theta}, \quad u_\theta = -\frac{1}{B}\cdot\frac{1}{\rho}\frac{\partial \psi}{\partial r} \quad (19)$$

where ρ is the density and $B = \left|\frac{dz}{d\sigma}\right|$ is the modulus of the conformal transformation.

The two equations to be solved numerically, for the unknown ψ and ρ, are: (a) the equation for irrotational flow

$$\frac{\partial}{\partial r}\left(\frac{r}{\rho}\frac{\partial \psi}{\partial r}\right) + \frac{\partial}{\partial \theta}\left(\frac{1}{r\rho}\frac{\partial \psi}{\partial \theta}\right) = 0 \quad (20)$$

which is an elliptic second order partial differential equation for ψ; and (b) Bernoulli's equation, which may be expressed as a relation between the local mass flow rate $f = \rho q$ (where q is the magnitude of the local velocity) and the density ratio $\tau = \rho/\rho_0$ (where ρ_0 is the stagnation density), namely

$$F \equiv \tfrac{1}{2}(\gamma - 1)f^2/(\rho_0^2 a_0^2) = \tau^2(1 - \tau^{\gamma-1}) \quad (21)$$

with f given by

$$f = \frac{1}{B}\left\{\frac{1}{r^2}\left(\frac{\partial \psi}{\partial \theta}\right)^2 + \left(\frac{\partial \psi}{\partial r}\right)^2\right\}^{\frac{1}{2}} \quad (22)$$

The form of equation (21) is shown in Fig. 25.

The boundary condition on the circle in the computing plane is simply $\psi = 0$ on $r = 1$. At the centre of this circle, $r = 0$ (which corresponds to "infinity" in the physical plane) two singularities appear in ψ:

as $r \to 0$, $\psi \sim -\frac{1}{r}\sin(\theta + \alpha) - \frac{\Gamma}{2\pi}\sqrt{1 - M_\infty^2}\ln r \quad (23)$

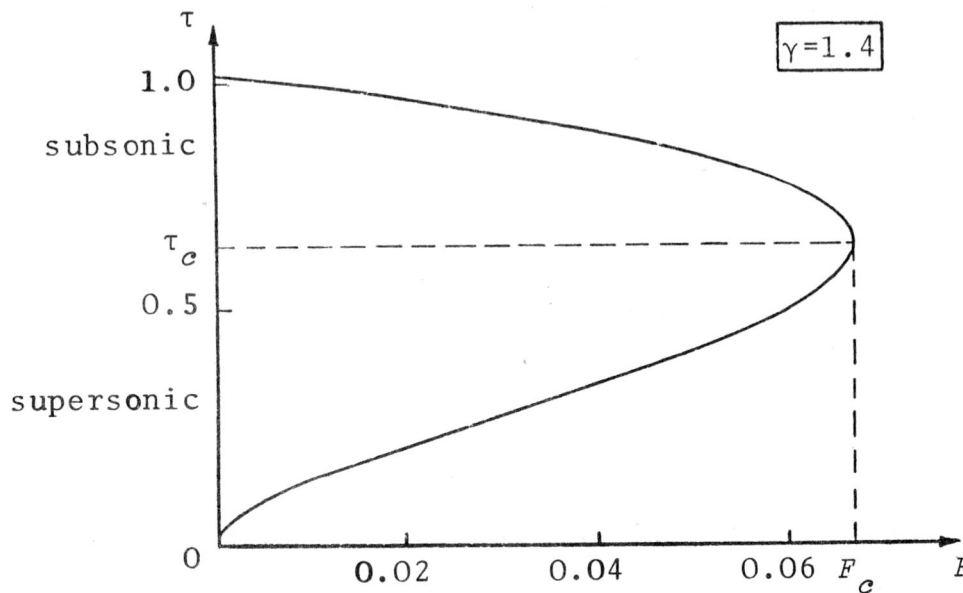

Fig. 25. Density as function of mass flow

The first of these singularities is a dipole of known strength, corresponding to the uniform free stream; while the second is a vortex, of unknown strength, which provides the circulation and hence the lift. The strength Γ of the circulation has to be determined so that the Kutta condition is satisfied at the trailing edge of the aerofoil: in the computing plane this condition implies that $\frac{\partial \psi}{\partial r} = 0$ at the point P, $r = 1$, $\theta = 0$, which corresponds to the trailing edge. These two singularities at the origin ($r = 0$) are subtracted from the physical stream function (ψ), leaving a regular modified stream function ($\bar{\psi}$), with which the subsequent calculations are performed.

In the numerical treatment of the problem, all partial derivatives are replaced by their standard central difference approximations. The method of solution is an iterative one. At the beginning of each basic cycle of the iteration, the partial differential equation for the modified stream function ($\bar{\psi}$) (corresponding to equation (20)) is first solved, by block Gauss-Seidel relaxation. Next, a revised value is obtained for the circulation Γ to satisfy the Kutta condition; and finally the latest values of $\overline{\Psi}$ and Γ are used to calculate new values of the mass flow rate (f) (equation (22)) and then of the density (ρ), by solving equation (21) for τ. This last step is a source of slow convergence when the maximum local velocity is close to the speed of sound, because of the infinite slope of the curve of τ against F at the source point (see Fig. 25). It has been found recently that a simple iterative solution of equation (21), expressed by the relation

$$\tau^{(n+1)} = \left\{ 1 - \frac{F^{(n)}}{\tau^{(n)2}} \right\}^{1/\gamma - 1} \quad (24)$$

as suggested by Albone [20], leads to much more rapid convergence in such cases than with the explicit approximate solution used originally by Sells.

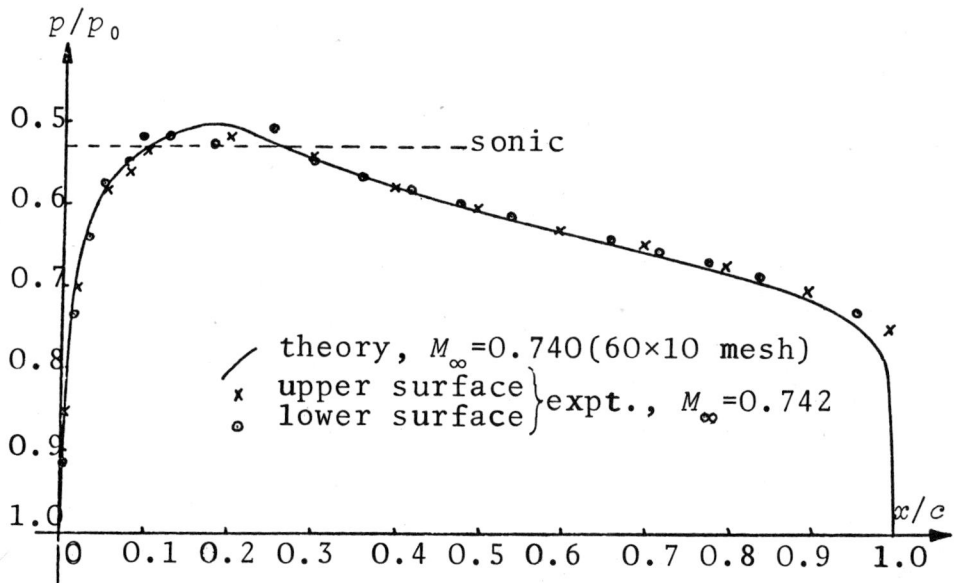

Fig. 26. Comparison of calculated and measured pressure distributions over a NACA 0012 section at zero angle of incidence

In Fig. 26 an example is shown in which this latest method has been used. The aerofoil is a symmetrical NACA 0012 section, 12% thick, and is at zero incidence at a Mach number of 0.74. The comparison with experiment - which in this case would be expected to be little affected by viscosity - is excellent, although the flow is just supercritical.

Building on the foundations which Sells had laid, Garabedian and Korn [21] have recently developed an analagous method, using the same coordinate system, but replacing the stream function by the velocity potential. In this way the difficulties caused by the double valued nature of the density function in Sells' method (see Fig. 25) are avoided; and the method has been successfully extended to deal with supercritical flows - (see pages 242-269).

So far as I am aware, the only analagous work on the corresponding three-dimensional problem has been done by Jameson [22] at the Courant Institute, New York, who has written a computer program for wings of near rectangular planform and constant section shape; and that of Professor Mangler's group at Southampton, who are applying the basic ideas described in a recent paper by Mangler and Murray [23], in which the equations of motion are derived in a general, not necessarily orthogonal, coordinate system.

The first application has been to the problem of flow past an ellipsoid (see [24]), whose equation is

$$\frac{x^2}{a^2} + \frac{y^2}{b^2} + \frac{z^2}{c^2} = 1, \quad \text{with} \quad a > b > c. \quad (25)$$

An ellipsoidal coordinate system (ζ_1, ζ_2, ζ_3) is used, such that the coordinate surfaces, ζ_1, ζ_2, ζ_3 = constant, form families of ellipsoids, hyperboloids of one sheet and hyperboloids of two sheets, respectively, the basic ellipsoid (25) being given by $\zeta_1 = \zeta_{10}$. This coordinate system is of course fully orthogonal. A transformation $\zeta_1 \to \tau_1$ is applied to the first of these coordinates (the one directed outwards from the ellipsoid), so that the infinite domain of the physical space is transformed into a finite domain in the computing (τ_1, ζ_2, ζ_3) space. A uniform computing grid in this coordinate system then has the desirable property that the corresponding grid in physical space is relatively fine in regions where the variables are changing rapidly - just as in Sells' method [19] in two dimensions.

In order to avoid a singularity in the computing space on the plane corresponding to "infinity" in physical space, the independent variable is taken to be the *perturbation* potential (ϕ). By combining the equation of continuity with Bernoulli's equation, a single partial differential equation for ϕ is obtained, which is expressed in finite difference form using standard central difference expressions for all derivatives. To solve the resulting system of finite difference equations a block Gauss-Seidel relaxation scheme was used, solving first in the direction of one particular coordinate by direct inversion of the corresponding tri-diagonal matrix. Rapid convergence was obtained with the aid of overrelaxation.

Results have been obtained for a number of examples with maximum local Mach numbers close to 1. Some of these are shown in Figs. 27 and 28, both for an ellipsoid with axis ratio 1:0.2:0.02. Fig. 27 shows contours of constant local Mach number for the flow when the free stream, of Mach number 0.95, is in the direction of the longest axis; there is nothing remarkable about this, since the ellipsoid in this position is so slender that the maximum local Mach number (0.963) is only 0.013 higher than that of the free stream. A more interesting example is shown in Fig. 28; here the free stream, of Mach number 0.80, is parallel to the medium axis, so the flow is like that over an unswept wing at zero incidence, the aspect ratio being 6.4 and thickness/chord ratio 0.10. The

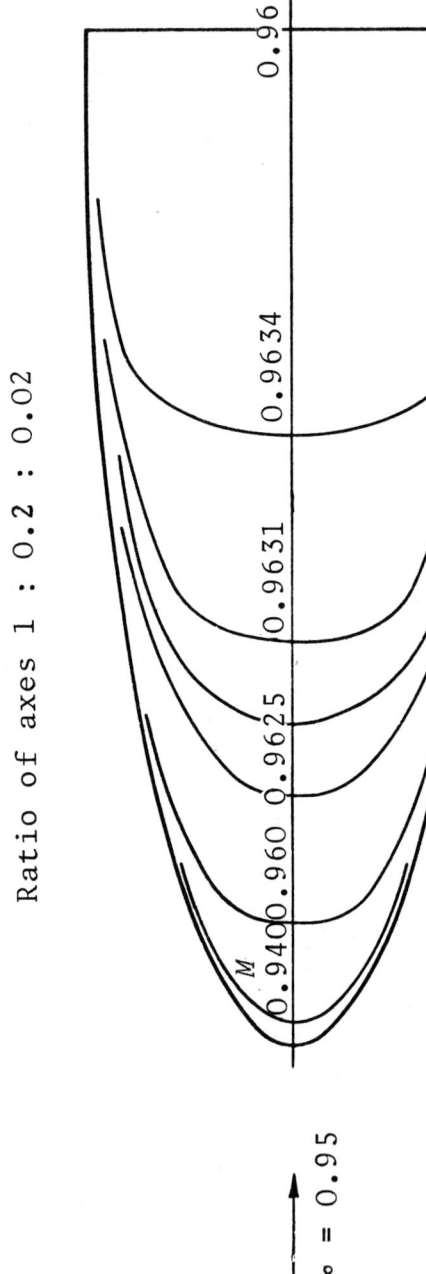

Fig. 27. Lines of constant Mach number

Elliptic Problems

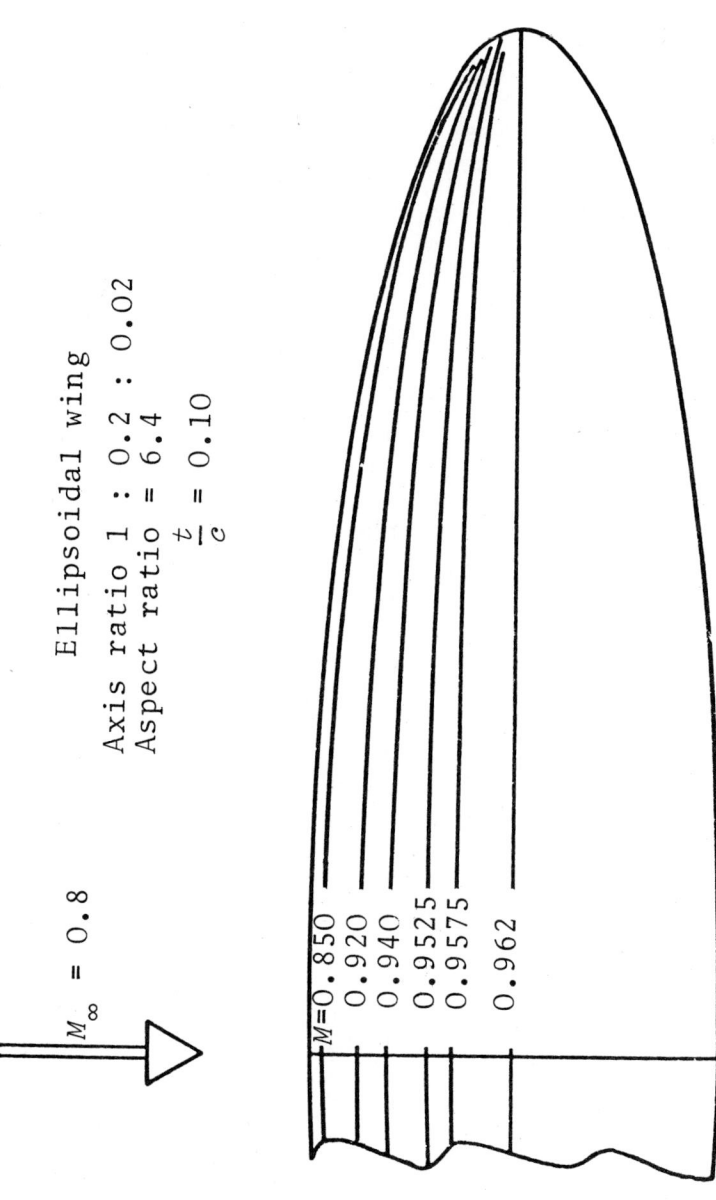

Fig. 28. Contours of constant Mach number

figure again shows contours of local Mach number, the maximum value being 0.962. An interesting property of both solutions is that the maximum local Mach number on any streamwise section (which occurs along the maximum thickness line) appears to be invariant across the "span," at least within the limits of the numerical accuracy of the method just as it is, precisely, for the corresponding incompressible flow.

CONCLUDING REMARKS

In conclusion, I shall outline the current state of the subject and possible future developments. For incompressible or low speed problems, the "panel" methods, in one form or another, seem the most efficient means of calculating the inviscid flow, because of their versatility and the relative ease with which they can be applied to complex configurations. Apart from minor improvements they are probably close to the limit of development; but it remains to exploit them further, particularly with regard to the inclusion of the effects of the boundary layer and wake, and of flow separations and engine flows.

None of these methods, however, is likely to be adequate for the practical problems encountered at high subsonic speeds, where a proper treatment of compressibility is essential. I am sure the future lies here with methods for solving the full compressible flow equations with the proper boundary conditions: and finite difference methods, although not the only possible way of tackling the problem, have the great advantage that they can now - thanks to the developments in numerical techniques Dr. Hall describes (pp. 242 - 269) - be extended fairly readily to deal also with transonic flows.

The two main difficulties are, first, that the equations are nonlinear and, second, the complicated shape of the objects with which we must deal. As a result, there is an inevitable incompatibility between the desire, on the one hand, to use the simplest possible coordinate system, to simplify the equations and their finite difference representation: and on the other hand to devise an efficient coordinate system, in which the solid boundaries form a coordinate surface, to provide the greatest possible accuracy for a given size of computing grid. The finite difference methods I have described certainly come under the second category; but then they are only applicable to relatively simple shapes (in three-dimensional problems). So I am by no means certain where the most efficient methods of solution will be found in the future.

REFERENCES

1. Loeve, W. and Slooff, J.W., "On the use of panel methods for predicting subsonic flow about aerofoils and aircraft configurations," NLR MP 71018U (1971).

2. Weber, J., "Second order small perturbation theory for finite wings in incompressible flow," RAE TR72171 (1972).

3. Sells, C.C.L., "Iterative solution for thick symmetrical wings at incidence in incompressible flow," RAE TR73047 (1973).

4. Sells, C.C.L., "Iterative method for thick cambered wings in subcritical flow," RAE TR74044 (1974).

5. Sells, C.C.L., "Calculation of the induced velocity field on and off the wing plane for a swept wing with a given load distribution," ARC R & M 3725 (1973).

6. Ledger, J.A., "Computation of the velocity field induced by a planar source distribution approximating a symmetrical non-lifting wing in subsonic flow," ARC R & M 3751 (1972).

7. Lighthill, M.J., "A new approach to thin aerofoil theory," *Aero. Q.*, **3**, 193 (1951).

8. "Method for predicting the pressure distribution on swept wings with subsonic attached flow," ESDU Transonic Data Memo. 6312 (Revised version 73012) (1973).

9. Hedman, S.G., "Vortex lattice method for calculation of quasi-steady state loadings on thin elastic wings in subsonic flow," FFA Rep. 105 (1966).

10. Roberts, A. and Rundle, K., "The computation of first order compressible flow about wing-body combinations," BAC (Weybridge) Rep. Aero. MA 20 (1973).

11. Hess, J.L. and Smith, A.M.O., "Calculation of potential flow about arbitrary bodies," *in* "Progress in Aero Sciences," 8, Pergamon (1966).

12. Hess, J.L., "Calculation of potential flow about arbitrary three-dimensional lifting bodies: Final Technical Report," McDonnell Douglas Co. Rep. No. MDC J5679-01 (1972).

13. Rubbert, P.E. and Saaris, G.R., "Review and evaluation of a three-dimensional lifting potential flow analysis method for arbitrary configurations," AIAA Paper No. 72-188 (1972).

14. Labrujere, T.E., Loeve, W. and Slooff, J.W., "An approximate method for calculation of the pressure distribution on wing-body combinations at subcritical speeds," AGARD CP-71-71, Paper 11 (1970).

15. Mangler, K.W. and Smith, J.H.B., "Behaviour of a vortex sheet at the trailing edge of a lifting wing," *Aero. J. Roy. Aero. Soc.* (November 1970).

16. Labrujere, T.E., "A numerical method for the determination of the vortex sheet location behind a wing in incompressible flow," NLR TR 72091U (1972).

17. Roberts, A. and Rundle, K., "Computation of incompressible flow about bodies and thick wings using the spline mode system," BAC

(Weybridge) Rep. Aero. MA19 (1972).

18. Craggs, J.W., Mangler, K.W. and Zamir, M., "Some remarks on the behaviour of surface source distributions near the edge of a body", *Aero. Q.*, **24**, Part 1 (February 1973).

19. Sells, C.C.L., "Plane subcritical flow past a lifting aerofoil," *Proc. Roy. Soc. A.*, **308**, 377 (1968).

20. Albone, C.M., "Plane subcritical flow past a lifting aerofoil: an improvement on Sells' method of calculation," RAE Tech. Rep. 71230 (1971).

21. Garabedian, P.R. and Korn, D.G., "Analysis of transonic aerofoils," *Com. Pure Appl. Math.*, **24**, 841 (1971).

22. Jameson, A., "Iterative solution of transonic flows over airfoils and wings including flows at Mach 1," *Com. Pure Appl. Math.*, *in the press*.

23. Mangler, K.W. and Murray, J.C., "Systems of coordinates suitable for the numerical calculation of three-dimensional flow fields," RAE Tech. Rep. 73074 (1973).

24. Duck, P.W., "The numerical calculation of subcritical steady potential flow round an unguarded ellipsoid," RAE Tech. Rep., *to be issued*.

COMPRESSIBLE SUBCRITICAL FLOW THROUGH AXIALLY SYMMETRIC SHARP-LIPPED ORIFICES AND NOZZLES

G.M. Alder

(University of Edinburgh)

INTRODUCTION

This paper is concerned with the irrotational, axially symmetric flow of an ideal gas through a conical duct. The gas accelerates from negligible velocity through the duct and discharges into a still atmosphere to form a jet bounded by a constant velocity streamline. A special case of this problem is the flow through a circular orifice in a plane wall. Attention is restricted to the range of subcritical flows defined as those in which the gas velocity never exceeds the local speed of sound. The range is characterised by the velocity on the jet boundary and at one end we have incompressible flow and at the other "critical" flow, when the gas on the jet surface just reaches sonic velocity.

Solutions to the corresponding two-dimensional problem (*i.e.*, flow between planes converging in the stream direction) have been well established for many years. Methods of conformal mapping have been used for incompressible flows [1], [2] and the hodograph transformation has been used for compressible flows [3], [4]. By contrast, the axially symmetric problem has received little attention and there appear to be no published solutions for compressible flows. Several workers have examined the incompressible flow through a circular orifice. The nozzle contraction coefficient (C_c) is a convenient parameter to use in a discussion of their results. We may define C_c as the ratio A_J/A_E, where A_E is the nozzle throat area and A_J is the jet cross-section area a long way downstream where all streamlines approach the asymptotic condition of uniform parallel flow. Trefftz [5] studied the problem using integral equations and obtained the result $C_c = 0.61$. Southwell and Vaisy [6] and Rouse and Abul-Fetouh [7] used relaxation methods and obtained the values 0.608 and 0.611, respectively. Garabedian [8], by considering related mathematical problems with simpler solutions, was able to determine by interpolation the value $C_c = 0.579$. This result was also obtained by Jeppson [9] with a finite difference method based on stream function and velocity potential as independent variables. Hunt [10] used integral equations derived from surface vortex distributions and obtained the result $C_c = 0.578$. The disagreement between these results was cleared up by Bloch [11] who used a conformal mapping technique and obtained the result $0.59131 < C_c < 0.59139$.

This paper describes a finite difference method for the solution of axially symmetric subcritical flows through convergent conical nozzles. Transformation to hodograph coordinates maps the whole flowfield into a rectangle and allows a rectangular finite difference mesh to be used. The error in the solution may be estimated by considering different mesh node densities. Typical results for both axially symmetric and two-dimensional flows are given.

FORMULATION OF PROBLEM

An axial cross-section of the physical flowfield is shown in Fig. 1. The flow is determined by the following three parameters.

(a) β, the nozzle wall angle. The case $\beta = 90°$ gives the flow through a circular orifice.

(b) V_J, the velocity on the jet boundary LC. Incompressible flow is given when $V_J \to 0$.

Critical flow is given when V_J is equal to the local sonic velocity and it is convenient to non-dimensionalise velocities so that $V_J = 1$ at the critical flow condition.

(c) γ, the ratio of the specific heats of the ideal gas.

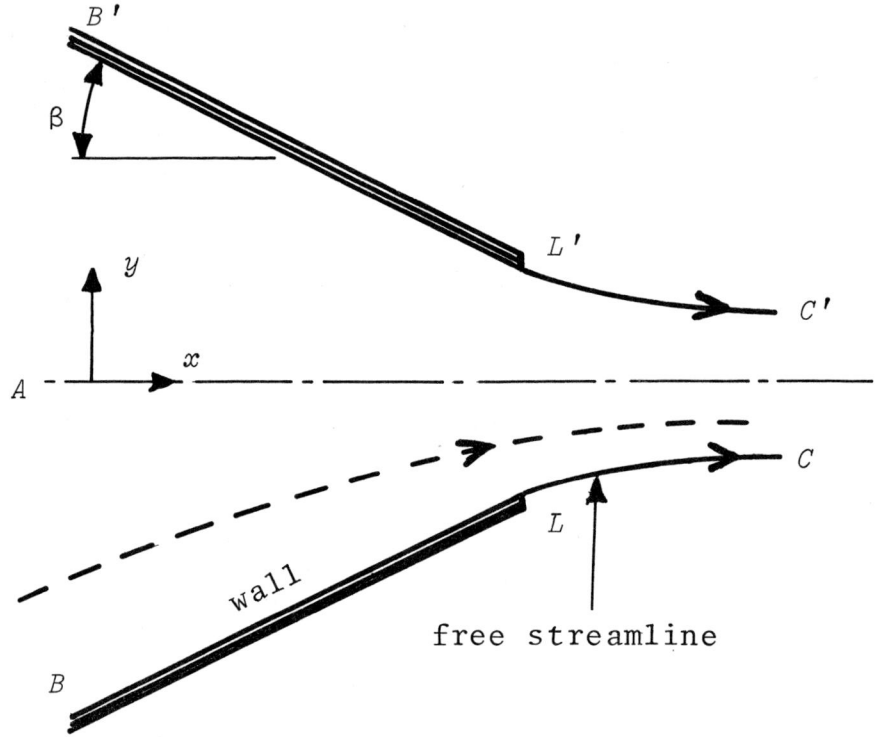

Fig. 1. The physical flowfield (broken line shows typical streamline)

The continuity equation for the flow is:

$$\frac{\partial}{\partial x}(\rho u y^k) + \frac{\partial}{\partial y}(\rho v y^k) = 0 \qquad (1)$$

where x and y are the physical coordinates in the axial and radial directions, respectively, u and v are the non-dimensional velocity components in the x and y directions, and ρ is the non-dimensional density scaled by the stagnation density. The parameter k takes the values $k = 1$ for axially symmetric flow and $k = 0$ for plane, two-dimensional flow.

The condition for irrotational flow is:

$$\frac{\partial u}{\partial y} - \frac{\partial v}{\partial x} = 0 \qquad (2)$$

Since the flow is isentropic, the density may be related to the velocity by means of the usual ideal gas relationships as follows:

$$\rho = [1 - \frac{\gamma-1}{\gamma+1}(u^2 + v^2)]^{1/\gamma-1} \qquad (3)$$

Equations (1), (2) and (3) are sufficient to determine the flow through the nozzle shown in Fig. 1 subject to the following boundary conditions:

(a) $-v/u = \tan \beta$ along the nozzle wall BL
(b) $v = 0$ along the axis
(c) $u^2 + v^2 = V_J^2$ along the free streamline LC
(d) $u \to 0$ and $v \to 0$ when $x \to -\infty$
(e) $v \to 0$ and $u \to V_J$ when $x \to +\infty$

In the case of two-dimensional, critical flow it has been shown [12] that uniform flow in the jet (boundary condition (e)) is achieved within a finite distance of the nozzle exit plane LL'. This is probably also true of axially symmetric critical flow.

It is convenient to transform the problem into the hodograph coordinates, V and θ, defined as follows:

$$V^2 = u^2 + v^2$$
$$\theta = \tan^{-1}(v/u) \qquad (4)$$

This has the effect of removing some of the difficulties associated with the above boundary conditions, where the location of the free streamline in the physical coordinate system is not known initially, and the upstream and downstream boundaries lie at infinity. When transformed, the lower half of the physical flowfield shown in Fig. 1 maps into the rectangle shown in Fig. 2.

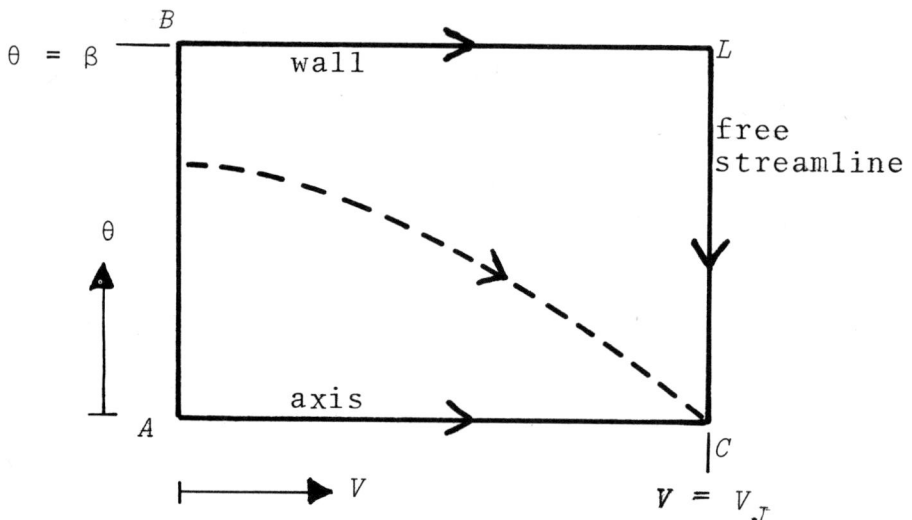

Fig. 2. The hodograph flowfield (broken line shows typical streamline)

It is convenient here to use rectangular coordinates to represent the hodograph plane, rather than the more usual polar coordinates. The derivation of the hodograph equations given below follows the conventional lines for two-dimensional flow (e.g., [13]) except that here the additional terms associated with axially symmetric flow are included.

The existence of the continuity equation (1) allows us to define the stream function, ψ, by the following:

$$u = \frac{1}{\rho y^k} \cdot \frac{\partial \psi}{\partial y}$$
$$v = -\frac{1}{\rho y^k} \frac{\partial \psi}{\partial x}$$
(5)

We define the velocity potential, ϕ, in the usual way:

$$u = \frac{\partial \phi}{\partial x}$$
$$v = \frac{\partial \phi}{\partial y} \qquad (6)$$

Thus, combining (4), (5) and (6) we write:

$$u = V\cos\theta = \phi_x = \alpha \cdot \psi_y$$
$$v = V\sin\theta = \phi_y = -\alpha \cdot \psi_x \qquad (7)$$

where the subscripts on ϕ and ψ denote partial differentiation and $\alpha = 1/\rho y^k$. We now have three alternative sets of coordinates — (x,y), (V,θ) and (ψ,ϕ) — and the following expressions can be derived from (7) by the usual rules of partial differentiation.

$$x_V = \frac{\cos\theta}{V} \phi_V - \alpha \frac{\sin\theta}{V} \psi_V \qquad (8a)$$

$$x_\theta = \frac{\cos\theta}{V} \phi_\theta - \alpha \frac{\sin\theta}{V} \psi_\theta \qquad (8b)$$

$$y_V = \frac{\sin\theta}{V} \phi_V + \alpha \frac{\cos\theta}{V} \psi_V \qquad (8c)$$

$$y_\theta = \frac{\sin\theta}{V} \phi_\theta + \alpha \frac{\cos\theta}{V} \psi_\theta \qquad (8d)$$

Since the order of partial differentiation is immaterial, x and y can be eliminated from equations (8) to give:

$$\phi_\theta = V\alpha\psi_V \qquad (9a)$$

$$\phi_V = \alpha_V \cdot \psi_\theta - \frac{\alpha}{V} \psi_\theta - \alpha_\theta \psi_V \qquad (9b)$$

Elimination of ϕ from (9) by further differentiation leads to:

$$V\psi_{VV} + \left[1 + \frac{V\alpha_V}{\alpha} + \frac{\alpha_{\theta\theta}}{\alpha}\right]\psi_V + \frac{\alpha_\theta}{\alpha}\psi_{V\theta}$$

$$+ \left[\frac{1}{V}\frac{\alpha_\theta}{\alpha} - \frac{\alpha_{V\theta}}{\alpha}\right]\psi_\theta + \left[\frac{1}{V} - \frac{\alpha_V}{\alpha}\right]\psi_{\theta\theta} = 0 \quad (10)$$

Since $\alpha = 1/\rho y^k$, we may rewrite (10) as follows:

$$V\psi_{VV} + \left[(1 - \frac{V\rho_V}{\rho}) - k(V\frac{y_V}{y} + \frac{y_{\theta\theta}}{y} - 2\frac{y_\theta^2}{y^2})\right]\psi_V$$

$$-k\left(\frac{y_\theta}{y}\right)\psi_{V\theta} + k\left[\frac{y_{V\theta}}{y} - \frac{2y_V y_\theta}{y^2} - \frac{1}{V}(1 + \frac{V\rho_V}{\rho})\frac{y_\theta}{y}\right]\psi_\theta$$

$$+ \left[\frac{1}{V}(1 + \frac{V\rho_V}{\rho}) + k\frac{y_V}{y}\right]\psi_{\theta\theta} = 0 \quad (11)$$

Expressions in terms of ψ, θ and V for the various derivatives of y contained in (11) may be obtained from (8c), (8d), (9a) and (9b). After substitution of these into (11) and some rearrangement we finally obtain

$$\left[V^2\psi_{VV} + V\psi_V(1 - \frac{V\rho_V}{\rho}) + \psi_{\theta\theta}(1 + \frac{V\rho_V}{\rho})\right]$$

$$+ \frac{2k}{\rho y^2}\left\{\sin\theta\left[V\psi_{VV}\psi_\theta + (1 - \frac{V}{\rho}\rho_V)\psi_V\psi_\theta - V\psi_V\psi_{V\theta}\right]\right.$$

$$\left. - \cos\theta\left[\frac{1}{V}(1 + \frac{V}{\rho}\rho_V)\psi_\theta^2 + V\psi_V^2\right]\right\}$$

$$+ \frac{k}{\rho^2 y^4}\left\{\sin^2\theta\left[\frac{1}{V}(3 + \frac{V\rho_V}{\rho})\psi_\theta^2\psi_V + 2V\psi_V^3 - 2\psi_{V\theta}\psi_\theta\psi_V\right.\right.$$

$$\left.\left. + \psi_V^2\psi_{\theta\theta} + \psi_\theta^2\psi_{VV}\right]\right.$$

$$\left. + \sin\theta\cos\theta\left[\frac{1}{V^2}(1 + \frac{V}{\rho}\rho_V)\psi_\theta^3 + \psi_V^2\psi_\theta\right]\right\} = 0 \quad (12)$$

We require the solution of (12) over the rectangular region shown in Fig. 2. The boundary values of the dependent variable may be assigned as follows.

(a) $\psi = 0$ on the axial streamline AC

(b) $\psi = \psi_{max}$ on the outer streamline BLC

(c) $\begin{cases} \psi/\psi_{max} = \theta/\beta \text{ (two dimensional)} \\ \psi/\psi_{max} = \dfrac{1-\cos\theta}{1-\cos\beta} \text{ (axially symmetric)} \end{cases}$ on $V = 0$

It is convenient to choose $\psi_{max} = 1$. The expressions in boundary condition (c) are obtained by examining the form of (12) in the limit $V \to 0$.

Equation (12) is linear for two-dimensional flow (when $k = 0$), and its solution may be expressed analytically as a series of hypergeometric functions [3]. For axially symmetric problems, when the nonlinear terms involving the physical coordinate y are present, the equation must be solved numerically. The incompressible solution is given when $\rho = 1$ and $\rho_V = 0$. For compressible flow we have, from (3):

$$\rho_V/\rho = -2V/[(\gamma+1) - (\gamma-1)V^2] \qquad (13)$$

The solution of (12) is obtained by an iterative process and, at any stage, the values of y required when $k = 1$ are given by integration of the current approximation. It is most convenient to integrate along lines of constant V. Combining (8d) and (9a) and noting that $\alpha = 1/\rho y$ we have:

$$y_\theta = \frac{1}{V\rho y}(V\sin\theta\psi_V + \cos\theta\psi_\theta) \qquad (14)$$

Equation (14) may be integrated along lines of constant V to give:

$$y^2 = \frac{1}{V\rho}\left[\int_0^\theta (\psi + V\psi_V)\sin\theta\, d\theta + \psi\cos\theta\right]_{V=\text{const}} \qquad (15)$$

where $y = 0$ and $\theta = 0$ on the axis.

NUMERICAL SOLUTION

A finite difference method was adopted for the solution of (12). The rectangular hodograph region was divided into N strips in each direction forming a finite difference mesh containing $(N-1)^2$ interior nodes. The mesh nodes were concentrated near the line $\theta = 0$ by taking equal intervals in the $V - \sigma$ plane, where σ is an auxiliary independent variable replacing θ and defined as follows:

$$\theta = \beta \cdot \sigma^t \qquad (16)$$

The range $0 \leq \sigma \leq 1$ covers the range $0 \leq \theta \leq \beta$, and the degree to which mesh points are concentrated near the axis is controlled by the value of the exponent t. The derivatives ψ_θ, $\psi_{\theta\theta}$ and $\psi_{V\theta}$ in (12) were replaced by the following equivalent expressions in terms of σ:

$$\psi_\theta = \frac{\sigma^{1-t}}{\beta t} \cdot \psi_\sigma$$

$$\psi_{\theta\theta} = \left(\frac{\sigma^{1-t}}{\beta t}\right)^2 \cdot \psi_{\sigma\sigma} + \frac{1-t}{t^2 \beta^2} \cdot \sigma^{1-2t} \psi_\sigma \qquad (17)$$

$$\psi_{V\theta} = \frac{\sigma^{1-t}}{\beta t} \psi_{V\sigma}$$

The conventional three-point central-difference approximations for the derivatives of ψ were substituted into (12) leading to a set of $(N-1)^2$ nonlinear algebraic equations to be solved for the values of ψ at the interior mesh nodes. Each of these finite difference equations was of the form:

$$f_i = F_i(\underline{\psi}) = 0 \; ; \; i = 1, (N-1)^2 \qquad (18)$$

where $\underline{\psi}$ is the vector containing the nodal values of ψ. The iterative solution started from an initial estimate of $\underline{\psi}$, and the values f_i were finite in general but approached zero as the iteration progressed.

The method of "Non-Linear Over-Relaxation" (NLOR), described by Ames [14], was used to solve the finite difference equations. In this method the nodes were scanned sequentially and at each step the value of ψ at the node in question was updated as follows:

$$(\psi_i)^{new} = \psi_i - \omega f_i \left(\frac{\partial f_i}{\partial \psi_i}\right)^{-1} \qquad (19)$$

where ω is a constant acceleration factor in the range $1 \leq \omega \leq 2$. The derivative $\partial f_i/\partial \psi_i$ was estimated numerically by perturbing ψ_i. The NLOR process was continued until successive corrections were acceptably small.

The solution of the linear two-dimensional problem ($k = 0$) was obtained first, and this was used to start the full, nonlinear solution. When $k = 0$ the NLOR method reduced to the well known linear "Successive Over-Relaxation" (SOR) method [15]. The SOR method converged in this case from an arbitrary choice of ψ to the two-dimensional solution. No convergence problems were encountered in the NLOR process, if the values of ψ were restrained from becoming zero or negative. The nodes were scanned along lines of constant V, starting at the axis, which allowed the integration for y by equation (15) to be carried out. The value of the acceleration factor ω had a considerable effect on the computing time necessary for a given NLOR solution. In the absence of a better criterion, the optimum acceleration factor for the initial SOR solution was used, and appeared always to give satisfactory results. The optimum SOR factor was estimated by the method of Carré [16].

Truncation errors in the solution are likely to be most serious in the region of the singularity at the corner ($V_J, 0$) in the hodograph plane. For this reason, the integration of the hodograph solution using equations (8) to obtain the physical coordinates of the nozzle wall and free streamline followed the path shown in Fig. 3 which avoids the singularity.

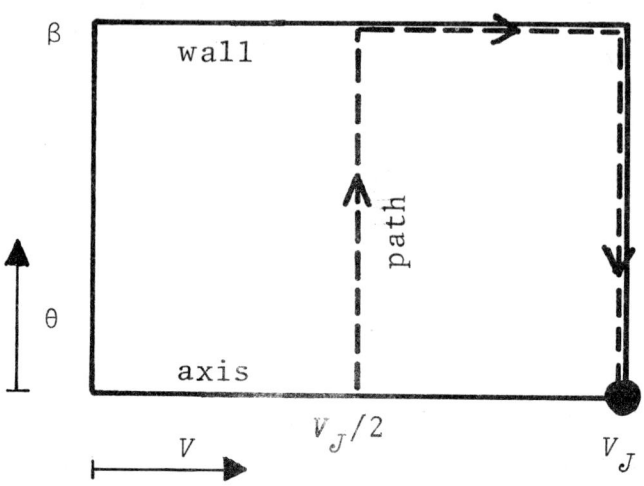

Fig. 3. The hodograph plane, showing integration path which avoids singularity at $(V_J, 0)$

The magnitude of the truncation errors in the results can be estimated by plotting calculated values against $1/N$ and extrapolating to the limit $1/N \to 0$. The calculated contraction coefficients for critical flow for the case $\beta = 90°$ are shown in Fig. 4. Results for both two-dimensional and axially symmetric calculations are given. In each case the truncation error is proportional to $1/N^2$, which is to be expected when three-point central-difference approximations are used for the derivatives of ψ in (12). The marked reduction in truncation error as a result of a suitable choice of exponent t in (16) is shown in Fig. 4. For axially symmetric critical flow the optimum value (found by numerical experiment) is $t = 2.0$, and accurate results are given by a relatively coarse finite difference mesh. Similar improvements in truncation error are shown also for other calculated features of the physical flowfield. For axially symmetric incompressible flow the optimum value of t was found to be $t = 1.5$, and a linear relationship between t and V_J was assumed over the subcri-

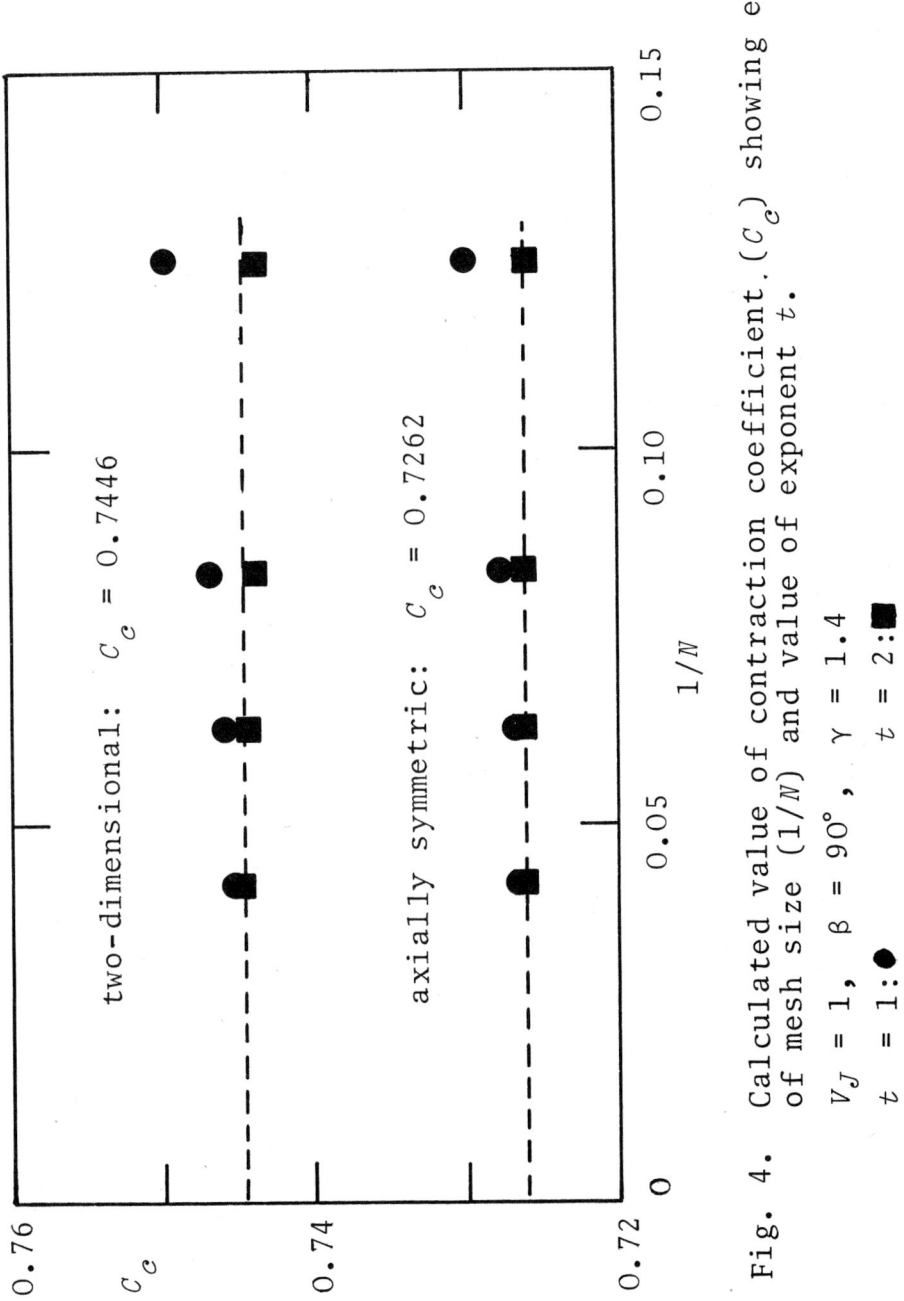

Fig. 4. Calculated value of contraction coefficient.(C_c) showing effect of mesh size (1/N) and value of exponent t.
$V_J = 1$, $\beta = 90°$, $\gamma = 1.4$
$t = 1$: ● $t = 2$: ■

tical range when evaluating the results presented in the following section.

RESULTS

Computed results for both axially symmetric and two-dimensional nozzles are shown in Table I and in Figs. 5 and 6.

Table I

Values of Contraction Coefficient (C_c)

β	incompressible		critical (γ=1.4)	
	2D	axi.	2D	axi.
90°	0.611	0.591	0.745	0.726
60°	0.692	0.668	0.814	0.794
30°	0.814	0.793	0.906	0.892

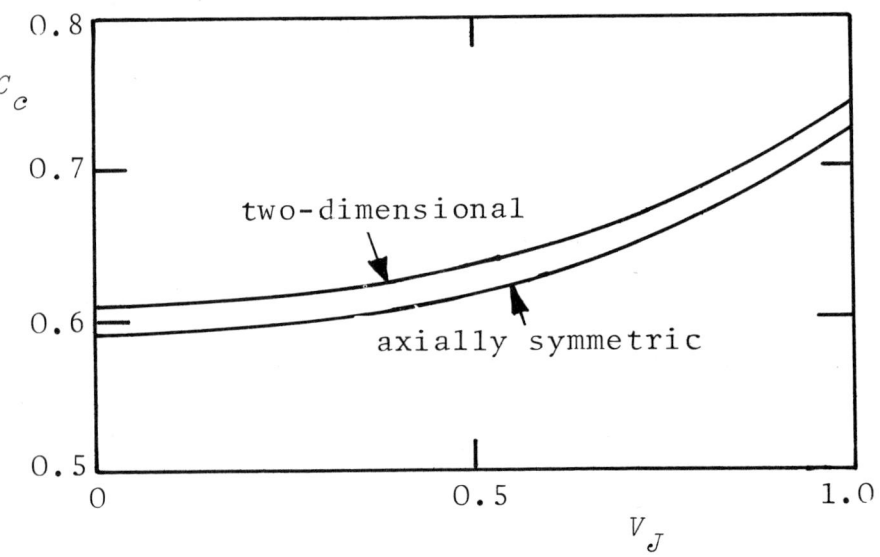

Fig. 5. Subcritical contraction coefficient as function of jet velocity (β=90°, γ=1.4)

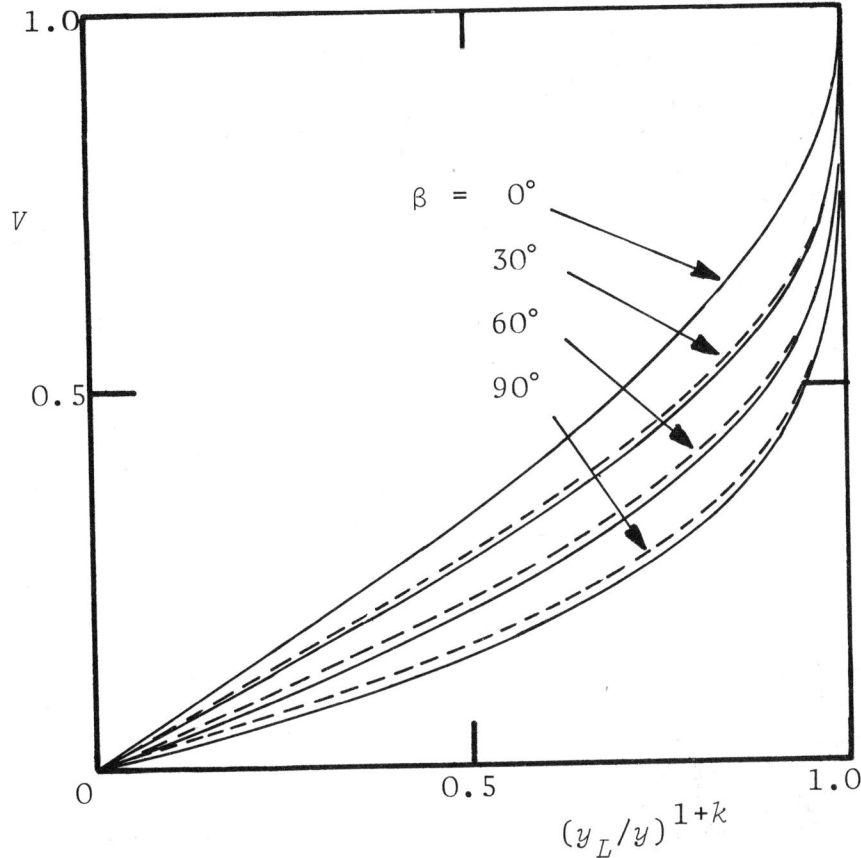

Fig. 6. Velocity on nozzle wall (V) as function of the ratio (lip radius/wall radius) for critical flow ($\gamma = 1.4$)

——————— axially symmetric ($k = 1$)
- - - - - - - two-dimensional ($k = 0$)

Contraction coefficients for incompressible and critical flow over a range of nozzle angles are given in Table I. The two-dimensional values are in agreement with the analytical results of Lord Rayleigh [1] for incompressible flow and of Chaplygin [3] for critical flow. The axially symmetric, incompressible value of C_c agrees with the result of Bloch [11]. The present method of solution is superior to Bloch's; first because it also covers compressible flow and second because it requires considerably less computation time. To achieve 3 figure accuracy, Bloch used a mesh containing 6370 points and required 4300 iterations. The method described in this paper could achieve the same accuracy using 256 mesh points and requiring 40 iterations.

Computed contraction coefficients for $\beta = 90°$ are shown in Fig. 5 as a function of jet velocity. The results show that the difference between the values of C_c for two-dimensional and axially symmetric flow is virtually constant over the whole subcritical range. The distribution of velocity in critical flow along the nozzle wall is shown for a range of values of β in Fig. 6. Here, the velocity on the wall is plotted against $(y_L/y)^{1+k}$, where y_L is the lip radius. When presented in this way, there is little difference between the results for two-dimensional ($k = 0$) and axially symmetric ($k = 1$) flow. Also shown in Fig. 6 are the corresponding graphs for $\beta = 0$ (*i.e.*, one-dimensional flow).

CONCLUSIONS

The numerical solution of two-dimensional and axially symmetric subcritical convergent nozzle flows has been described in this paper. A finite difference method is used to solve the problem in the hodograph plane. It has been shown that even relatively coarse finite difference meshes lead to small truncation errors in the results. There appear to be no other published solutions to the compressible axially symmetric problem and in the incompressible case the present method shows advantages over that of Bloch [11] in terms of computing efficiency.

I have used the hodograph approach to axially symmetric flows as the subsonic part of a solution of the supercritical nozzle problem, which involves matching subsonic and supersonic calculations along the line $V = 1$. The basic computational method can also be applied to other axially symmetric problems, such as the design of wind tunnel contractions and of subsonic intake cowls.

REFERENCES

1. Lord Rayleigh, "Notes on Hydrodynamics," *Phil. Mag.*, **2**, (5th Series) 441 (1876).

2. von Mises, R., *"Berechnung von Ausfluss - und Uberfal-Zahlen,"* ZVDI, **61**, 447-498 (1917).

3. Chaplygin, S., "Gas Jets," Scientific Memoirs, Moscow Univ., (1902). [*Trans: NACA TM* 1063 (1944)]

4. Jacob, C., *"Sur un Jet Gazeux,"* Comptes Rendus, **203**, 423 (1936).

5. Trefftz, E., *"Über die Kontraktion Kreisformiger Flussigkeitsstrahlen,"* Z. Math. Phys., **64**, 34 (1917).

6. Southwell, R. and Vaisy, G., "Relaxation Methods Applied to Engineering Problems. XII. Fluid motions characterised by free stream lines," *Phil. Trans. R. Soc. Lond. A*, **240**, 117 (1948).

7. Rouse, H. and Abul-Fetouh, A., "Characteristics of Irrotational Flow through Axially Symmetric Orifices," *J. Appl. Mech.*, **17**, 421 (1950).

8. Garabedian, P., "Calculation of Axially Symmetric Cavities and Jets," *Pacif. J. Math.*, **6**, 611 (1956).

9. Jeppson, R.M., "Free Streamline Problems Solved by Inverse Formulation and Finite Differences," *Developments in Mechanics* **5**: Midwestern Mech. Conf. Publication (1969).

10. Hunt, B.M., "Numerical Solution of an Integral Equation for Flow from a Circular Orifice," *J. Fluid Mech.*, **31**, 361 (1968).

11. Bloch, E., "Numerical Solution of Free Boundary Problems by the Method of Steepest Descent," *The Physics of Fluids Supplement II* (1969).

12. Ovsiannikov, L.V., "Gas Flow with Straight Transition Line," *Prikl. Mat. Mekh.*, **13** (1949) [*Trans:* NACA TM **1295** (1951)]

13. Shapiro, A.H., "The Dynamics and Thermodynamics of Compressible Fluid Flow," New York, Ronald Press (1953).

14. Ames, W.F., "Numerical Methods for Partial Differential Equations," London, Thomas Nelson and Sons Ltd. (1969).

15. Young, D.M., "Iterative Methods for Solving Partial Difference Equations of Elliptic Type," *Trans. Amer. Math. Soc.*, **76**, 92 (1954)

16. Carré, B.A., "The Determination of the Optimum Accelerating Factor for Successive Over-relaxation," *Comput. J.*, **4**, 73 (1961).

SUBSONIC FLOWS IN TURBOMACHINES
H. Marsh

(University of Durham)

INTRODUCTION

The prediction of the flow in a multistage compressor or turbine is an extremely difficult mathematical problem. The objective is to provide the design engineer with a method for predicting the performance of a turbomachine, so that he can ensure an efficient machine with the minimum time for development. During the past 50 years, the methods for calculating the flow in turbomachines have progressed from the basic mean line analysis with velocity triangles at the mean radius, first to the method of simple radial equilibrium [1], which was necessary to deal with machines of low hub to tip ratio. In the 1950's, actuator disc theory [2] was developed in which each blade row was represented by a plane discontinuity. This model contributed greatly to our understanding of the interaction of blade rows. By 1965, digital computers were sufficiently large and fast to allow numerical solutions of the governing equations and two separate methods of analysis were developed, namely streamline curvature [3] and matrix through-flow [4]. These two numerical techniques are now widely used in the analysis of turbomachinery flows and the flow through ducts and nozzles.

Although this paper concentrates on the problem of predicting the flow pattern in a multistage machine, it must be remembered that the machine consists of many blade rows whose individual performance must be specified. The prediction of the flow through a cascade of blades has progressed from the incompressible flow solutions and transformation methods to Martensen's singularity method [5] and, more recently, to numerical solutions for compressible flow using the methods of streamline curvature and finite differences [6]. Although there is still discussion about the details of the flow at a rounded trailing edge, the calculations do lead to a good estimate for the unstalled behaviour of a cascade of blades.

Until the early 1960's, the mathematical model for the flow in a turbomachine was more advanced than the methods of computation. When Wu [7] published the general theory for flow in an arbitrary turbomachine in 1952, computers were not sufficiently large or fast to obtain solutions to the governing equations. It was not until 1964 that computers became available with sufficient speed and storage capacity to deal economically with these problems of flow calculation. The techniques now exist for calculating the flow in turbomachines on the basis of a flow model which includes the effects of compressibility, radial variations of lift, a consistent loss model, blade row interaction, secondary flows and the development of the wall boundary layers. This flow model is based on our understanding of the flow through cascades, isolated blade rows and multistage machines; it is a model that is continually being questioned and revised.

METHODS OF FLOW CALCULATION

The flow in a multistage turbomachine is a complex three-dimensional time varying flow. It is only for an isolated blade row, or a cascade, that the flow is steady. In 1952, Wu [7] published the general theory for the flow through a turbomachine. In Wu's analysis, the governing equations are developed for the flow on two intersecting families of stream surfaces, Fig. 1, and the complete three-dimensional flow is obtained by iteration between solutions for the flow on the two sets of surfaces. Throughout the analysis, it is assumed that the flow is steady relative to the blade row. However, at exit from a row of blades

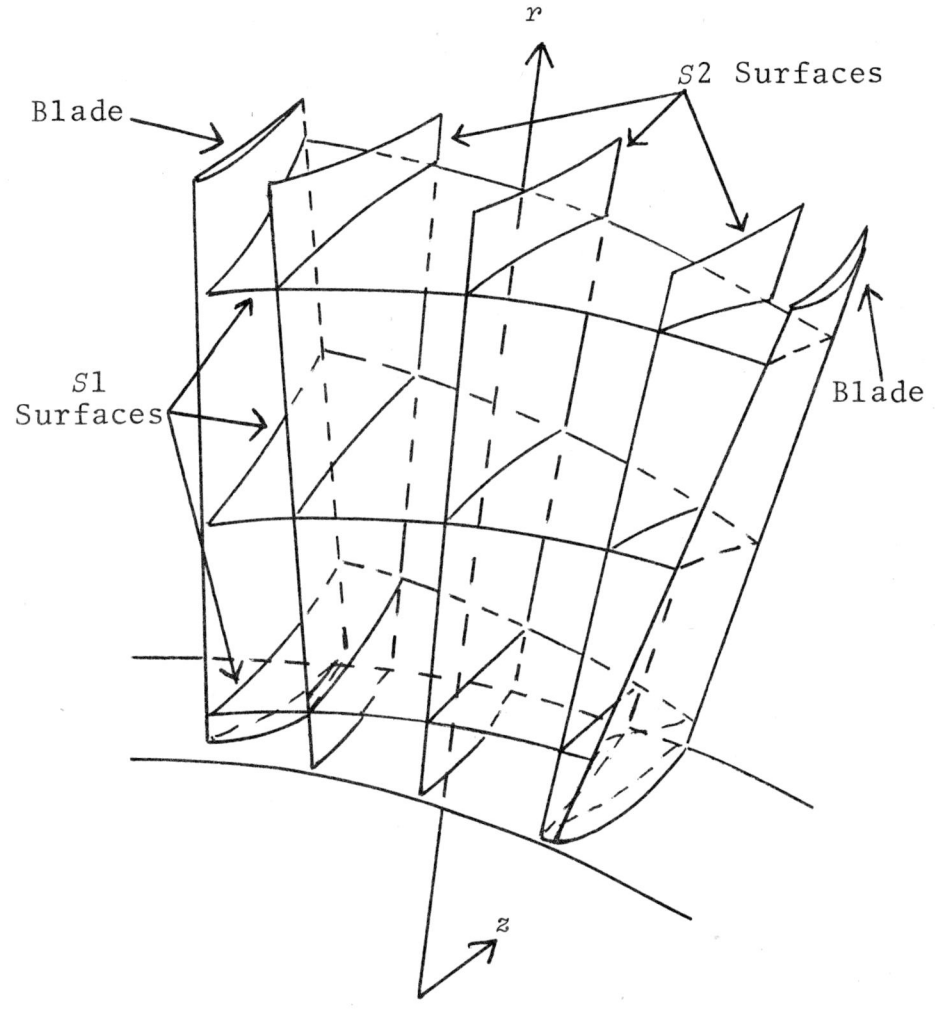

Fig. 1. $S1$ and $S2$ stream surfaces

with lift the flow and gas state vary across the pitch and if the following blade row has a motion relative to the first it is subjected to a time varying inlet flow. Wu's general theory is, therefore, only applicable to the flow in an isolated blade row. The theory in general in that it is a method for estimating a steady three-dimensional flow by calculating the flow on two intersecting sets of stream surfaces, the $S1$ blade-to-blade surfaces and the $S2$ blade-like surfaces.

Flow calculations for turbomachines are usually based on calculating the flow on the mean $S2$ stream surface of Fig. 1, it being assumed that the flow on the blade-to-blade $S1$ stream surfaces is already known from numerical solutions, or experiments. The flow on the mean $S2$ surface within a blade row is assumed to be steady and the solution is taken to represent average values of the thermodynamic properties and velocity components for the flow in the blade passage. In a multistage machine, the time dependence of the flow is removed by treating the $S2$ through-flow solution as an axially symmetric flow in the duct region between each pair of blade rows. This approach neglects the effect of time dependence, but experience has shown that the predictions provide a good estimate for the flow in turbomachines. When the solution is known for the flow on the mean $S2$ stream surface it is possible to redefine the $S1$ surfaces and obtain a better estimate for the blade-to-blade flow pattern.

This approach to turbomachinery flow calculations requires that the mean $S2$ stream surface is specified and the calculated flow pattern is assumed to be an average for the blade passage. The flow models for a blade row have been examined [8] and it has been shown that it is not possible to obtain complete correspondence between the averaged actual flow and the predicted flow. In general, the mathematical models cannot fully represent the averaged flow within a cascade. However, [8] shows that the over-all behaviour of the cascade can be correctly represented by the model of flow on a mean stream surface, although the local variations of the flow properties are not fully matched in the averaged flow and the model. The flow model can predict the correct over-all changes of the flow across a blade row, but the calculated flow on the mean $S2$ stream surface may not provide a good local representation for the averaged actual flow. The main conclusion of [8] is that although it is not possible to give a precise definition of the mean $S2$ stream surface, a reasonable approximation will lead to a mathematical model with an over-all behaviour which is a good approximation to the real cascade. This problem of mathematical models for cascade flows deserves more attention in that there is little point in developing sophisticated numerical techniques if the model is not a good approximation for the real flow.

THE GOVERNING EQUATIONS

There are two widely used methods of flow calculation for turbomachines and both can be used for calculating the flow on the $S1$ and $S2$ stream surfaces. For the flow on the mean $S2$ stream surface, these methods are usually known as streamline curvature and through-flow analysis. The two methods of flow calculation were developed independently, but it has been shown [9] that they are merely two different techniques for solving the same set of equations for the same mathematical model. The continuing use of both methods of analysis suggests that neither has yet shown sufficient advantages to become the accepted method for turbomachinery flow calculations.

Notation is defined on p. 136. In an r, θ, z coordinate system, which rotates with the blade row at an angular velocity ω, the equations governing the steady, reversible inviscid flow are:

(a) <u>continuity</u> $\nabla \cdot (\rho \bar{W}) = 0$ (1)

(b) <u>motion</u> $2\omega \bar{W} - \bar{W}(\nabla \bar{W}) = T\nabla s - \nabla I$ (2)

(c) <u>energy</u> $\dfrac{DI}{Dt} = 0$ (adiabatic flow)   (3)

(d) <u>state</u> $\rho = f(h, s)$ (4a)

or for a perfect gas,

$$\rho \propto h^{\frac{1}{\gamma-1}} \cdot e^{\frac{-s}{R}} \qquad (4b)$$

where h is the enthalpy per unit mass and

$$I = h - \omega r V_\theta$$

is the rothalpy. In the analysis, these equations are only solved for flow on a prescribed mean stream surface, defined by

$$S(r,\theta,z) = 0 \qquad (5)$$

or
$$\theta = \theta(r,z)$$

where it is assumed that the surface is single valued in θ.

If $\overline{\partial q/\partial r}$ and $\overline{\partial q/\partial z}$ are partial derivatives taken along the stream surface, then

$$\left.\begin{aligned}\overline{\frac{\partial q}{\partial r}} &= \frac{\partial q}{\partial r} - \frac{n_r}{n_\theta} \cdot \frac{1}{r}\frac{\partial q}{\partial \theta} \\ \overline{\frac{\partial q}{\partial z}} &= \frac{\partial q}{\partial z} - \frac{n_z}{n_\theta} \cdot \frac{1}{r}\frac{\partial q}{\partial \theta}\end{aligned}\right\} \qquad (6)$$

where n_r, n_θ and n_z are the components of the unit vector normal to the surface. The special derivative $\overline{\partial q/\partial r}$ is the rate of change of q with r on the surface at a given value of z, whereas $\partial q/\partial r$ is the rate of change of q with r at given values of z and θ. For an axially symmetric flow, there is no difference between the two partial derivatives $\partial q/\partial r$ and $\overline{\partial q/\partial r}$.

By introducing the special derivatives, Wu was able to express the governing equations in the form:

(a) <u>continuity</u>

$$\frac{1}{r}\overline{\frac{\partial}{\partial r}}(r\rho W_r) + \overline{\frac{\partial}{\partial z}}(\rho W_z) = -\frac{\rho}{rn_\theta}\left[n_r\frac{\partial W_r}{r\partial\theta} + n_\theta\frac{\partial W_\theta}{\partial\theta} + n_z\frac{\partial W_z}{z\partial\theta}\right]$$
$$= \rho C(r,z) \qquad (7)$$

(b) motion

$$-\frac{W_\theta}{r} \cdot \frac{\overline{\partial}}{\partial r}(rV_\theta) + W_z\left[\frac{\overline{\partial W_r}}{\partial z} - \frac{\overline{\partial W_z}}{\partial r}\right] = T\frac{\overline{\partial s}}{\partial r} - \frac{\overline{\partial I}}{\partial r} + F_r \qquad (8)$$

$$\frac{W_r}{r}\frac{\overline{\partial}}{\partial r}(rV_\theta) + \frac{W_z}{r}\frac{\overline{\partial}}{\partial z}(rV_\theta) = F_\theta \qquad (9)$$

$$-W_r\left[\frac{\overline{\partial W_r}}{\partial z} - \frac{\overline{\partial W_z}}{\partial r}\right] - \frac{W_\theta}{r}\frac{\overline{\partial}}{\partial z}(rV_\theta) = T\frac{\overline{\partial s}}{\partial z} - \frac{\overline{\partial I}}{\partial z} + F_z \qquad (10)$$

where
$$\overline{F} = -\frac{1}{r\rho n_\theta} \cdot \frac{\partial p}{\partial \theta} \cdot \underset{\sim}{n}$$

(c) energy

$$W_r\frac{\overline{\partial I}}{\partial r} + W_z\frac{\overline{\partial I}}{\partial z} = 0 \qquad (11)$$

In an inviscid flow, the force vector $\underset{\sim}{\overline{F}}$ is normal to the stream surface S and is therefore normal to the relative velocity vector $\underset{\sim}{\overline{W}}$, so that

$$W_r F_r + W_\theta F_\theta + W_z F_z = 0 \qquad (12)$$

It is convenient to define the local form of the stream surface by two angles λ and μ where

$$\left.\begin{array}{c} \tan \lambda = \dfrac{n_r}{n_\theta} = \dfrac{F_r}{F_\theta} \\[1em] \tan \mu = \dfrac{n_z}{n_\theta} = \dfrac{F_z}{F_\theta} \end{array}\right\} \qquad (13)$$

The three velocity components are then related by

$$W_\theta = -W_z \tan \mu - W_r \tan \lambda \qquad (14)$$

which is the geometrical condition that the flow follows the mean stream surface.

Following Wu's analysis, an integrating factor B is introduced such that the equation of continuity becomes

$$\overline{\frac{\partial}{\partial r}}(rB\rho W_r) + \overline{\frac{\partial}{\partial z}}(rB\rho W_z) = 0 \qquad (15)$$

This equation indicates that B is proportional to the local angular thickness of the stream surface and, to a first approximation, B is taken as proportional to the local width of the blade passage.

THE METHOD OF STREAMLINE CURVATURE

In the method of streamline curvature, the radial equation of motion is combined with the equations of continuity and energy to obtain a single equation which involves the local slope and curvature of the streamlines. A detailed derivation is given by Smith [3]; the final equation is quoted here

$$T\overline{\frac{\partial s}{\partial r}} - \overline{\frac{\partial I}{\partial r}} + F_r = -\frac{W_\theta}{r}\overline{\frac{\partial}{\partial r}}(rV_\theta) - W_m\overline{\frac{\partial W_m}{\partial r}} - \left(\frac{M_r^2}{1-M_m^2}\right)\frac{V_\theta^2}{r}$$

$$- \left(\frac{1-M_z^2}{1-M_m^2}\right)\frac{W_m^2}{R_m\cos\phi} + \left(\frac{M_r M_\theta}{1-M_m^2}\right)F_\theta$$

$$+ \left(\frac{W_r}{1-M_m^2}\right)\left[\frac{W_m}{R}\overline{\frac{\partial s}{\partial m}} - \frac{W_z}{r}\overline{\frac{\partial}{\partial r}}(r\tan\phi) - \frac{W_m}{B}\overline{\frac{\partial B}{\partial m}}\right] \qquad (16)$$

where M is the Mach number and ϕ and R_m are the slope and curvature of meridional streamlines.

Equation (16) can be expressed as

$$W_m\overline{\frac{\partial W_m}{\partial r}} + W_m^2 K(r,z) + L(r,z) = 0 \qquad (17)$$

The method of solution assumes that an estimate for the flow pattern is available from a previous iteration, so that the functions $K(r,z)$ and $L(r,z)$ can be evaluated. At any position within the flow field, a value for W_m is chosen at some radial position, such as the mid-annulus, and the equation for W_m is integrated radially to obtain the axial velocity profile. The mass flow rate across the calculating plane is then computed and compared with the upstream mass flow rate. If necessary, a new value for W_m is chosen and the calculations repeated until the required mass flow rate is obtained. When the axial velocity profiles are known at each calculating plane, the new streamline pattern can be found and new values obtained for the functions $K(r,z)$ and $L(r,z)$. The complete cycle of calculations is repeated until a convergence criterion is satisfied. The numerical procedure can become highly unstable and it is usually necessary to restrict the calculation to small changes of the flow pattern between successive iterations.

A major difficulty with the method of streamline curvature is that it is necessary to estimate the local slope and curvature of the streamlines. A widely used approximation for the streamlines is a spline fit through points of equal stream function, the spline being differentiated once to obtain the slope and twice to obtain the curvature. The net result is that the values for the slope and curvature may not be good estimates for the flow. Shaalan and Daneshyar [10] suggest that a more accurate value for the curvature can be obtained by fitting a second spline to the variation of slope and then differentiating to obtain the second derivative and curvature. Even with this double spline technique, a good estimate for the curvature of an oscillating streamline requires five to ten points per wavelength.

In a multistage turbomachine, the wavelength of the oscillating streamlines may correspond to the length of a stage, a rotor row followed by a stator row. It follows that if a good estimate for the curvature requires five to ten points per wavelength, it is necessary to place calculating planes within the blade rows. This is relatively simple for subsonic flows, but for transonic flows with shocks the mathematical model for the flow in the blade passage is not adequate. When calculating the performance of transonic machines, the calculating planes are often placed outside the blade rows and the blade rows are treated as devices with a specified over-all behaviour. The wide spacing of the calculating planes, typically one half wavelength, reduces the accuracy of the calculation, but experience has shown that there is still good agreement between the calculated performance and that found experimentally.

THROUGH-FLOW ANALYSIS

The alternative method for solving the equations governing the flow on the mean stream surface is to define a stream function ψ where

$$\left. \begin{array}{c} \dfrac{\partial \psi}{\partial r} = rB\rho W_z \\[6pt] \text{and} \quad \dfrac{\partial \psi}{\partial z} = - rB\rho W_r \end{array} \right\} \qquad (18)$$

The radial equation of motion can then be expressed as

$$\frac{\partial^2 \psi}{\partial r^2} + \frac{\partial^2 \psi}{\partial z^2} = \frac{\partial \psi}{\partial r} \cdot \frac{\partial}{\partial r}\left[\ln(rB\rho)\right] + \frac{\partial \psi}{\partial z} \cdot \frac{\partial}{\partial z}\left[\ln(rB\rho)\right]$$

$$+ \frac{r\rho B}{W_z}\left[\frac{\partial I}{\partial r} - T\frac{\partial s}{\partial r} - F_r - \frac{W_\theta}{r}\frac{\partial}{\partial r}(rV_\theta)\right] \qquad (19)$$

or

$$\frac{\partial^2 \psi}{\partial r^2} + \frac{\partial^2 \psi}{\partial z^2} = q(r,z) \qquad (20)$$

Equation (19) is often referred to as Wu's principal equation; it is a nonlinear equation, but can be solved by the repeated solution and correction of the quasilinear equation (20). A solution for the stream function ψ is obtained for a given distribution $q(r,z)$, the function $q(r,z)$ is then corrected using the improved solution for ψ and the process repeated until a convergence criterion is satisfied, as described in reference [4].

Mathematically the through-flow analysis is extremely simple, the difficulty lies in obtaining numerical solutions for the stream function. Many finite difference approximations use a rectangular grid of points, since this leads to simple expressions for the approximations to derivatives. However, for calculating the flow through an axial flow compressor, a more suitable form of grid is a distorted or non-rectangular mesh which is straight in only one direction.

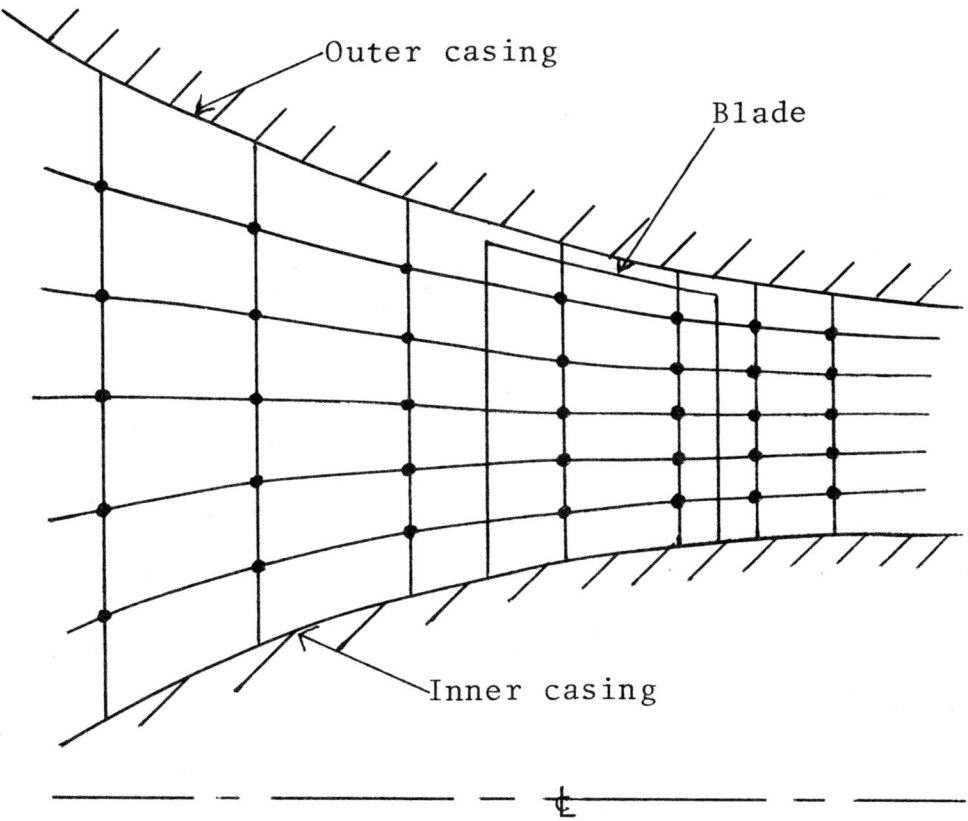

Fig. 2. Distorted Grid, reference [4]

Fig. 2 shows a distorted grid with parallel radial lines having equally spaced points between the inner and outer casings. By definition, the machine casings form curved grid lines and there are no additional difficulties for grid points that lie close to or on the boundaries. There is an automatic refining of the grid as the annulus height is reduced.

There is clearly no difficulty in forming finite difference approximations for the radial derivatives, but there are no simple expressions for derivatives with respect to z. The finite difference approximation for the Laplacian operator at the point (i,j) is obtained by choosing the coefficients a_{pq} such that

$$\sum_{p,q} a_{pq} \psi(p,q) = \overline{\frac{\partial^2 \psi}{\partial r^2}} + \overline{\frac{\partial^2 \psi}{\partial z^2}} + O(\Delta r^2) \qquad (21)$$

where $\psi(p,q)$ is the stream function at a grid point (p,q) which lies close to (i,j) and Δr is the local radial spacing of the grid points. In general, the Laplacian operator can be approximated by a combination of values for the stream function at ten neighbouring grid points, the ten conditions being that, in equation (21), the coefficients of ψ, $\overline{\partial \psi/\partial r}$, $\overline{\partial \psi/\partial z}$, $\overline{\partial^2 \psi/\partial r \partial z}$, $\overline{\partial^3 \psi/\partial r^3}$, $\overline{\partial^3 \psi/\partial r^2 \partial z}$, $\overline{\partial^3 \psi/\partial r \partial z^2}$ and $\overline{\partial^3 \psi/\partial z^3}$ are all zero and the coefficients of $\overline{\partial^2 \psi/\partial r^2}$ and $\overline{\partial^2 \psi/\partial z^2}$ are unity. For certain conditions, such as a square or rectangular grid, there is more than one solution for the coefficients a_{pq} and the method for determining the finite difference approximation becomes ill conditioned with singular matrices. Examination of the 10 × 10 matrices shows that this difficulty can be overcome by relaxing the condition that the coefficient of $\overline{\partial \psi/\partial z}$ is zero and replacing this by setting the coefficient of $\overline{\partial^4 \psi/\partial r^2 \partial z^2}$ to zero. This is equivalent to rephrasing the principal equation in the form

$$\overline{\frac{\partial^2 \psi}{\partial r^2}} + \overline{\frac{\partial^2 \psi}{\partial z^2}} + E\overline{\frac{\partial \psi}{\partial z}} = q(r,z) + E\overline{\frac{\partial \psi}{\partial z}}$$

$$= Q(r,z) \qquad (22)$$

where

$$E = 2\left[\frac{1}{(z_{i+1} - z_i)} - \frac{1}{(z_i - z_{i-1})}\right]$$

and z_i is the z coordinate of the radial grid line. With the governing equation for ψ expressed in this form, there is no difficulty in evaluating the coefficients a_{pq}. The formation of the grid and the finite difference approximations can then become part of the computer program, as described in reference [4].

The finite difference approximation for equation (22), together with the upstream and downstream boundary conditions of parallel flow, can be expressed in the matrix form

$$[M][\psi] = [Q] \qquad (23)$$

where $[\psi]$ and $[Q]$ are column vectors and $[M]$ is a band matrix which remains unchanged throughout the calculation. The method is to solve for the stream function with a given value of $[Q]$, to correct $[Q]$ using the new flow pattern and then to repeat the cycle of calculation until convergence is obtained. The numerical stability of the through-flow method is dependent on the problem and although under-relaxation of the stream function is often required, in examples such as nozzle and duct flows, it has been possible to use over-relaxation. The radial grid lines are usually positioned in the duct regions and within the blade rows and there appears to be no difficulty in obtaining numerical solutions, provided the machine boundaries form smooth curves.

LIMITATIONS ON THE MACH NUMBER

The through-flow analysis has certain limitations on the Mach number which occur in the calculation of the density and are caused by the coupling of density and velocity on the current iteration. At each point in the mesh, there are two possible flows, one corresponding to a meridional or relative Mach number less than unity and the other to a Mach number greater than unity. To avoid ambiguity and to ensure that the governing equation remained elliptic, the computer program of [4] was restricted to flows where

(a)   in a duct flow, $M_m \leq 1$

(b)   within or immediately downstream of a blade row, $M_{rel} \leq 1$

Following a suggestion by D. Gelder (private communication, 1969), it has been found that these Mach number restrictions can be relaxed if the velocity on the current iteration is determined by using the density from the previous iteration. The calculation for density then lags behind that for velocity, but this should not be important provided the process converges. By using this technique, numerical solutions can be obtained with regions where the Mach number is greater than unity, but the difficulty then lies in determining whether or not these solutions represent real flows.

The Mach number limitations of the streamline curvature method are less easily defined, but in [11] the uniqueness of the solution is examined. It is shown that if the density and velocity are both determined for the current iteration, then at each calculating plane the requirements for uniqueness are:

(a)   duct flow, $M_m \leq 1$ at all radii,

(b)   within or immediately downstream of a blade row, $M_{rel} \leq 1$ at all radii
or $M_{rel} \geq 1$ at all radii.

Within the blade row, uniqueness of the solution cannot be established unless the flow is entirely subsonic or entirely supersonic.

In the streamline curvature method, a value for the meridional velocity, W_m', is chosen at some radial position and equation (17) is integrated radially to obtain the velocity profile. The mass flow rate $\dot{m}$ is then calculated and compared with the specified value.

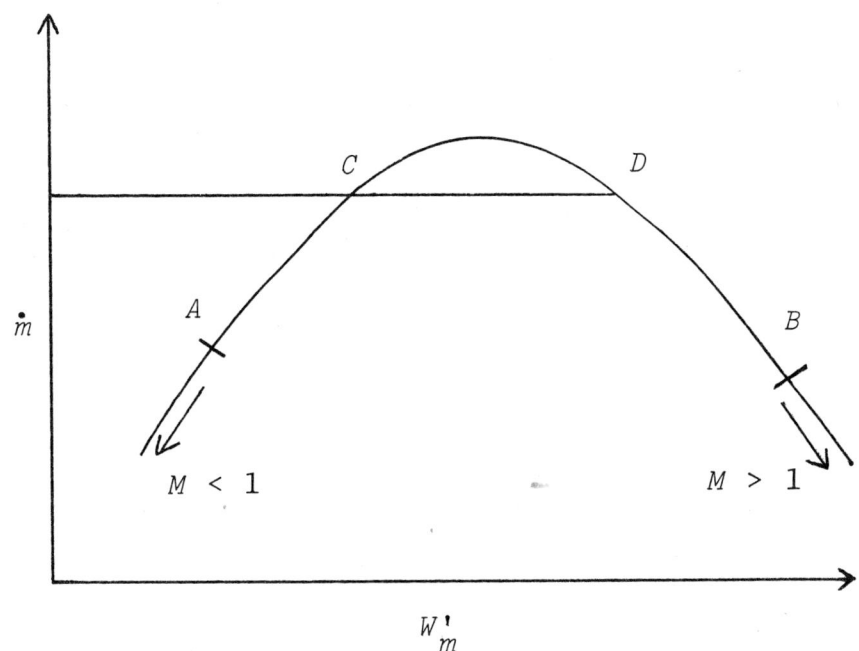

A $M = 1$ at one radius
B $M \geq 1$ at all radii

Fig. 3. Variation of $\dot{m}$ with W_m'

Fig. 3 shows the typical variation of theoretical mass flow rate with the chosen value of W_m'. When the flow is known to be entirely supersonic or entirely subsonic, there is no ambiguity about the correct value for W_m'. However, if the specified mass flow rate leads to solutions which are partly subsonic and partly supersonic, points C and D of Fig. 3, the problem is to decide which of these two is correct. A decision to take the lower value of W_m', as in many streamline curvature programs, may allow the computer to proceed to a numerical solution, but this, by itself, does not necessarily imply that the computer has calculated the true flow pattern. It would be interesting to know whether there is any rigorous justification for taking the solution corresponding to the lower value of W_m', since without this there is no certainty that the computer has calculated the correct flow pattern.

The flow through a convergent-divergent nozzle is an interesting test case for the computer programs, the nozzle being treated as a turbomachine with no blades and no hub. A calculation method is required that can predict the flow in the nozzle for any pressure ratio without the need to specify the subsonic and supersonic regions or the location of any shock. At present there are no published solutions to this problem using either the streamline curvature method or the through-flow analysis, but solutions have been obtained by considering the steady flow as the asymptotic solution for a time dependent flow [12], [13]. The introduction of time as a variable leads to stable numerical schemes for predicting nozzle flows and for determining the location and strength of any shock.

THE LOSS MODEL

There is a serious defect in the early analyses of turbomachinery flows in that the governing equations are those for a reversible, adiabatic and therefore isentropic flow. There is no mechanism in the mathematical model that allows for losses or changes of entropy. The use of experimental data for the loss of stagnation pressure in the flow through a cascade therefore appears to be inconsistent with the method of flow calculation. This problem is discussed in [9] where it is shown that for an axial flow machine the through-flow analysis is based on the assumption that the losses are caused by a dissipative force which acts in the axial direction. More recently, it has been shown [14] that it is possible to rephrase the problem such that the calculations are consistent with a dissipative force that opposes the velocity vector. The magnitude of this dissipative force is not required in the analysis, but it can be derived from the loss data. This new form of the through-flow analysis is based on the equation of motion normal to the streamlines and it is consistent with the use of loss data. Existing computer programs, usually based on equation (19), can easily be modified to include the consistent loss model. The loss data used in these calculations can be obtained by experiment or by applying the loss correlations published by Lieblein [15] and Swan [16].

WALL BOUNDARY LAYERS AND SECONDARY FLOWS

As the flow passes through a compressor, boundary layers form on the inner and outer walls of the annulus and reduce the effective flow area. A method is required for calculating the development of these wall boundary layers. As the flow passes through a blade row, the pressure and velocity gradients in the mainstream flow are extremely large and doubts are expressed about the validity of using conventional boundary layer theory to deal with the wall boundary layers in turbomachines [17].

If the pressure does not vary across the boundary layer then, for a cascade flow, there is no variation of the blade force within the boundary layer. For a many bladed cascade this would lead to a variation of the exit flow angle across the boundary layer and this is not consistent with the prescribed exit angle of the many bladed cascade. It has been suggested [17] that a better model is obtained by assuming that the mean pressure gradient along the blade passage is constant in the boundary layer. This model leads to the conclusion that the blade force varies in the boundary layer region and this has been observed by Smith [18]. Comparisons have been made [17] between several theories for predicting the wall boundary layers and the experimental results obtained [19] with isolated blade rows. The main conclusion is that the development of the wall boundary layers can only be predicted with reasonable accuracy for lightly loaded blades, a flow condition where there is little change in the wall boundary layers.

Although a blade row may turn the mainstream flow from an inlet angle α_1 to an exit angle α_2, the flow in a shear layer, such as a wall boundary layer, behaves differently and the exit angle varies along the blades. This variation of flow angle at exit from a blade row is caused by the turning of the shear flow and it is usually referred to as secondary flow. If the inlet flow has a normal component of vorticity ξ_n, then at exit from a cascade the distributed secondary vorticity in the direction of flow, ξ_{sec}, is given by

$$\xi_{sec} = \frac{\xi_n}{\cos\alpha_1 \cos\alpha_2}[\tfrac{1}{2}(\sin 2\alpha_2 - \sin 2\alpha_1) + \alpha_2 - \alpha_1] \quad (24)$$

The secondary flow associated with this streamwise component of vorticity can be calculated and the predicted variation in flow angle is in good agreement with that measured in cascade tests. Two reviews of secondary flow theory have been given by Hawthorne and Novak [20] and Horlock and Lakshminarayana [21]. Further research is in progress to combine secondary flow and wall boundary layers into a single theory and to include the effect of compressibility.

CASCADE FLOWS

Although I have concentrated on the flow on the $S2$ surface in a turbomachine, the two methods of calculation can also be applied to the related problem of the blade-to-blade flow on the $S1$ surface. Smith and Frost [6] have described the use of both finite difference and streamline curvature methods for computing the compressible flow through a cascade and have obtained good agreement between the predicted and measured pressure distribution on the blades. A recent development in the calculation of cascade flows is the use of the finite element approach [22], but it is not yet clear whether this will offer any significant advantage over the other two methods of calculation.

CONCLUSIONS

The mathematical model for the flow in turbomachines was developed in 1952 by Wu, but it was not until the mid-1960's that the speed and storage capacity of computers became adequate to allow numerical solutions to be obtained. Two methods of analysis were developed for solving the governing equations, streamline curvature and finite differences. The relationship between these two methods has been discussed in detail together with their limitations on Mach number. Both the streamline curvature and finite difference method can be used for calculating subsonic flows in turbomachines, axisymmetric ducts and nozzles, but there remain five major problems that require more attention:

1. transonic flows,
2. unsteady flows,
3. the use of a consistent loss model,
4. wall boundary layers,
5. secondary flows.

In addition to these specific practical problems, there is a need for more research on the mathematical models that form the basis for the calculation methods. If there are to be further improvements in turbomachinery fluid dynamics we must improve our understanding of the flow, the mathematical models and the methods of computation. The flow calculations depend on forming the correct mathematical model and the model is dependent on our physical understanding of the fluid dynamics.

NOTATION

- B angular thickness of the mean $S2$ stream surface
- $\bar{F}$ force normal to the stream surface,
- h enthalpy,
- I rothalpy, $I = h - \omega r V_\theta$,
- m direction of the meridional streamline,
- $\dot{m}$ mass flow rate,
- $\tilde{n}$ normal to the stream surface,
- r radius
- R_m radius of curvature of the meridional streamline,
- s entropy,
- t time,
- T temperature,
- $\bar{V}$ velocity,
- $\bar{W}$ velocity relative to the blade row,
- z axial coordinate,
- γ ratio of specific heats,
- θ angular coordinate,
- $\left.\begin{array}{l}\lambda\\ \mu\end{array}\right\}$ angles defining the stream surface,
- ρ density,
- ϕ slope of the meridional streamline,
- ψ stream function,
- ω angular velocity.

REFERENCES

1. Cohen, H. and White, E.M., "The theoretical determination of the three dimensional flow in an axial compressor, with special reference to constant reaction blading," ARC Report No. 6842 (1943).

2. Horlock, J.H., "Some actuator disc theories for the flow of air through axial turbomachines," ARC R & M 3030 (1952).

3. Smith, L.H., "The radial equilibrium equation of turbomachinery," *Trans. ASME*, Series A, **88** (1966).

4. Marsh, H., "A digital computer program for the through-flow fluid mechanics in an arbitrary turbomachine using a matrix method," NGTE Report R282, (1966); ARC R & M 3509 (1968).

5. Martensen, E., "Calculation of the pressure distribution over profiles in cascade in two-dimensional potential flow, by means of a Fredholm integral equation," *Archs ration. Mech. Analysis*, **3**, no. 3 (1959).

6. Smith, D.J.L. and Frost, D.H., "Calculation of the flow past turbomachine blades," Thermo and Fluid Mech. Conv., I. Mech. E., Paper 27 (1970).

7. Wu, C.H., "A general theory of three dimensional flow in subsonic and supersonic turbomachines of axial, radial and mixed flow types," NACA, TN 2604 (1952).

8. Horlock, J.H. and Marsh, H., "Flow models for turbomachines," *J. Mech. Eng. Sci.*, **13**, no. 5 (1971).

9. Marsh, H., "The through-flow analysis of axial flow compressors," *in* "AGARD Lecture series No. 39: Advanced Compressors," (1970)

10. Shaalan, M.R.A. and Daneshyar, H., "A critical assessment of methods of calculating slope and curvature of streamlines in fluid flow problems," *Proc. I. Mech. E.*, Part I, **186** (1972).

11. Marsh, H., "The uniqueness of turbomachinery flow calculations using the streamline curvature and matrix through-flow methods," *J. Mech. Eng. Sci.*, **13**, No. 6 (1971).

12. Marsh, H. and Merryweather, H., "The calculation of subsonic and supersonic flow in nozzles," Symp. on Internal Flows, I. Mech. E. (1971).

13. Daneshyar, H. and Glynn, D.R., "The calculation of flow in nozzles using a time-marching technique based on the method of characteristics," *Int. J. Mech. Sci.*, **15**, p921 (1973).

14. Bosman, C. and Marsh, H., "An improved method for calculating the flow in turbomachines, including a consistent loss model," *J. Mech. Eng. Sci.*, **16**, no. 1 (1974).

15. Lieblein, S., "Loss and stall analysis of compressor cascades," *Trans. ASME, J. Basic Eng.*, Series D, **81** (1959).

16. Swan, W., "A practical method of predicting transonic compressor performance," *Trans. ASME, J. Eng. Power*, Series A, **83** (1961).

17. Marsh, H. and Horlock, J.H., "Wall boundary layers in turbomachines," *J. Mech. Eng. Sci.*, **14**, No. 6 (1972).

18. Smith, L.H., "Casing boundary layers in multi-stage axial flow compressors," *in* "Flow Research in Blading," Brown Boveri Symposium, Elsevier (1969).

19. Gregory-Smith, D.G., "An investigation of annulus wall boundary layers in axial flow turbomachines," ASME, Paper No. 70-GT-92 (1970).

20. Hawthorne, W.R. and Novak, R.A., "The aerodynamics of turbomachinery," *in* "Annual Review of Fluid Mechanics," No. 1, Annual Reviews Inc. (1969).

21. Horlock, J.H. and Lakshminarayana, B., "Secondary flows," *in* "Annual Review of Fluid Mechanics," No. 5, Annual Reviews Inc. (1973).

22. Thompson, D.S., "Finite element analysis of the flow through a cascade of aerofoils," Cambridge University Engineering Department, Report No. Turbo/TR 45 (1973).

FINITE ELEMENT AND DIFFERENCE METHODS FOR CASCADES

M.J. O'Carroll and L.A. Morgan

(Lanchester Polytechnic)

INTRODUCTION

Origin

The flow in turbomachines is generally complex, three-dimensional and unsteady. Design considerations for such flows begin with global parameters such as work rate and flow rate. Many machines require strong local variations of flow characteristics, for example spanwise variation of flow angle. This calls for an understanding of the flow pattern and a first description is expressed in terms of circumferential averages. Analysis of this type, essentially meridional flow analysis, is described by Professor Marsh. Such a representation depends on information about how the flow turns in the circumferential direction so that a mean (circumferential average) stream surface can be defined. For thin closely spaced blades this may be approximated fairly well. But in general it is necessary to solve a blade-to-blade problem for the information. Further, the outlet flow angle usually deviates from the direction of the blade trailing edge so there is some uncertainty in the effectiveness of the blades in turning the flow.

More complete discussions of design processes and methods are given by Gostelow, Horlock and Marsh [1], Serovy [2] and Horlock [3]. Conventional methods of analysis combine two two-dimensional representations: (*a*) blade-to-blade flow on assumed axisymmetric stream surfaces and (*b*) circumferentially averaged "meridional" flows. These are interdependent and analysed iteratively, because the meridional analysis is needed to locate the axisymmetric stream surface for blade-to-blade analysis and the resulting blade-to-blade flow is needed to locate the meridional stream surface. However, stream surfaces will generally twist away from surfaces of revolution. In a potential flow model, the design requires some variation of lift or circulation along the span of the blades. This causes vortex sheets to be shed from trailing edges. Consequently a secondary flow effect is generated and the model (*a*) is not strictly valid.

A complete three-dimensional analysis is needed but this is only practicable for a single blade row because of the unsteady nature of the coupling of adjacent stationary and rotating blade rows. The meridional analysis for the whole machine can then be combined with three-dimensional blade-to-blade analysis for separate blade rows. The former is needed to present upstream and downstream boundary conditions for the latter, in a sense time averaged and simplified, defining a steady three-dimensional flow problem.

Flow Considerations

It is commonly assumed that viscosity effects in turbomachines are confined to boundary layers and wakes. This means that three-dimensional effects and secondary flows may arise from three causes: (*a*) transport of vorticity from annulus boundary layers; (*b*) transport of vorticity in the incoming flow; and (*c*) twisting of potential flow by the blade configuration. We are particularly interested in (*c*). It may be an important aspect of the flow in many machines where high efficiency and loading are prime considerations so a greater understanding of the flow is valuable. For example, it is likely to be a factor in flow in leading fans of aircraft engines producing conditions for separation as well as downstream vortices and irregularities. Another case is the highly flared low pressure stages of large steam turbine generators with highly twisted blades.

To investigate the three-dimensional blade-to-blade problem we make use of velocity potential [4]. This easily embraces both three-dimensional and compressible (but isentropic) flow. Other approaches have been tried but appear to be computationally exorbitant (see [5]). The main limitation of velocity potential is that it represents only irrotational flows, or irrotational components of a flow. A scheme for extending the representation by adding a solenoidal component was presented [6] which may be practicable when the vorticity distribution is simple (e.g., primarily streamwise). A more general approach along these lines has been made by Wu [7] for viscous flow problems using explicit integral relations. Here, however, we restrict ourselves to purely potential flows, or relative potential flows in rotors.

Computational Considerations

The three-dimensional analysis is computationally formidable and the major difficulty in producing a useful analytic tool is to keep down the computing time. Even with irrotational flow there are three obstacles to rapid solution: (*a*) algebraic equations arising with net methods are large and in linearised form have large bandwidth as a result of three-dimensional coupling; (*b*) for compressible flows the equations are nonlinear; and (*c*) the wake is generally a surface of discontinuity and its location is unknown. The last two features call for iterative processes and this generally creates a succession of linearised problems with varying coefficient matrices. (It is possible to use formulations with a fixed coefficient matrix for the linear problems but then the convergence may be more questionable; we have not tried this.) Together with the large bandwidth, these conditions make direct solution of the linearised problems prohibitively long. Block iteration mixed with iterations for the nonlinear features (*b*) and (*c*) proves to be much better.

Discretisation methods now must provide a formulation well suited to rapid iterative solution. The geometry of the problem demands either a skew net or else an irregular matrix structure and a complicated formulation of the boundary conditions. Consider nets of the type shown in Fig. 1 for a plane cascade. Similarly constructed nets in three dimensions [4] become highly skewed for typical cascades of large outlet and sweep angles. This has the effect of spoiling the property of diagonal dominance for linearised finite difference equations. A guarantee of Successive Over-Relaxation (SOR) convergence is then lost and high SOR parameters may cause divergence. It is difficult to select net space ratios and acceleration parameters for convergence and in some cases the best results are prohibitively slow [4].

A variational finite element method on the other hand leads to symmetric semidefinite matrices and an alternative guarantee of SOR convergence regardless of the skewness of the net. Initial comparisons of difference and element methods for plane ducts involving point SOR demonstrate this convergence [5]. But apart from extreme cases of skewness the finite difference method gives a faster optimal convergence. Nevertheless the results, together with those reported here, suggest that finite elements may be effective for the general three-dimensional problems, and this is reinforced since the conditions of classical optimal convergence apply for block SOR iterations.

Both methods are capable of admitting local net refinements in order to represent the flow more accurately in regions of significance, and both represent the continuous problem to the same order of accuracy. The element method does have an advantage of simplicity in coping with local refinements and especially with locally higher order approximations by using more complex ele-

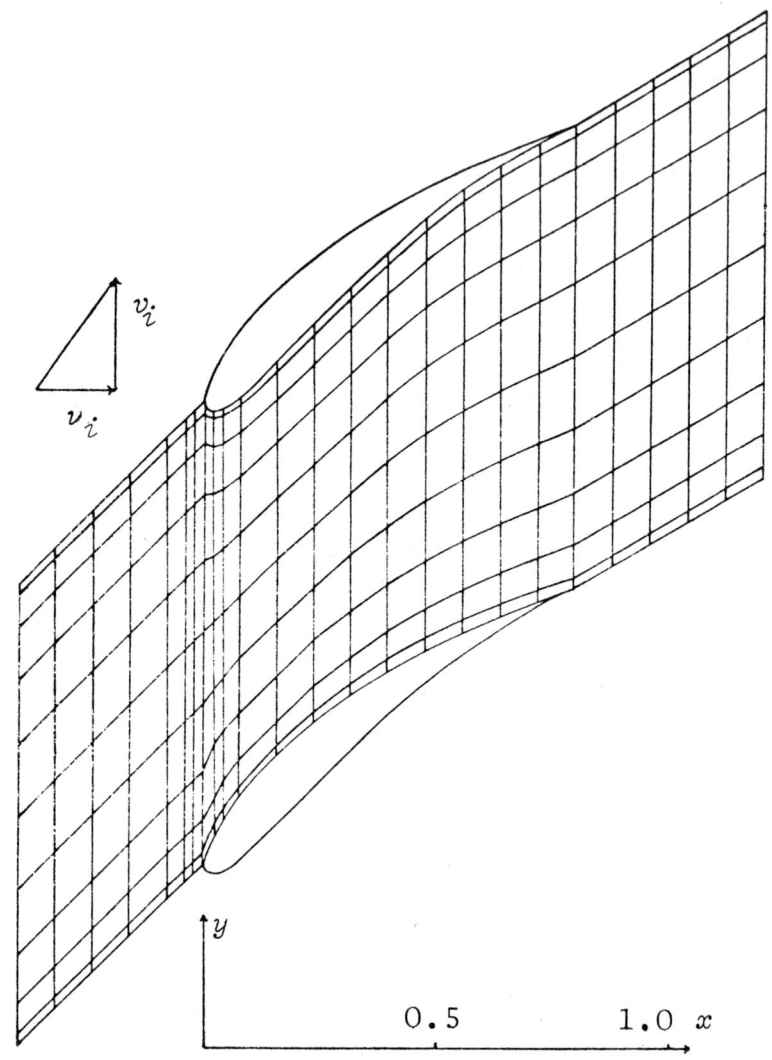

Fig. 1. Finite element net for compressor cascade of Gostelow with inlet velocity components u_i = 0.594823, v_i = 0.803857

ments. The stock of element models developed in recent years should be a valuable contribution to the "geometric tool kit" needed for the configuration sensitive problems of aerodynamics.

Scope of Present Paper

Some comparisons of element and difference methods for steady nonviscous potential flow in plane cascades are given here. A test cascade is chosen for which an exact solution is known so the discretisation accuracy of the methods may be exposed. The flow is incompressible although the methods apply to compressible potential flows; this does not detract from the main purpose - to find rapid iterative techniques for the linearised problems on skew nets. Our experience has been that maximum local Mach numbers up to 1.2 add no special difficulties beyond slowing down iteration convergence slightly [4]. The two methods are compared for block SOR performance on a moderately skewed net. Our results follow on from those published earlier [5] for point SOR for flow in plane ducts.

DISCRETISATION METHODS

Interior points and elements

In the finite difference method interior point equations are local approximations to the mass conservation equation div ρq = 0 where ρ is density and $q = \nabla \phi$ is velocity. For plane isentropic flow this has the quasilinear form

$$(1-u^2/c^2)\phi_{xx} - 2uv/c^2 \phi_{xy} + (1-v^2/c^2)\phi_{yy} = 0 \quad (1)$$

with subscripts denoting partial differentiation, (u,v) the velocity components and c the local speed of sound. We use a transformation to net coordinates to construct a second order 7-point difference formula for this equation in a manner favourable to maintaining diagonal dominance. Details are given elsewhere [5], [8].

The finite element approach seeks to maximise the functional

$$J(\phi) = \int_D p \, dxdy + \int_{\partial D} \phi m ds \qquad (2)$$

over a bounded flow domain D with boundary ∂D, where p is the pressure related to ϕ by Bernoulli's equation in the form

$$dp/d(\tfrac{1}{2}q^2) = -\rho. \qquad (3)$$

The Euler-Lagrange equation for this functional is the mass conservation equation div $\rho \underline{q} = 0$.

We seek to maximise $J(\phi)$ among piecewise linear functions on a triangulation such as obtained from the net of Fig. 1 by introducing the shorter diagonals of the quadrilateral cells. We showed that selecting these diagonals is favourable to diagonal dominance [5] and have since found that this leads to slightly better computational performance.

Consider a typical triangle with nodal values ϕ_1, ϕ_2, ϕ_3 for an approximation ϕapprox. interpolated linearly in the form

$$\phi_{\text{approx.}} = a_1 \phi_1 + a_2 \phi_2 + a_3 \phi_3$$

where the coefficients a_1, a_2, a_3 are linear in the coordinates x, y. This gives a constant velocity $q_{\text{approx.}} = \nabla \phi_{\text{approx.}}$ in the triangle and this velocity is linear in the nodal values ϕ_1, ϕ_2, ϕ_3. Now the stationary condition $\partial J / \partial \phi_1 = 0$ includes contributions from the integrals over various triangles. The contribution from the triangle under consideration is

$$\partial J_\Delta / \partial \phi_1 = -\rho A \, \partial (\tfrac{1}{2} q^2_{\text{approx.}}) / \partial \phi_1 \qquad (4)$$

where ρ is the density corresponding to $q$approx., A the area of the triangle and q^2approx. is a quadratic form in ϕ_1, ϕ_2, ϕ_3. Together with contributions $\partial J_\Delta / \partial \phi_2$ and $\partial J_\Delta / \partial \phi_3$ to the other stationary conditions this involves an element coupling matrix $\rho A M$ where M is the matrix of the quadratic form $q^2_{\text{approx.}}$. (The factor $\tfrac{1}{2}$ disappears in the differentiation). The complete expression $\partial J/\partial \phi_1$ involves terms in other nodal values from other triangles involving ϕ_1. Commonly in finite element work a global coupling matrix is assembled by adding in element matrices from each triangle. Here we compose equations locally in an efficient manner suitable for block SOR solution. The equations are linearised at each iterative step by freezing the density factor in (4).

The piecewise linear representation is simple and in much finite element work it has given way to more elaborate elements, with piecewise cubics among the favourites. For the present purpose, however, the simpler form offers the rapid reassembly of local equations which will be needed with the iterative approaches it is necessary to adopt. The integrand in J_Δ is then constant so that no numerical integration procedure is required to arrive at equation (4). Moreover the approximation to the differential equation achieved is of second order as with the finite differences; for Laplace's equation on a rectangular net the two methods coincide.

Boundary Conditions

On blade and wake surfaces the tangential flow condition is that the normal derivative $\partial \phi/\partial n$ is zero. The natural condition for the maximisation of J is

$$\rho \, \partial\phi/\partial n = m \qquad (5)$$

and so m is taken as zero on these boundaries. Then no special treatment is needed with the element method. We can show that we automatically achieve a second order representation of this boundary condition by the equations made up from triangle contributions of the form (4) above. Finite differences do require special treatment of the boundary condition. A second order representation spoils diagonal dominance and in these computations we use a first order formulation which gives results of similar accuracy and superior iteration performance. Details are described elsewhere [8]. A special formula is adopted for leading edge cells [4].

Inlet and outlet boundaries carry conditions that specify the total mass flow and the boundary distribution of one scalar variable. In the difference calculations it is convenient to specify the velocity component along the grid lines to be a fixed value across the boundary. The element method uses the natural condition for $J(\phi)$ with m uniformly specified across the boundary. The second integral in J is approximated by terms of the form $\phi_i \int m ds$ where ϕ_i is a nodal value and the integral extends half a net space either side of the node. When ϕ_i is at a corner of the region the integral is taken on one side only. In the element assembly process a constant term $\int m ds$ appears on the right of equation (4).

Upstream Periodicity

Irrotationality of the flow, with its periodicity, implies that the jump in ϕ across the flow region is constant upstream of the blades. Thus

$$\phi_A - \phi_B = s V_i \qquad (6)$$

where s is the space between consecutive blades, V_i is a given average inlet swirl velocity and ϕ_A, ϕ_B are taken at nodes at the opposite ends of a net transversal. The difference net is not the same as the element net of Fig. 1 and equation (6) is applied directly to upstream boundary nodal values relating them to opposite interior nodal values [4]. In the element method it is important to use equations of the form (6) as constraints in the maximising of J, thus preserving the symmetry of the coefficient matrix for the final linearised system. Loss of symmetry by imposing periodicity conditions as extra equations in terms of a stream function led to convergence difficulties as encountered by Thompson [9].

The constraint (6) may be applied by substitution. Eliminating, say, ϕ_B from the approximate expression for $J(\phi)$ leads to a new form of the stationary equation $\partial J/\partial \phi_A = 0$, as

$$(\partial J/\partial \phi_A)_{\text{constr.}} = (\partial J/\partial \phi_A)_{\text{free}} + (\partial J/\partial \phi_B)_{\text{free}} (\partial \phi_B/\partial \phi_A)$$

Now $\partial \phi_B/\partial \phi_A = 1$ so we add the "old" equations for ϕ_A and ϕ_B to produce the single "new" equation. Also by substituting $\phi_B = \phi_A - sV_i$ we have an easily implemented condensation technique for dealing with the constraints and, of course, preserving the symmetry. This applies to linear constraints in general and is appealing when the constraints strongly involve particular nodes for elimination.

Wake Location

In both of these methods an initial guess is made of the wake position, and the net is fitted to this wake as a boundary. After the flow problem has been partly solved the pressure jump across the guessed wake is inspected and the wake moved to unload it. This wake shifting process proceeds iteratively along with the flow calculation. A similar net shifting approach is used by Wilkinson [10] with his streamline curvature method. It is possible to consider a thick wake with some momentum and corresponding transverse loading consistent with its curvature but this is not done in the present work.

Let Δp be the calculated pressure jump across the wake in the negative y direction at a given x station: that is, Δp is found as the pressure jump across the flow region of Fig. 1 in the positive y direction. Thus when Δp is positive an upward force acts on the assumed wake. To unload this the downward reaction on the flow must be removed so the flow gains momentum in the y direction. The required effect on the total flow is to increase the downstream swirl velocity by δv given by

$$m \, \delta v = h \, \Delta p, \qquad (7)$$

where m is the total mass flow rate per blade and h is the x component of distance over which Δp acts. Such extra swirl is added for each net cell along the wake. The extra velocity δv is imposed uniformly with respect to y; this is not exact but, as the wake equilibrium position is approached, $\delta v \to 0$ and this inaccuracy does not affect the limiting result. A piecewise linear variation of δv with respect to x is introduced so the momentum is gradually assumed across each cell on which Δp acts. The net position is then moved to match the flow angle by a piecewise quadratic displacement $\int \delta v / u \, dx$, where u is the velocity in the x direction. To first order the net movement does not affect the measured pressure discrepancy Δp but the additional swirl δv removes it precisely. Details are given in O'Carroll [4] together with a description of the treatment to unload the wake immediately behind the trailing edge. When the trailing edge is rounded the pressure is made to balance symmetrically across the projected camber line. This condition is important and fixes the circulation as does the Kutta condition for a sharp trailing edge. Basically it agrees with Preston's results on vorticity shedding but ultimately the potential flow computation should be coupled with a boundary layer analysis (e.g., [11]).

COMPUTATION

Algebraic Iterations

The linearised discrete equations can be grouped for nodes on the same transversal and a typical group written in the form

$$A_i \underline{\phi}_{i-1} + B_i \underline{\phi}_i + C_i \underline{\phi}_{i+1} = \underline{b}_i \tag{8}$$

where ϕ_i is the vector of nodal values on the ith transversal, A_i, B_i and C_i are coefficient matrices (with frozen density factors) and b_i is a constant vector. These are solved directly for ϕ_i and the new values are used in a block SOR with relaxation factor ω:

$$\phi_i^* = B_i^{-1}(b_i - A_i \phi_{i-1}^{(n+1)} - C_i \phi_{i+1}^{(n)}),$$

$$\phi_i^{(n+1)} = \phi_i^{(n)} + \omega(\phi_i^* - \phi_i^{(n)}),$$

the superscript n denoting iteration number. The matrix B_i is tridiagonal apart from extra coupling of outermost nodes by the upstream periodicity condition. The block equations are thus efficiently solved by a suitable algorithm [4]. In the finite element approach we assemble contributions from triangles between the ith and $i+1$th consecutive transversals, collecting C_i and part of B_i, "remembering" A_i and the rest of B_i from the previous strip. The global coupling matrix is not stored so relatively little computing space is required.

Convergence theorems for block SOR (e.g., Varga [12]) do not go far for the difference method. If the coefficient matrix is irreducibly diagonally dominant and of positive type, the block Gauss-Seidel method (i.e., with $\omega = 1$) converges faster than the point Gauss-Seidel method, which in turn is faster than point SOR, with $\omega < 1$. However, little can be said for $\omega > 1$ and even the conditions of diagonal dominance and positivity can be lost when dealing with highly skewed nets. On the other hand the element method provides a symmetric semidefinite matrix with definite block diagonal and the triple transversal coupling of equation (8) gives the block property "A." Under these conditions and with sweeping iterations through the transversals from left to right, Young's optimal theory applies. An optimal acceleration factor is

$$\omega_b = 2/(1 + \sqrt{1 - \mu^2})$$

where μ is the block Jacobi spectral radius. The corresponding SOR spectral radius is $\omega_b - 1$, giving a sharp optimum at $\omega = \omega_b$ much better than the Gauss-Seidel case for $\omega = 1$.

These theorems, in fact, require a definite rather than semidefinite coefficient matrix but this problem can be avoided by projecting out null space components corresponding to the singularity of the matrix for a Neumann problem [13]. The idea is used to measure increment seminorms of successive approximations simply by taking their standard deviations. The ratio of successive increment norms asymptotically averages to the spectral radius of the iteration matrix. We study convergence behaviour of our methods by considering these ratios λ for various values of ω in the same way for block SOR here as for point SOR in O'Carroll and Morgan [5]. Estimates for λ were taken after 20 to 50 iterations, depending on the speed of settling.

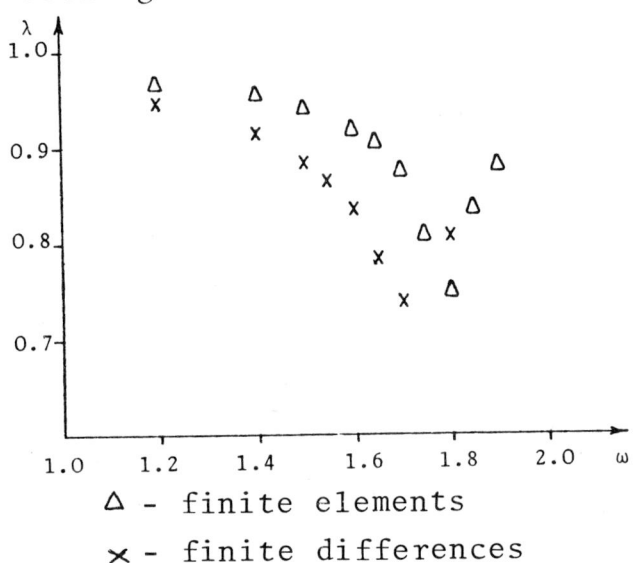

$\triangle$ - finite elements

$\times$ - finite differences

Fig. 2. Averaged block SOR increment norm ratios *versus* acceleration parameter ω for a net of 15 × 5 nodes for a cascade with 70° outlet angle

Fig. 2 shows comparative results for difference and element methods for a cascade with outlet angle approximately 70° so the net is moderately skewed. The net used was composed of 15 × 5 nodes with equal spacing in the y direction, and the wake shifting iteration was suppressed to expose the behaviour of the block SOR process. The results show some similarity to those for point SOR [5] in that the optimal convergence for finite elements occurs at a higher relaxation factor than for finite differences. There is little to choose here in optimal speed.

Discretisation accuracy

The compressor cascade of Fig. 1 was obtained by conformal transformation so the exact incompressible flow through it is known [14]. The outlet slope for the exact solution is 0.5779 and the net settles close to this in the wake shifting process.

Fig. 3 shows the calculated pressure distributions around the blade using the net of Fig. 1, together with the exact solution, plotted against chordwise position. The inclusion of refining transversals near the leading edge improves the resolution but shows an exaggerated peak velocity. This exaggeration is diminished when the net is equally spaced in the y direction but otherwise alteration of the y spacing makes little difference. Finite difference results are also shown and give much the same discretisation accuracy as with finite elements.

The addition of refining transversals carries a penalty of reducing convergence speed, but this is fairly small in optimal speeds allowing for the upward shift of optimal relaxation factor with the refined net. For example, for the problem of Fig. 1 the addition of the two extra transversals both upstream and downstream of the leading edge in a net previously of 21 × 10 cells raised the optimum error multiplier from about 0.80 to 0.83.

CONCLUSIONS

Computational methods for three-dimensional blade-to-blade cascade flows using velocity potential on a skew net have been investigated. Rea-

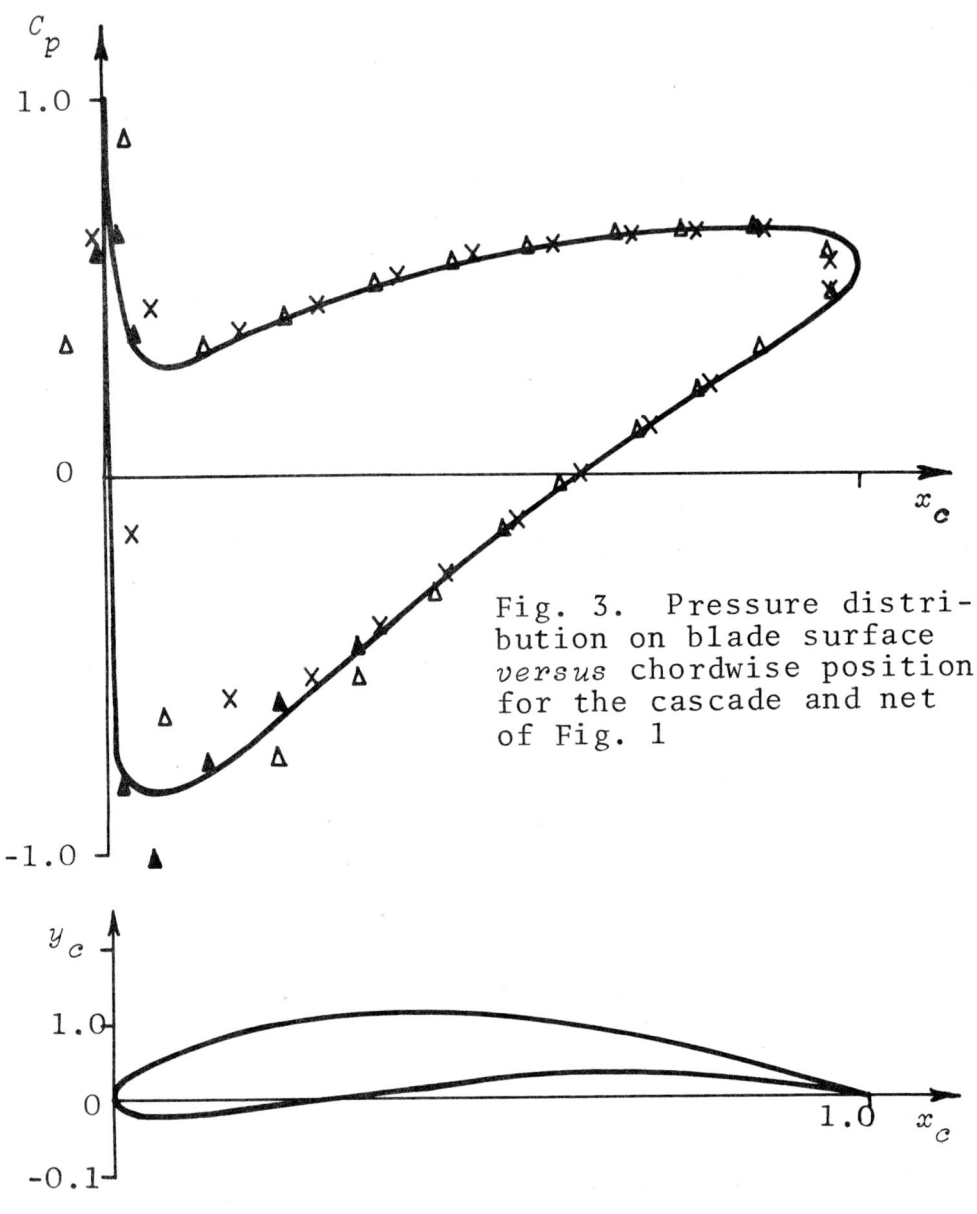

Fig. 3. Pressure distribution on blade surface *versus* chordwise position for the cascade and net of Fig. 1

× - finite difference calculation
△ - finite element calculation (without refinement near leading edge)
▲ - finite element calculation (with refinement near leading edge)
——— - exact solution

sonable limitation of computing time requires block iterative techniques for which finite difference equations are ill suited when the net is highly skewed.

The finite element method as constructed satisfies the necessary conditions for optimal block SOR theory and this is borne out by the two-dimensional computations presented. In the plane the element equations take about three times longer to assemble than the comparable difference equations, and convergence of iterations is only of similar speed. Optimal convergence is guaranteed for the more severe problems in three dimensions which is not so with difference equations.

To solve the problem of Fig. 1 (omitting the refining transversals) with an algebraic error of less than 1% in velocities and allowing for wake location takes 45 seconds on Atlas by finite differences and about three times as long with finite elements. Extending the same problem to three dimensions with only three cells in the third direction increases the finite difference time to 750 seconds. Comparable three-dimensional finite element calculations have yet to be made.

The two methods give results of similar discretisation accuracy but the element method lends itself more easily to local net refinement. Indications of results so far are that finite elements offer a practicable way of computing three-dimensional blade-to-blade flows although there remain uncertainties in the speed of the location of twisted wakes and in the achievement of good resolution in regions of intricate geometry.

REFERENCES

1. Gostelow, J.P., Horlock, J.H. and Marsh, H., "Recent developments in the aerodynamic design of axial flow compressors," *Proc. I. Mech. E.*, **183**, part 3N, paper 6 (1969).

2. Serovy, G.K., "Recent progress in the aerodynamic design of axial flow compressors in the United States," *J. Eng. Power, Trans. ASME.*, **88** (series A), 251 (1966).

3. Horlock, J.H., "Axial flow turbines," Butterworth (1966).

4. O'Carroll, M.J., PhD Thesis, Lanchester Polytechnic (1972).

5. O'Carroll, M.J. and Morgan, L.A., "Difference and element methods for potential flow," *in* Oden, J.H., *et alia, Editors,* "Finite Element Methods in Flow Problems," UAH Press, Huntsville (1974).

6. O'Carroll, M.J., "Computation of rotational flows," *Q. Appl. Math.*, **29**(4), 547-550 (1972).

7. Wu, J.C., "Integral representation of field variables for the finite element solution of viscous flow problems," *in* Pulmano, V.A. and Kabaila, A.P., *Editors,* "Proceedings of the 1974 International Conference on Finite Element Methods in Engineering," University of New South Wales, 827-840 (1974).

8. O'Carroll, M.J., "Velocity potential computation by finite differences for compressible flow in ducts," *Int. J. Num. Meth. Eng.*, *in the press.*

9. Thompson, D.S., "Finite element analysis of the flow through a cascade of aerofoils," Cambridge University Engineering Department, Turbo/TR 45 (1973).

10. Wilkinson, D.H., "Calculation of blade-to-blade flow in a turbomachine by streamline curvature," ARC R & M 3704 (1972).

11. Gostelow, J.P., Lewkowicz, A.K. and Shaalan, M.R.A., "Viscosity effects on the 2-dimensional flow in cascades," ARC, C.P.872 (1967).

12. Varga, R.S., "Matrix Iterative Analysis," Prentice Hall (1962).
13. O'Carroll, M.J., "Inconsistencies and SOR convergence for the discrete Neumann problem," *JIMA*, **11**, 343-350 (1973).
14. Gostelow, J.P., "Potential flow through cascades - a comparison between exact and approximate solutions," ARC, C.P.807 (1963).

THE FINITE ELEMENT METHOD APPLIED TO FLUID MECHANICS

J.H. Argyris and P.C. Dunne

(University of Stuttgart)

ORIGINS OF THE FINITE ELEMENT METHOD

Depending on the point of view the finite element method may be traced to mathematical [1] or engineering sources. There is no doubt that engineers used the method successfully long before its mathematical basis was firmly established. Intuitive physical lumping of complex plate and shell structures was used by aeronautical engineers before the war [2]. Aeronautical engineers were also the first to realise the enormous potential of the electronic computer to organise and calculate. In fact, in the structural field a theoretical foundation was given through the principle of virtual work more than 20 years ago [3].

On the other hand, mathematicians now consider the method as a special application of weighted residual or extremum methods, in which the interpolation functions are defined separately in a large number of small domains with certain necessary continuities between them at their boundaries [4], [5], [6].

For many years the method was applied very little outside the field of solid mechanics. In this field a vast store of computer "software" exists. Much of this could be applied to other fields such as fluid mechanics.

Most applications have been to linear problems and especially to those problems for which a definite extremum formulation exists.

More recently it has been realised that the weighted residual methods (for example, Galerkin's method) make it possible to apply the method to any problem which may be stated as one or more differential equations. Existing routines are applicable to the linear parts of the equations. Nonlinear parts greatly increase the difficulties and cost of solution.

Definition of a finite element

A finite element is a portion of a continuum. The shape is generally simple - triangle, tetrahedron, quadrilateral, etc. - but may also be curved. The most important property is that the values of certain field variables defined at nodes on the boundary define these variables all over the part of the boundary defined by the same nodes. Interior nodes may also exist but the value of a field variable at these nodes does not influence boundary values.

The above property ensures that the specified field variables are continuous through the boundary with an adjacent element. Field variables may be scalars or vectors or gradients of these.

Interpolation functions in finite elements

If ϱ_1 is vector of values of a field variable ψ at nodal points then ψ at any point in or on the boundary of the element is defined by

$$\psi = \omega \varrho_1 \qquad (1)$$

where ω is a row matrix of interpolation functions. The interpolation functions may be given in terms of the most convenient coordinates for the element such as natural (area) coordinates in triangles, curvilinear coordinates in curved elements.

(a) Linear ψ over a triangle

TRIM 3

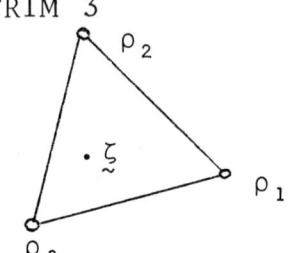

$$\underset{\sim}{\rho}_I = \{\rho_1 \;\; \rho_2 \;\; \rho_3\}$$
$$\underset{\sim}{\omega} = [\zeta^1 \;\; \zeta^2 \;\; \zeta^3]$$

(b) Parabolic over a line

internal node

$$\underset{\sim}{\rho}_I = \{\rho_1 \;\; \rho_2 \;\; \rho_3\}$$
$$\underset{\sim}{\omega} = [(1-\xi)(1-2\xi) \;\; 4\xi(1-\xi) \;\; \xi(2\xi-1)]$$

$\xi = x/h$

Fig. 1. Simplest cases of interpolation functions

Simplex Elements
$$\psi = \sum_{r,s=0}^{n} a_{rs} x^r y^s \quad r+s \leq n \quad (2)$$

These are perhaps the most useful of all elements. They use complete polynomial interpolation of any degree n. The first member is given in Fig. 1. Higher members are shown in Fig. 2.

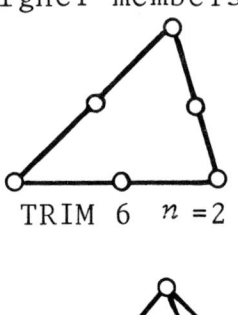
TRIM 6 $n=2$

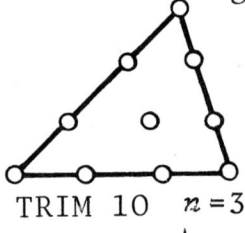
TRIM 10 $n=3$

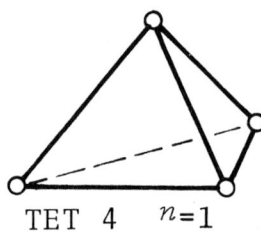
TET 4 $n=1$

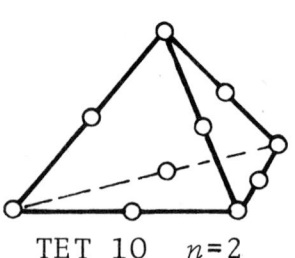
TET 10 $n=2$

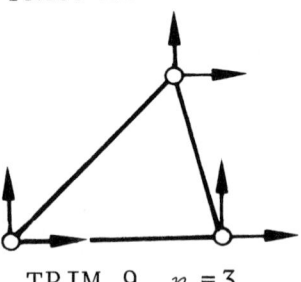
TRIM 9 $n=3$

o   field variable ψ

⟶   gradient of ψ

⟶⟶   gradient and second derivative

x   cross-derivative

In TRIM 9 the central node is eliminated so that the polynomial is not complete. Derivatives are continuous at nodes but not along sides.

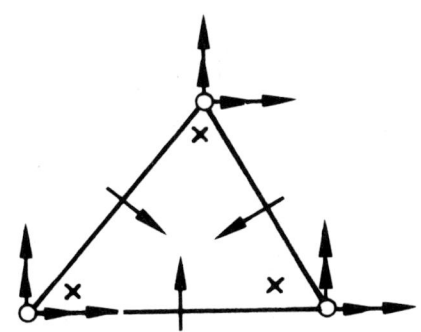

TUBA 6 is the simplest element giving continuous first derivatives. It gives also continuous second derivatives at vertices.

TUBA 6. $n=5$ (21 nodal parameters)

Fig. 2.   Various simplex elements

Rectangular elements (Fig. 3)

These elements use complete Lagrangian interpolation or a restricted version obtained by eliminating internal nodes. Thus the interpolating polynomial is of the form

$$\psi = \sum a_{rs} x^r y^s \qquad r,s = 0,1,2,\ldots n \qquad (3)$$

$\underset{\sim}{\rho}_I = \{\rho_1 \ \rho_2 \ \rho_3 \ \rho_4\}$

$\underset{\sim}{\omega} = [\tfrac{1}{4}(1+\xi)(1-\eta) \ \ \tfrac{1}{4}(1+\xi)(1+\eta)$

$\qquad \tfrac{1}{4}(1-\xi)(1+\eta) \ \ \tfrac{1}{4}(1-\xi)(1-\eta)]$

$\xi = x/a \quad \eta = y/b$

Simplest case $n=1$ QUAM 4

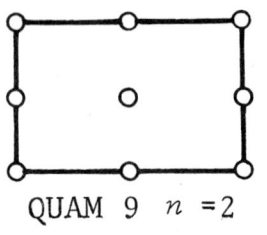
QUAM 9 $n=2$

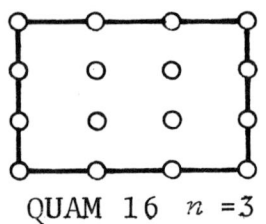
QUAM 16 $n=3$

Similar three-dimensional elements (LUMINA) may be formed.

Fig. 3. Rectangular elements - Lagrangian interpolation

From the case $n=3$ by allowing all nodes to pass to the nearest corner, Lagrangian interpolation on the sides becomes Hermitian and results in an element giving an interpolation complete up to $n=3$ and with continuous derivatives at the boundaries: 16 nodal parameters correspond to 16 coefficients in (3) (see Fig. 4).

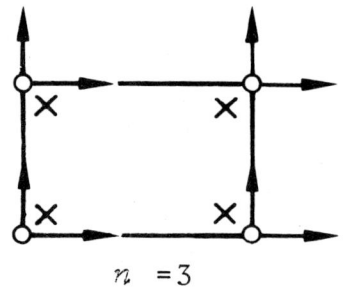
$n=3$

Fig. 4. Rectangular element - Hermitian interpolation

Curvilinear elements (Fig. 5)

Curved elements are obtained from straight-sided ones by a mapping procedure in which the mapping functions are the same as the interpolation functions for ψ. Some authors use the term isoparametric for the latter reason. Numerical integration is essential in these elements.

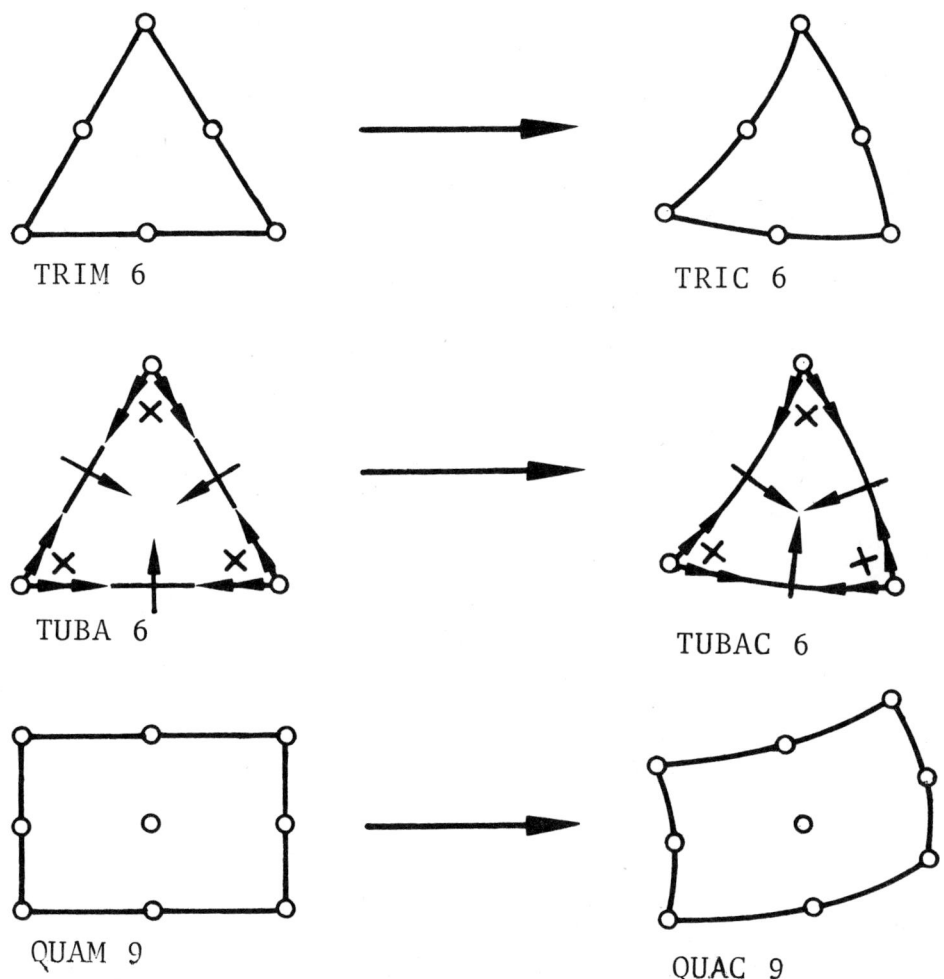

Fig. 5. Curvilinear elements

FORMULATION OF PROBLEM

Problems with positive definite quadratic forms

These are the simplest and are the type of problem for which the method was originally developed. Most of the common terms, such as stiffness matrix, mass matrix, damping matrix and consistent load matrix, derive from this type of problem.

As the simplest example in fluid mechanics we have plane incompressible irrotational flow in a duct (see Fig. 6).

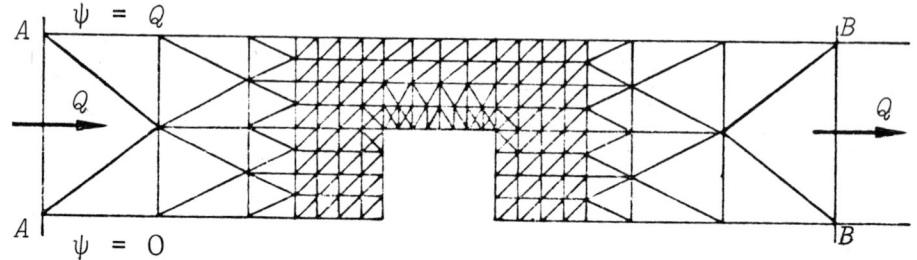

Fig. 6. Plane potential flow in duct

If we suppose the velocity to be uniform in the axial direction at sections AA and BB then ψ varies linearly over these sections and the problem reduces to minimising,

$$U = \tfrac{1}{2}\iint [(\tfrac{\partial \psi}{\partial x})^2 + (\tfrac{\partial \psi}{\partial y})^2] \, dx \, dy \qquad (4)$$

subject to given ψ on the boundaries.

Since $\psi = \omega \rho_{Ii}$ over an element i the contribution of the element to U is,

$$\tfrac{1}{2}\rho_{Ii}^t \iint [\omega_{,x}^t \omega_{,x} + \omega_{,y}^t \omega_{,y}] \, dx \, dy \cdot \rho_{Ii} = \tfrac{1}{2}\rho_{Ii}^t k \rho_{Ii} \qquad (5)$$

where k_i is a square matrix analogous with the stiffness matrix of a membrane under transverse deflection.

If r is the vector of all nodal values of ψ,

$$\rho_I = ar \qquad (6)$$

where a is a Boolean location matrix. Then the total U may be written

$$U = \tfrac{1}{2}r^t K r \qquad (7)$$

where

$$K = a^t k a \qquad (8)$$

and

$$\underset{\sim}{k} = \lceil \underset{\sim}{k}_1 \ \underset{\sim}{k}_2 \ldots \underset{\sim}{k}_i \ldots \underset{\sim}{k}_s \rfloor \qquad (9)$$

Minimisation of U with respect to nodal freedoms gives

$$\frac{\partial U}{\partial \underset{\sim}{r}_t} = \underset{\sim\sim}{K r} = \begin{bmatrix} 0 \\ \underset{\sim}{f} \end{bmatrix} \qquad (10)$$

where $\underset{\sim}{f}$ is the vector of unknown boundary "forces".

Note that the integration in (4) requires only that ψ be continuous at the boundaries. Thus the simplest elements such as TRIM 3 or QUAM 4 may be used. However, higher degree elements such as TRIM 6 will generally give a better result for a given number of freedoms.

Another way of treating this problem is to start with the differential equation,

$$\frac{\partial^2 \psi}{\partial x^2} + \frac{\partial^2 \psi}{\partial y^2} = - q(x,y) \qquad (11)$$

It is well known that treatment of this equation by the Galerkin method requires satisfaction of all boundary conditions. This means also that gradients must be continuous at finite element boundaries. Thus the Galerkin (or other weighted residual method) would require the use of higher order elements such as TUBA 6.

However, if the expression

$$- \int\int [\frac{\partial^2 \psi}{\partial x^2} + \frac{\partial^2 \psi}{\partial y^2}] \delta\psi \ dx \ dy \qquad (12)$$

is integrated by parts one obtains, *ignoring the boundary integrals*,

$$\tfrac{1}{2}\delta\int\int [(\frac{\partial \psi}{\partial x})^2 + (\frac{\partial \psi}{\partial y})^2] \ dx \ dy = \delta U \qquad (13)$$

Thus the Galerkin method then becomes equivalent to the extremum method. If

$$\frac{\partial \psi}{\partial n} = v_s$$

is specified on parts of the boundary this must be

regarded as part of the "loading" function $q(x,y)$. Then the total functional to be varied is

$$\Phi = \tfrac{1}{2}\iint [(\tfrac{\partial \psi}{\partial x})^2 + (\tfrac{\partial \psi}{\partial y})^2] \, dx \, dy - \iint q(x,y) \psi \, dx \, dy \\ - \int_C v_s \, \psi \, ds \quad (14)$$

where C is the part of the boundary with specified v_s.

In the present problem the vorticity q is supposed zero and v_s is not specified so that Φ is identical with U in (4).

Occasionally the expression for U obtained in the above manner does not correspond to the value obtained by direct calculation. For example, in plate bending theory the expression for U, in this case the strain energy, depends on Poisson's ratio whereas the latter does not enter, except in the multiplying constant D, in the differential equation,

$$D\Delta^4 \psi = q \quad (15)$$

The reason is that the expressions multiplying $\delta\psi$ and $\delta\tfrac{\partial \psi}{\partial n}$ in the neglected boundary integrals do not correspond to the boundary shears and bending moments except for Poisson's ratio $\nu = 1$. When the boundary integrals of the above expressions and the true expressions for the boundary shears and moments are included in U the correct expression is obtained. Nevertheless, in some cases the correct solution will be obtained without the latter correction - for example in simply supported or built-in plates.

These remarks emphasise the difficulties and pitfalls that may be expected in applying the method to non-positive definite problems and non-linear problems.

The contribution of the "loading" terms to the right hand side of an equation of the type (14) is most easily obtained by using interpolation functions for q and v_s which may be of equal or lower degree than $\underline{\omega}$. For example, in applying the TUBA 6 5th degree element a cubic, parabolic or even linear representation of the "loading" may be adequate. Thus if q is represented by

$$q_i = \bar{\underline{\omega}} \bar{\underline{\rho}}_i \tag{16}$$

then the contribution to Φ of an element i is

$$- \underline{\rho}_{Ii}^t \iint \underline{\omega}^t \underline{\bar{\omega}} \, dx \, dy \cdot \bar{\underline{\rho}}_i = - \underline{\rho}_{Ii}^t \bar{k}_i \bar{\underline{\rho}}_i \tag{17}$$

where $\bar{k}_i$ is a rectangular matrix.

Assembling $\bar{\underline{\rho}}$ as $\bar{a}\bar{\underline{r}}$ one obtains for the total contribution to Φ,

$$- \underline{r}^t \bar{\bar{K}} \underline{r} \tag{18}$$

where

$$\bar{\bar{K}} = \underline{a}^t \bar{\bar{k}} \bar{\underline{a}} \tag{19}$$

in which

$$\bar{\bar{k}} = \lceil \bar{k}_1 \bar{k}_2 \ldots \ldots \bar{k}_s \rfloor \tag{20}$$

Then the "load" vector resulting from q in the equation

$$K\underline{r} = \underline{f} \tag{21}$$

is

$$\underline{f}_q = \bar{\bar{K}} \underline{r} \tag{22}$$

A similar contribution f_s will arise from the boundary "loading" v_s where the interpolation will be only over the element boundary.

Problems with non-conservative properties

The simplest way of looking at problems of this type is to regard the non-conservative terms as a "loading" which may be entered in the right hand side of the differential equation. For example, in the supersonic flutter of plates by piston theory one requires a pressure proportional to the gradient $\frac{\partial \psi}{\partial y}$. Thus the equivalent loading is

$$q_a = \lambda \frac{\partial \psi}{\partial y} = \lambda \underset{\sim}{\omega}_{,y} \underset{\sim}{\rho}_I i \qquad (23)$$

It follows from (17) that the corresponding load vector is

$$\underset{\sim}{f}_a = \underset{\sim}{K}_a \underset{\sim}{r} \qquad (24)$$

where

$$\underset{\sim}{K}_a = \underset{\sim}{a}^t \underset{\sim}{k}_a \underset{\sim}{a} \qquad (25)$$

is a square non-symmetrical aerodynamic stiffness matrix

$$\underset{\sim}{k}_a = \lceil \underset{\sim}{k}_{a1} \; \underset{\sim}{k}_{a2} \ldots \underset{\sim}{k}_{as} \rfloor \qquad (26)$$

and

$$\underset{\sim}{k}_{ai} = \lambda \iint \underset{\sim}{\omega}^t \underset{\sim}{\omega}_{,y} \, dx \, dy \qquad (27)$$

An important consideration is the order of the highest derivative occurring in the simulated "loading" function. Ideally this should not be more than one greater than the order of the derivatives for which the interpolation function maintains continuity. Thus, for the TRIM series of elements the highest derivative would be the first. For the TUBA series it would be the second.

The above restriction would require very high order continuities in interpolation functions to deal with "loading" terms of order higher than 2. Such elements are possible but do not exist at present in any routine software package. Continuity in second derivatives would require a complete interpolation polynomial of degree nine over a triangle - a total of 55 nodal parameters. A rectangular element would require 36 parameters.

Because of the desire to avoid high degree interpolation functions, elements are often used which do not satisfy the full Galerkin boundary conditions. In such cases the boundary condition equations (in elasticity called the boundary equilibrium conditions) must be included in the total variational statement. Care must be taken that the correct sign and weighting is given to a contribution from the boundary. In elasticity this is easily done as all contributions represent virtual work. Similar considerations apply in other fields. Application of integration by parts is often used to reduce the order of the derivatives in the field equations to permit integration. As mentioned before the resultant boundary integrals will often cancel with those in the boundary contributions to the variational statement. Thus for a typical nodal displacement r_i one frequently obtains,

$$\frac{\partial \Phi}{\partial r_i} + \text{contribution of non-conservative terms} = 0 \qquad (28)$$

The first part of the equation is formed using routine software. The second part will require special programs, in general involving integration over the boundary as well as the interior of each element.

Another way of reducing the degree of the required interpolation function is to reduce the order of the differential equation by using several equivalent equations of lower order. In solid mechanics this procedure leads to the so called mixed elements which, for example, use displacements and bending moments in plates instead of displacements and rotations. In fluid mechanics one has the example of the plane incompressible Navier-Stokes equations treated in terms of three equations in the velocities and pressure, two equations in terms of the vorticity and stream function, or one equation in terms of the stream function. Taken to the extreme we obtain the situation in which the field variables are constant in each element [7].

However, in solid mechanics at least, one often pays for the apparent simplification in terms of an indefinite extremum and absence of energy bounds.

Although there is now a reasonable literature

on finite elements in fluid mechanics, including the beginnings of a more rigorous mathematical basis, finite difference methods have been much more used. It is fair to say that so far the method has been applied successfully only to problems for which the nature of the solution is known in advance. It is too early to predict whether the method will eventually contribute to a better understanding of the enormously difficult field of fluid mechanics.

POTENTIAL FLOW

The vast majority of finite element computer software existing today in applied mechanics is in the field of solids and structures. This arises from the historical reasons previously mentioned. Although it may be unwise to overdo the application of analogies because of the danger of pushing them beyond their physical validity, there is no doubt that they are of great value in suggesting solution procedures. In particular they enable rapid application of existing software to other fields by persons who are not themselves experts in those fields. In the following we consider a simple membrane (soap film) analogy and a more complex plate analogy.

Membrane analogy

The simple soap film analogy for plane incompressible irrotational flow may be given in terms of the stream function or the potential. Either may be made the basis for a finite element formulation using the TRIM or TRIC series of finite elements. The analogy is specially useful for confined flow with internal obstacles. Consider first the stream function method (see Fig. 7). We suppose the velocity at the entrance and exit of the duct to be purely axial so that

$$\frac{\partial \psi}{\partial x} = 0$$

Thus the nodes at section $x=0$ and L are unloaded.

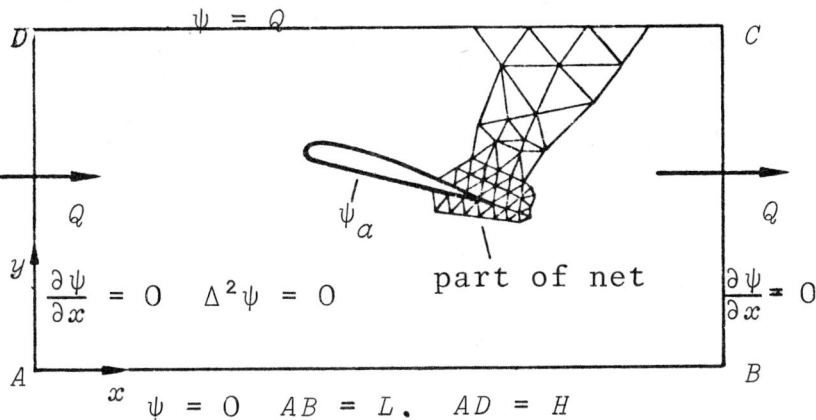

Fig. 7. Membrane analogy for incompressible inviscid flow

A suitable network of finite elements, TRIM 6 away from the aerofoil and TRIC 6 near the aerofoil, is constructed. If the value of ψ_1 is assumed at the aerofoil the membrane equilibrium problem may be solved by routine methods. In general this will give a streamline leaving the trailing edge not in accordance with the Kutta-Joukowsky condition. A repeat calculation with another value of $\psi = \psi_2$ is then made. From the two results an estimate of ψ to satisfy the Kutta-Joukowsky condition is easily made. The aerofoil lift may be found from the circulation

$$\Gamma = \oint v_s \, ds = \oint \frac{\partial \psi}{\partial n} \, ds \qquad (29)$$

= vertical force of membrane on aerofoil, which is equal to the total nodal forces on the aerofoil boundary. When there are several bodies with sharp trailing edges sufficient cases must be calculated to determine the various values of ψ at the bodies.

The second membrane analogy for plane incompressible irrotational flow is described by the equation,

$$\frac{\partial^2 \phi}{\partial x^2} + \frac{\partial^2 \phi}{\partial y^2} = 0 \qquad (30)$$

where

$$u = \frac{\partial \phi}{\partial x}, \quad v = \frac{\partial \phi}{\partial y} \qquad (31)$$

In this case the same problem as in Fig. 7 has the boundary conditions given in Fig. 8.

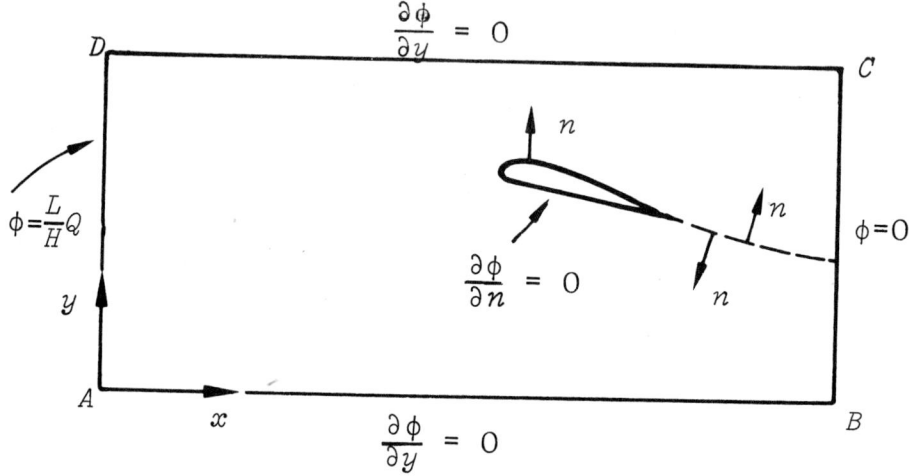

Fig. 8. Membrane analogy by potential method

The Kutta-Joukowsky condition now requires a discontinuity in ϕ along the streamline from the trailing edge which, however, must be constant to maintain equal velocity each side of the line. A trial and error process, which is not as simple as when using the stream function, must be used to fix this line. Note that the condition $\frac{\partial \phi}{\partial n} = 0$ implies zero loads on the corresponding boundary nodes. The above mentioned discontinuity in ϕ is in fact the circulation Γ.

Three-dimensional potential flow

The potential method has the advantage that it can be extended to three-dimensional flow and also may be modified to take account of subcritical compressibility. The resultant equation is nonlinear but in numerical experiments so far made no difficulties have been experienced with convergence of the iterative solution when the nonlinear part calculated from the values of the previous step is placed on the right hand side of the equation. By increasing v_0 in steps, free stream Mach numbers up to 0.8 have been calculated. Thus in two-dimensional flow, for example, one has to solve,

$$\Delta^2 \phi = (\phi_{,x}^2/a^2)\phi_{,xx} + (\phi_{,y}^2/a^2)\phi_{,yy} + (2\phi_{,x}\phi_{,y}/a^2)\phi_{,xy} \quad (32)$$

where a is the local speed of sound given for a perfect gas by,

$$a^2 = a_0^2 - \tfrac{1}{2}(\gamma-1)[\phi_{,x}^2 + \phi_{,y}^2 - v_0^2] \quad (33)$$

in which v_0 and a_0 are the free stream values of velocity and speed of sound. Since second derivatives occur on the right hand side, finite elements of the TUBA 6 or TUBAC 6 type are strictly required. However, cubic triangular elements with continuity in the first derivatives only at the vertices have given good results. This is an example where numerical integration is unavoidable even in the straight-sided elements, because of the term $\tfrac{1}{a^2}$. In the initial iterations one may make various approximations such as

$$a = a_0, \quad \phi_{,x} = v_0, \quad \phi_{,y} = 0.$$

An alternative approximate tangent stiffness method is to keep the nonlinear terms on the left hand side of the equation and regard the coefficients of $\phi_{,xx}$, etc. as constant within each element, the constant being calculated from suitable mean values of $\phi_{,x}$, etc. This method would be most suitable with a large number of simple elements such as TRIM 6 or even TRIM 3.

Figs. 9 to 15 show potential flow calculations made at ISD by Mareczek [8].

NAVIER-STOKES EQUATIONS: A PLATE ANALOGY

Finite element methods have been applied by various authors to the Navier-Stokes equations for incompressible viscous flow, particularly the plane case. Formulations in terms of velocity and pressure, vorticity and stream function, and

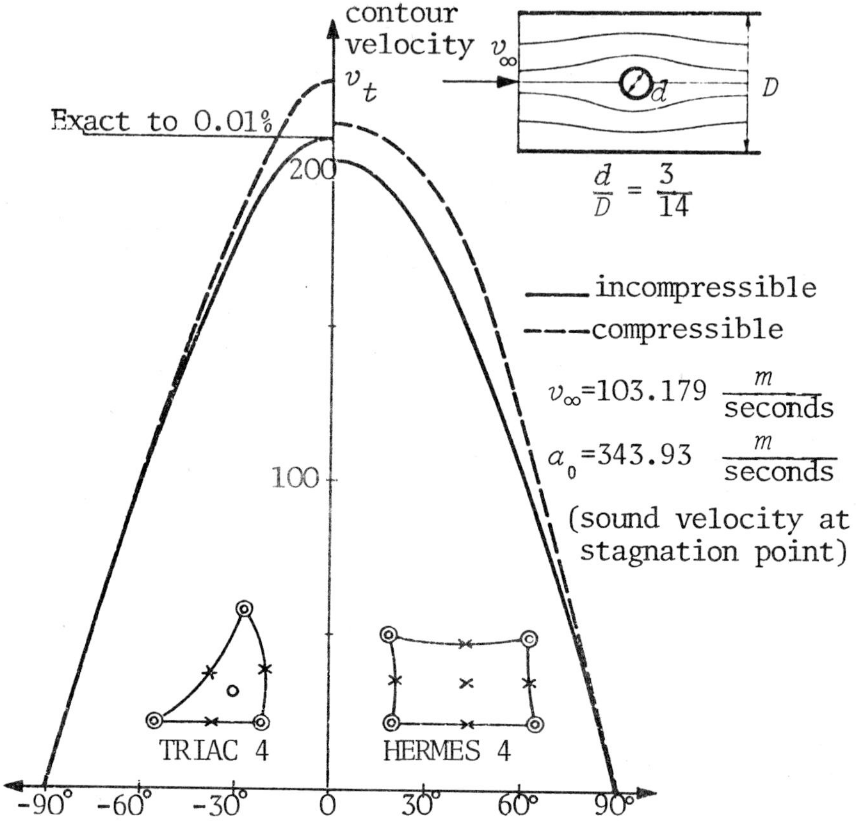

Fig. 9. Potential flow about cylinder in wind tunnel at $M_0 = \dfrac{v_\infty}{a_0} = 0.3$

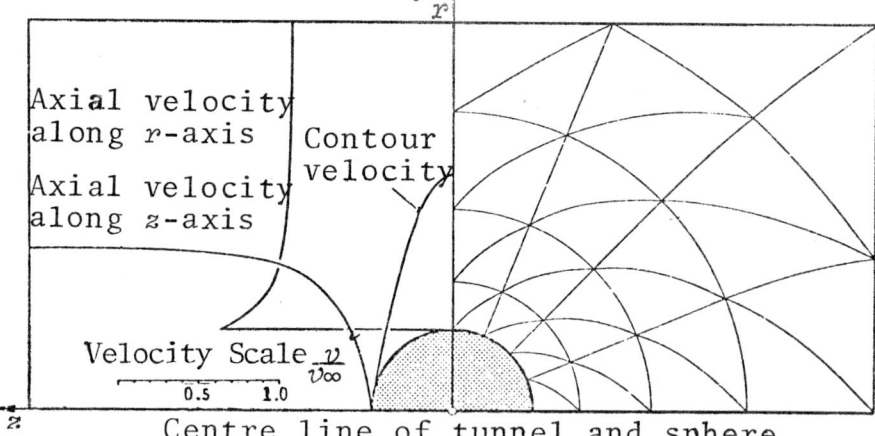

Fig. 10. Potential flow about a sphere in a wind tunnel

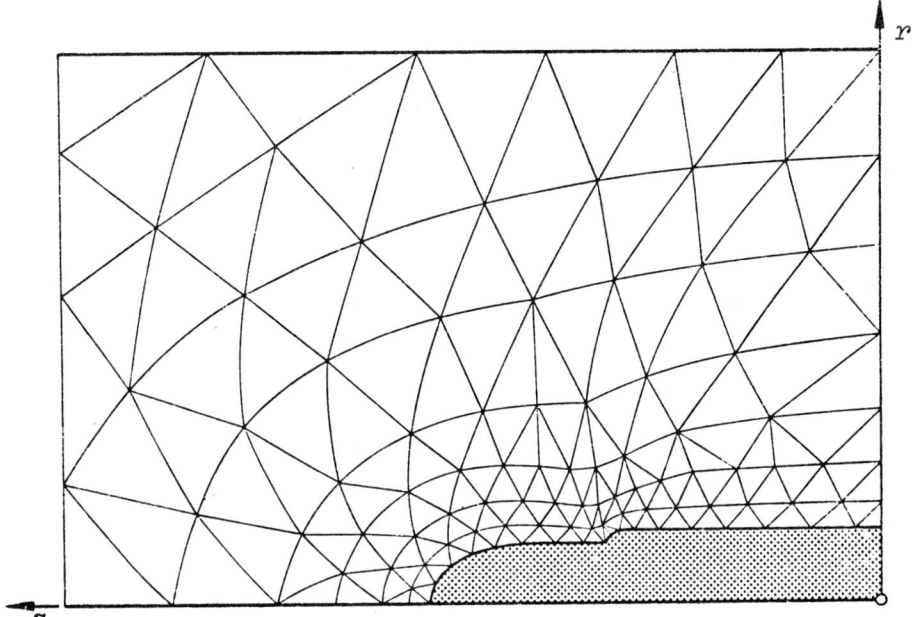

Fig. 11. Finite element mesh for potential flow about axisymmetric turbine hub

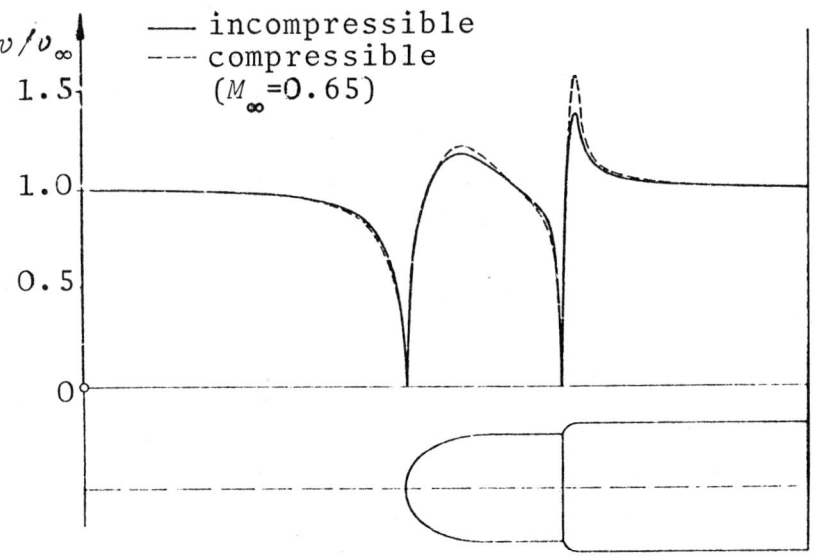

Fig. 12. Compressible potential flow about axisymmetric turbine hub

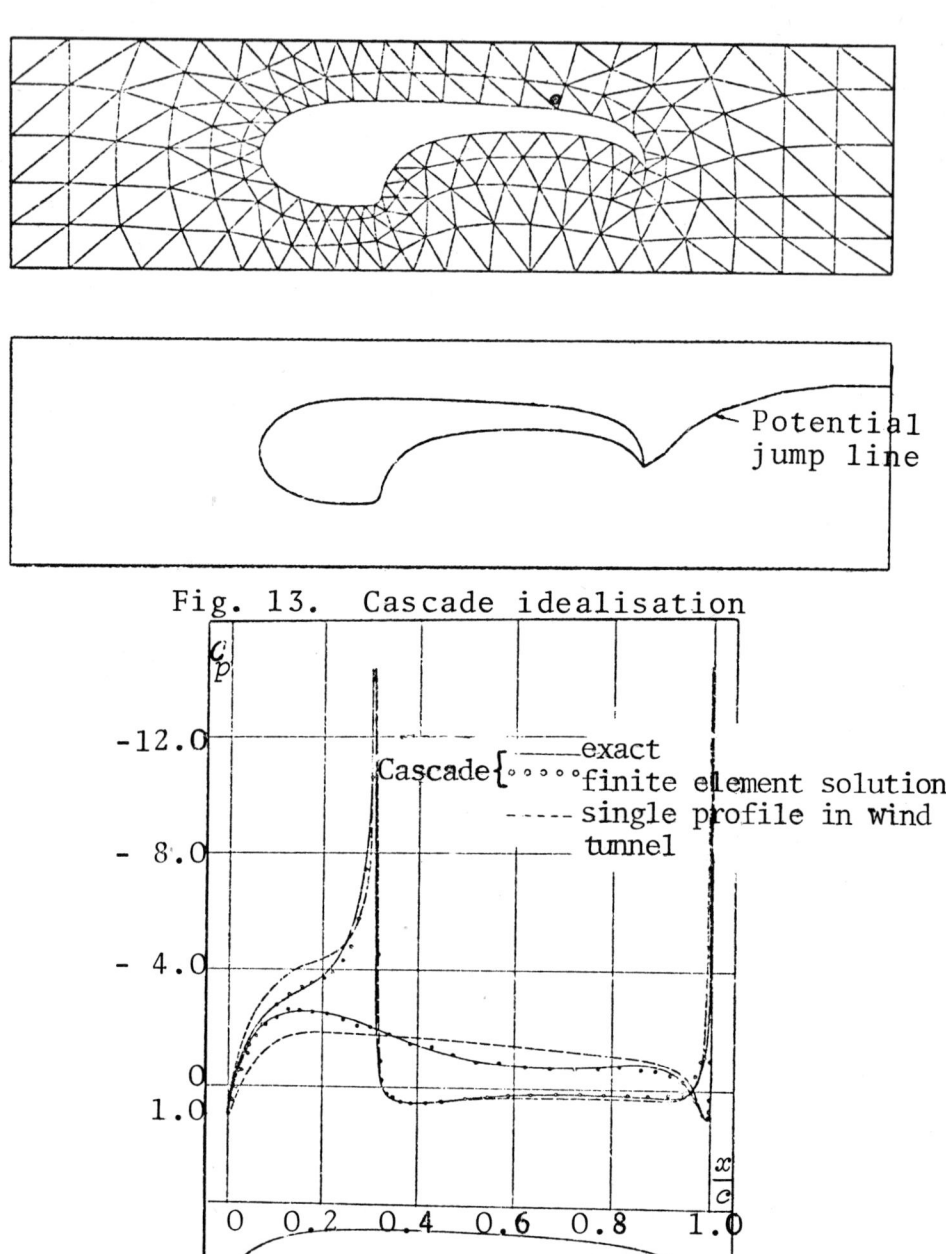

Fig. 13. Cascade idealisation

Fig. 14. Cascade at zero angle of attack $c_L = 0$

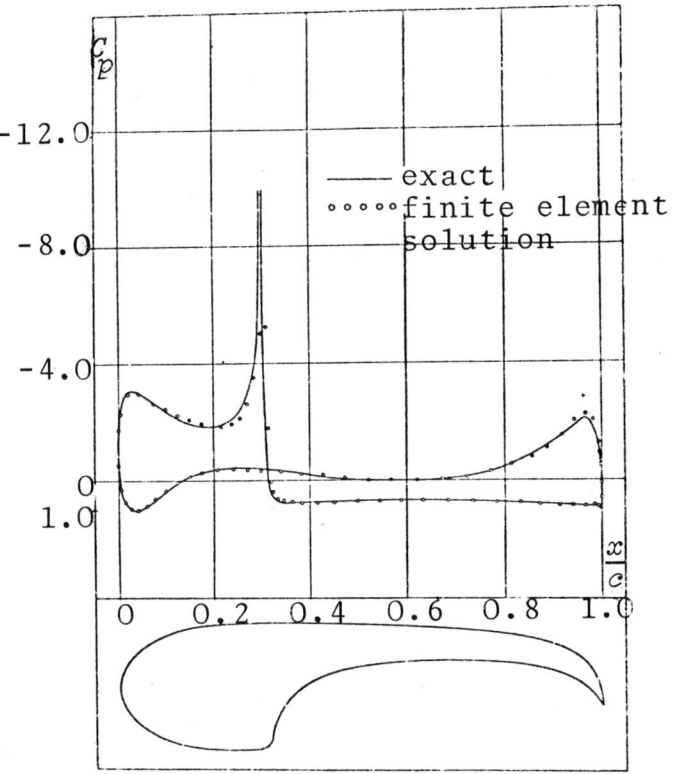

Fig. 15. Cascade under angle of attack $\alpha_I = -45°$, $c_L = 0$

stream function only, have been tried with varying success. Only the simplest of transient flows have been considered, most problems studied being steady state solutions at low Reynolds number. Whether the resultant steady state solutions are possible would appear to require a study of the variation of a perturbation of the steady state with time. It is clear that any numerical procedure which leads to steady state solutions under conditions where turbulence is known to be physically inevitable must be suspect. On the other hand if it proves impossible to establish a steady state solution whatever the numerical strategy employed one can be certain that no physical steady state exists.

One class of problem for which the stream function method is very suitable is Stokes flow - that is, slow viscous flow with neglect of inertia or $Re = 0$. The plate analogy for this case allows the use of the sophisticated software available for plate bending. This analogy may also be of use in the low Reynolds number regime.

The analogy derives from the Navier-Stokes equations in the form

$$\Delta^4 \psi = \frac{\rho}{\mu}\left[\frac{\partial \psi}{\partial y}\frac{\partial}{\partial x} - \frac{\partial \psi}{\partial x}\frac{\partial}{\partial y} + \frac{\partial}{\partial t}\right]\Delta^2 \psi \qquad (34)$$

Thus when $\rho = 0$ we have the equilibrium equation for an unloaded plate,

$$\Delta^4 \psi = 0 \qquad (35)$$

Stokes flow

Corresponding to the strain energy U of the plate we have the expression for half the energy dissipation rate resulting from fluid viscosity which is, for unit thickness,

$$U = \tfrac{1}{2}\mu \iint [(u_{,y} + v_{,x})^2 + 2u_{,x}^2 + 2v_{,y}^2]\, dx\, dy$$

$$= \tfrac{1}{2}\mu \iint [(\psi_{,xx} + \psi_{,yy})^2 - 4(\psi_{,xx}\psi_{,yy} - \psi_{,xy}^2)]\, dx\, dy \qquad (36)$$

Comparing with the plate energy equation we have the correspondence,

$$\mu \equiv D = \text{flexural rigidity}$$
$$4 \equiv 2\,(1-\text{Poisson's Ratio})$$
$$\text{Poisson's Ratio} = -1$$
$$\text{Youngs' modulus} = 0$$
$$\text{Shear modulus} = 12\mu t^{-3}$$

where t = plate thickness.

With this information we can construct Table I of analogous values.

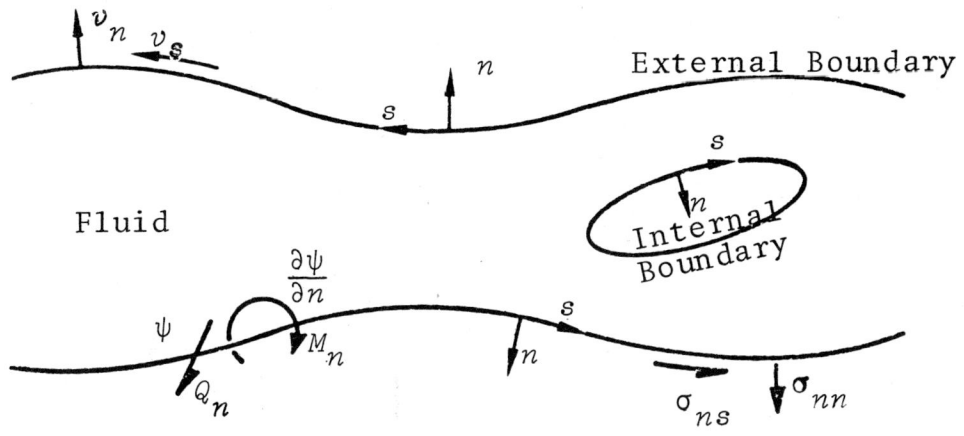

Table I
Stokes Flow Plate Analogy

Expression	Plate	Fluid
ψ	Displacement	Stream Function
$\dfrac{\partial \psi}{\partial n}$	Edge rotation	- Tangential velocity $(-v_s)$
$\mu\left(\dfrac{\partial^2 \psi}{\partial n^2} - \dfrac{\partial^2 \psi}{\partial s^2}\right)$	Edge moment M_n	- Edge shear stress $(-\sigma_{ns})$
$-\mu\dfrac{\partial}{\partial n}\Delta^2\psi - 2\mu\dfrac{d}{ds}\dfrac{\partial^2\psi}{\partial n \partial s}$	Edge shear Q_n	- $\dfrac{d\sigma_{nn}}{ds}$

The last correspondence is not obvious but, $\oint Q_n \delta\psi\, ds$ = virtual work of edge forces, and

$$-\oint \frac{d\sigma_{nn}}{ds}\delta\psi\, ds = \oint \sigma_{nn}\frac{d\delta\psi}{ds} ds = \oint \sigma_{nn}\delta v_n\, ds$$

= virtual increment of rate of work of edge direct stresses.

$\dfrac{d\sigma_{nn}}{ds}$ derives from $\sigma_{nn} = -p + 2\mu\dfrac{\partial v_n}{\partial n}$ \hfill (37)

and $\dfrac{dp}{ds} = \mu\Delta^2 v_s = -\mu\dfrac{\partial}{\partial n}\Delta^2\psi$ \hfill (38)

Boundary conditions in a duct flow problem will be easiest when velocities or ψ are specified. When both ψ and $\frac{\partial \psi}{\partial n}$ are known at the boundary ψ may be found from a plate program with any Poisson's ratio. Internal obstacles will have boundaries with ψ constant, $\frac{\partial \psi}{\partial n} = 0$ and since,

$$\oint \frac{d\sigma_{nn}}{ds} ds = 0 \qquad (39)$$

the total edge shear in the plate analogy is zero. Internal boundaries are therefore represented by floating rigid plates of constant undetermined displacement. This may be seen also from the fact that letting the plates float minimises the strain energy. Rotating internal cylinders will require only that $\frac{\partial \psi}{\partial n}$ is equal to $-v_s$ instead of zero.

When only the fluid stress σ_{nn} at inlets and outlets is given the ψ's on the fixed boundaries are to be determined. The most practically useful case is to suppose constant σ_{nn} and no tangential velocity at the inlet or outlet. In these circumstances the pressure at the inlet is also equal to the stress $-\sigma_{nn}$.

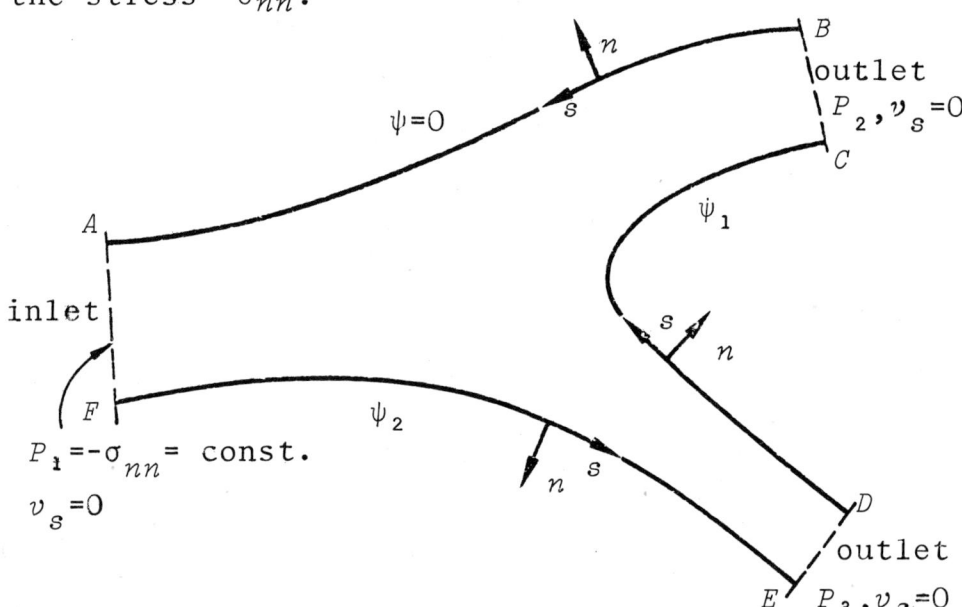

Fig. 16. Stokes flow in compound duct

Regarding boundary AB with $\psi = 0$ we have to determine ψ_1 on CD and ψ_2 on EF. It is supposed that $p_1 > p_2, p_3$. At F on FE $-\sigma_{nn}$ is also equal to p_1 and at E on FE it is equal to p_3. Then along FE,

$$-\int_F^E \frac{d\sigma_{nn}}{ds} ds = p_3 - p_1$$

Thus in the plate analogy,

$$\int_F^E Q_n\, ds = p_3 - p_1 \qquad (40)$$

and similarly

$$\int_D^C Q_n\, ds = p_2 - p_3$$

Thus the plate problem has the following boundary conditions: AB built-in; CD built-in to rigid body free to move vertically only under up load $(p_3 - p_2)$; FE as CD with up load $(p_1 - p_3)$; AF, BC, DE are built-in to frictionless walls.

Some examples calculated using TRICI 6 elements with velocity and pressure as field variables [9] are shown in Figs. 17 to 20. In Figs. 21 to 24 are examples calculated with standard ASKA plate programs using TUBA 6 elements.

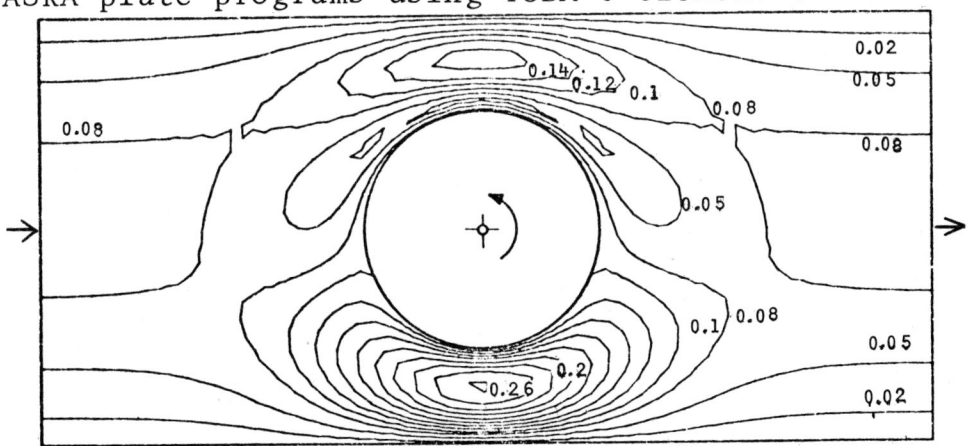

Fig. 17. Rotating cylinder in incompressible viscous channel flow at $Re=0$. Lines of constant velocity

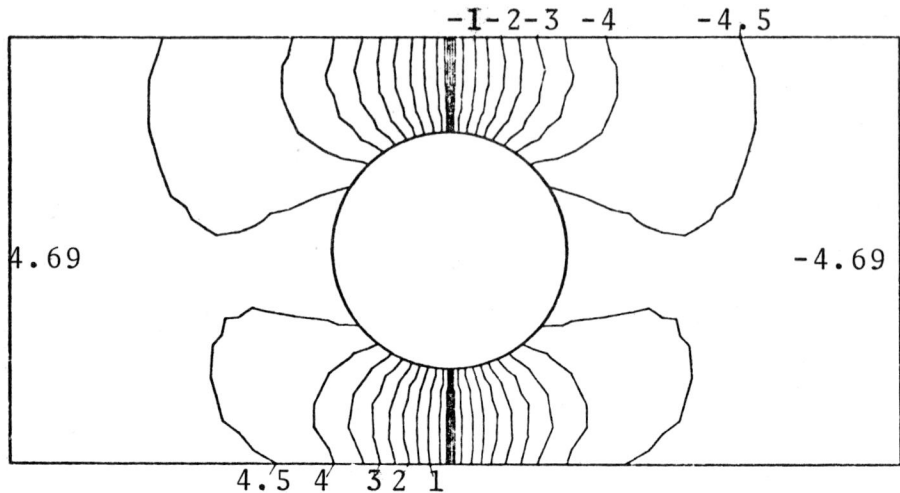

Fig. 18. Rotating cylinder in incompressible viscous channel flow at $Re=0$. Lines of constant hydrostatic pressure

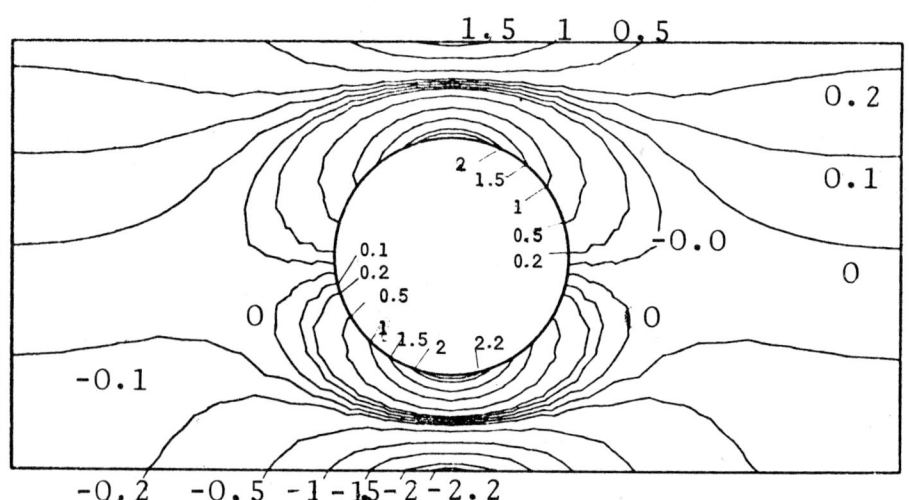

Fig. 19. Rotating cylinder in incompressible viscous channel flow at $Re=0$. Lines of constant vorticity

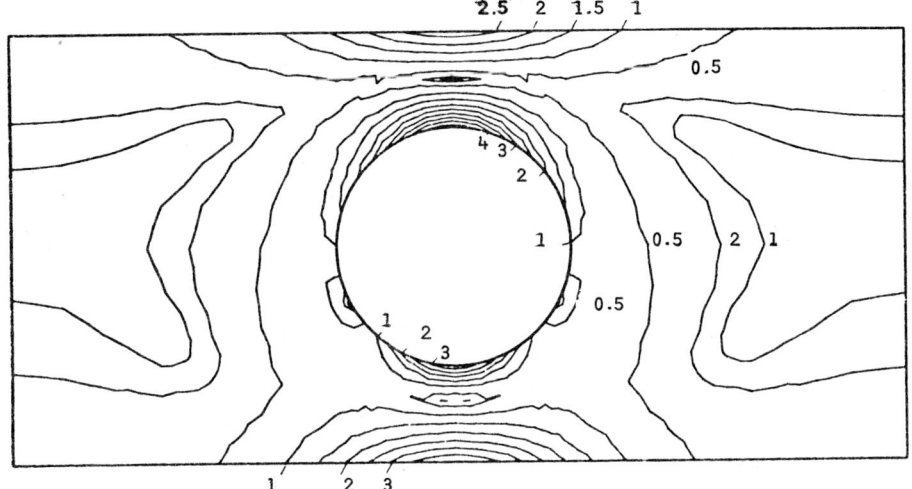

Fig. 20. Rotating cylinder in incompressible viscous channel flow at $Re=0$. Lines of constant equivalent stress

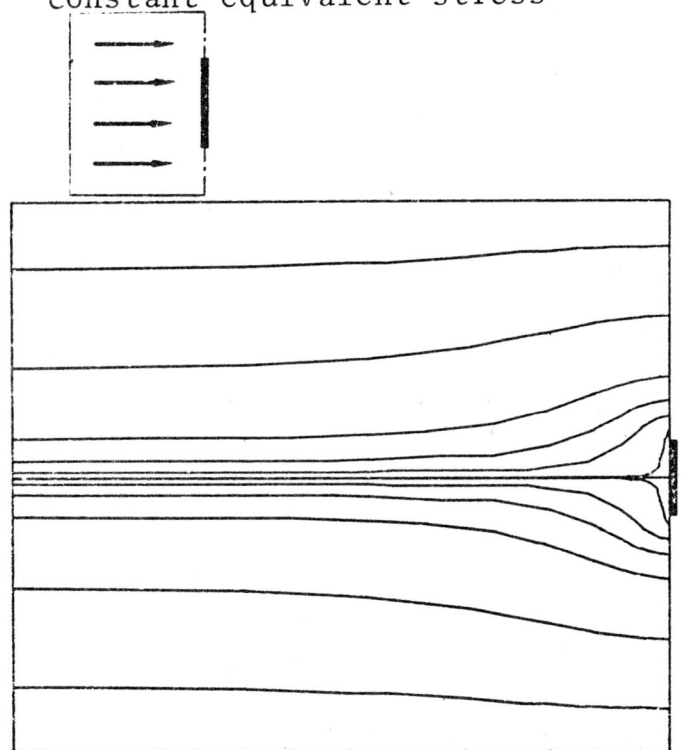

Fig. 21. Stokes flow around a flat plate in a frictionless duct

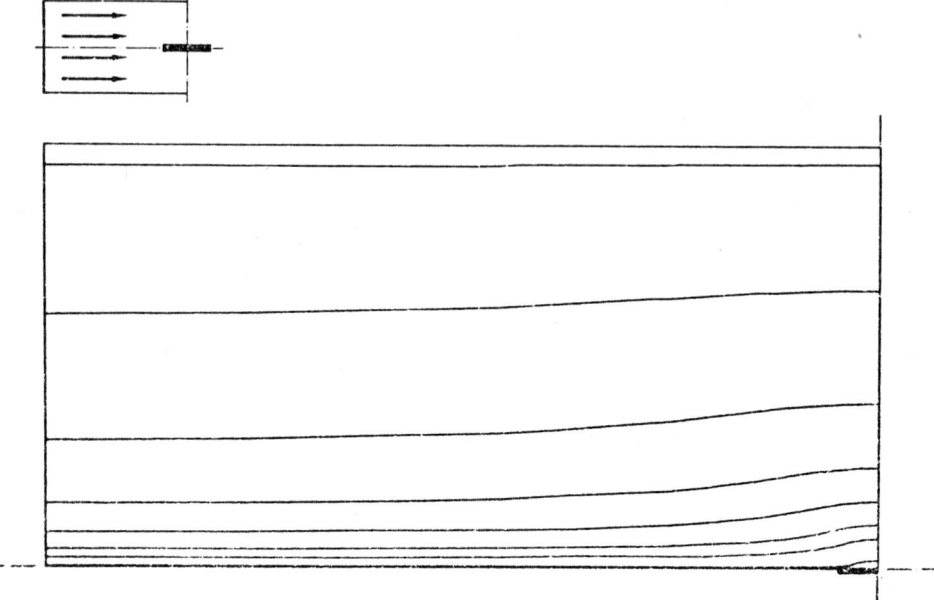

Fig. 22. Stokes flow: plate parallel to flow in a frictionless duct. Entry and exit are assumed to have constant u and $v=0$

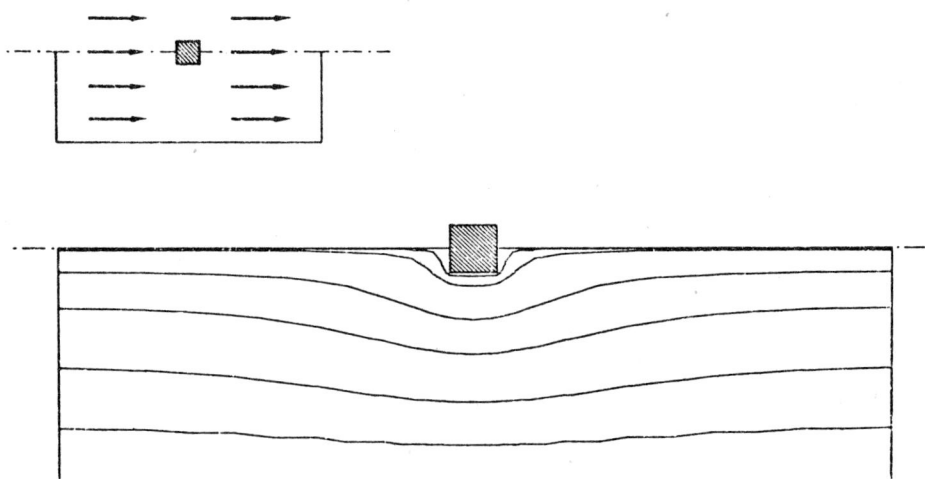

Fig. 23. Stokes flow around a square body in a frictionless duct. Entry and exit are assumed to have constant u and $v=0$

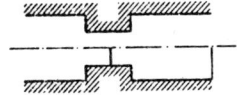
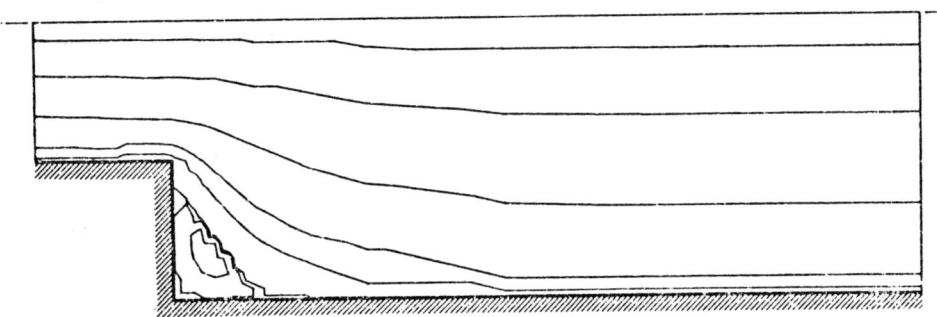

Fig. 24. Stokes flow: duct with Poiseuille flow at the entry and exit

Navier-Stokes flow

When the Reynolds number is not zero the full equation (34) must be considered. Assuming a steady state flow exists we may neglect the term in time and write,

$$\Delta^4 \psi = \frac{1}{\nu}\left[\frac{\partial \psi}{\partial y}\frac{\partial}{\partial x} - \frac{\partial \psi}{\partial x}\frac{\partial}{\partial y}\right]\Delta^2 \psi \tag{41}$$

This equation may be solved by regarding the right hand side as a "loading" and recalculating ψ. It is usual to start with the Stokes flow solution. Iteration by this method converges only for very low Reynolds numbers. The recalculation of the right hand side is an expensive operation as it consists of a vector of quadratic forms in the nodal freedoms $\underline{r}$.

An alternative method is to write the right hand side as an expression of the form

$$K(\underline{r}_0)\underline{r} \tag{42}$$

where $\underline{r}_0$ is the vector at the beginning of the iteration. Greater accuracy may result from using the implicit form

$$\tfrac{1}{2}[\underset{\sim}{K}(\underset{\sim}{r}_0) + \underset{\sim}{K}(\underset{\sim}{r}_1)]\underset{\sim}{r} \qquad (43)$$

where $\underset{\sim}{r}_1$ is the final vector.

$\underset{\sim}{K}(\underset{\sim}{r}_0)$ is a secant modulus and this method may converge to a stable solution when the ordinary iterative method does not. The Newton-Raphson method can also be used. However, it is easy to construct problems with unstable equilibrium configurations to which this method rapidly converges.

The boundary condition analogies remain as in the case for Stokes flow except for the edge shear force Q_n. This is now given by,

$$Q_n = -\frac{d\sigma_{nn}}{ds} + \frac{dq}{ds} - \rho v_n \Delta^2 \psi \qquad (44)$$

where q is the dynamic pressure.

At fixed boundaries $v_n = 0$ and $q = 0$ and the expression for the pressure differences between the ends of a duct remains as before (see equation (40)). If the boundary is frictionless v_n is still zero and on an internal obstacle

$$\oint Q_n ds = 0 \qquad (45)$$

Thus in the analogy the plate is still "floating." Along a frictionless duct boundary FE the pressure difference equation is modified to read:

$$\int_F^E Q_n ds = p_3 + q_3 - (p_1 + q_1) \qquad (46)$$

One assumes here that the conditions at the duct entrance and exit permit writing

$$\sigma_{nn} + p = 0$$

Otherwise p in (46) must be replaced by $-\sigma_{nn}$.

The results of attempts to calculate steady state solutions by the present method were perplexing. Only at very low Reynolds numbers (order of 10) was it possible to attain steady

state solutions starting from the Stokes flow.
Even for this case it was necessary to apply the
Reynolds number in increments. There was also
some evidence that with refinement of the finite
element net in problems with sharp edged boundaries the achievable Reynolds number would tend
to zero. On the other hand many authors have
obtained steady state solutions at quite high Reynolds numbers using other algorithms. At ISD the
formulation in terms of pressure and velocities
has produced steady state solutions for the cavity
flow problem up to Re = 100 at least [9] (see Figs.
25 to 28).

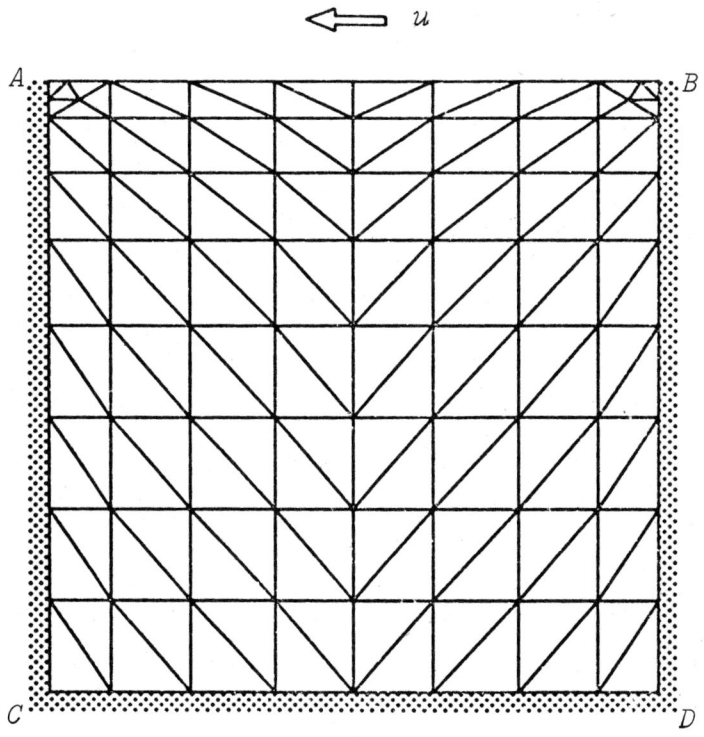

Fig. 25. Incompressible viscous cavity flow;
idealisation with 136 TRICI 6 or TRIACI
4 elements (583, 963 degrees of freedom,
respectively)

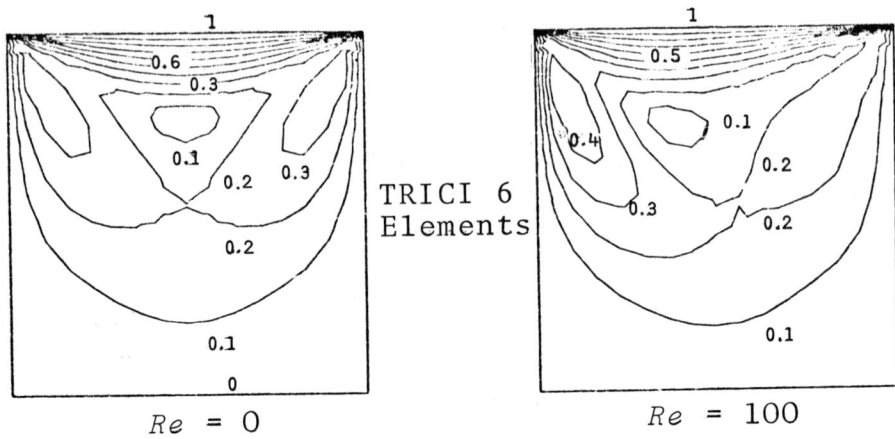

Fig. 26. Incompressible viscous cavity flow; lines of constant velocity

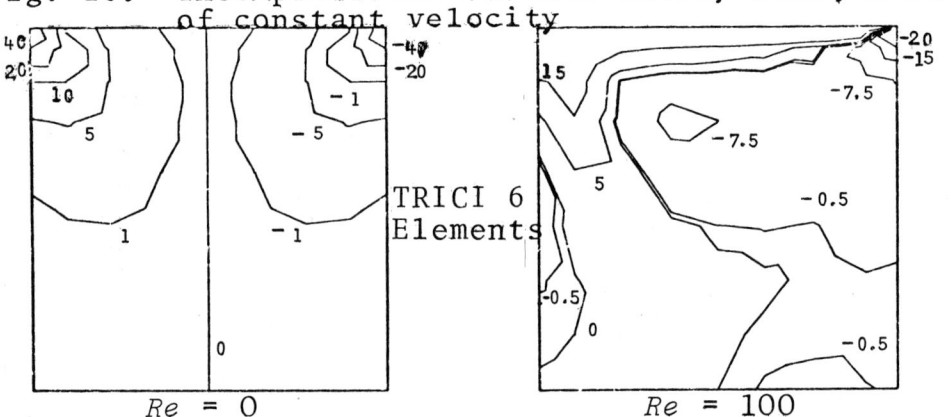

Fig. 27. Incompressible viscous cavity flow; lines of constant hydrodynamic pressure

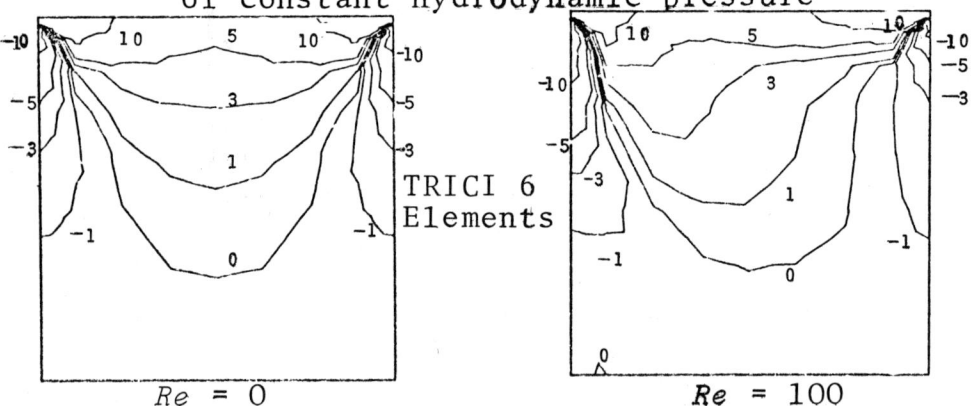

Fig. 28. Incompressible viscous cavity flow; lines of constant vorticity

Other examples of steady state solutions at Reynolds number up to 10^5 are given by Taylor and Hood [10].

It is evident that to obtain physically meaningful solutions in fluid mechanics much greater care and refinement is required than in most other applications. The TUBA and TUBAC elements are the most sophisticated elements available for fourth order elliptic problems and yet without special and expensive programming they are not entirely suitable for the Navier-Stokes incompressible plane flow problem. However, it is expected that either by making the necessary corrections to the TUBA program or by developing the higher order elements mentioned previously it will eventually be possible to give accurate estimates of the critical Reynolds number for Navier-Stokes flow in smoothly curved ducts or past smoothly curved objects.

In order to gain insight into the behaviour of the Galerkin method when applied to the Navier-Stokes equations we shall now apply the method to plane Poiseuille flow stability.

STABILITY OF PLANE POISEUILLE FLOW

So far as we know this problem has been studied only in relation to an infinite channel. The mathematical difficulties are considerable and since Heisenberg concluded in 1924 [11] that the flow would become unstable at Reynolds number a little less than 6000 other authors have agreed or disagreed with his conclusions. The question appears to have been settled by Thomas' machine calculations of 1953 [12] which were made following a suggestion of von Neumann. However, there are still many obscure points. Several authors have applied the Galerkin method [13], [14], [15], but although there is general agreement on the critical Reynolds number it has proved singularly difficult to establish with certainty even the first few members of the eigenspectrum of this linearised perturbation problem. For example, the higher eigenvalues of Grohne [16] have been disputed by Grosch and Salwen [14]. In finite element applications the inlet and outlet boundary conditions are generally assumed given so that the problem of stability is different from the infinite tube case. Recently we [17] examined the Poiseuille problem with fixed Stokes flow entry and exit conditions. Up to 36 polynomial degrees of freedom were used in a channel with length: width ratios of 1:1 and 10:1. No instability was found but the margin of stability decreased inversely as Re. However, when the constant axial velocity condition at the exit was replaced by the commonly assumed condition of zero vorticity gradient, instability occurred at the low Reynolds number of 20 when using only a cubic polynomial - that is ten freedoms.

Although most authors who have had success with the Galerkin method have used expansion functions forming a complete orthonormal set [18], and indeed warning has been given [19] of the numerical difficulties that arise when using non-orthogonal polynomials, it was decided to treat the infinite tube problem with a simple polynomial expansion. An orthodox finite element treatment is left for a later paper.

The calculations made in [17] on the infinite channel used a symmetrical stream function expansion of the form

$$\psi = e^{i\alpha x} e^{\lambda t} \phi(y) \qquad (47)$$

with

$$\phi(y) = \sum_{n=0}^{N-1} a_n (1 - y^2)^2 y^{2n} \qquad (48)$$

where N is the number of degrees of freedom.

The differential equation in ϕ is,

$$\lambda \left[\frac{d^2\phi}{dy^2} - \alpha^2\phi\right] = \nu\left[\frac{d^4\phi}{dy^4} - 2\alpha^2\frac{d^2\phi}{dy^2} + \alpha^4\phi\right]$$
$$- \tfrac{3}{2}\alpha i\left[2\phi + (1-y^2)(\frac{d^2\phi}{dy^2} - \alpha^2\phi)\right] \quad (49)$$

with
$$\phi = \frac{d\phi}{dy} = 0 \quad \text{at} \quad y \pm 1$$

Up to 40 terms were retained in equation (48). Substitution of equation (48) in (49) gives a complex matrix equation for the eigenvalue λ. This equation was solved by an algorithm developed by Eberlein [20]. The calculations proved to be exceptionally sensitive to numerical error. This was especially so at the higher Reynolds numbers. Table II gives the values of the eigenvalues of maximum and minimum modulus obtained for 12 degrees of freedom using 8, 14 or 28 digits in the computations. The wave number $\alpha = 1$ and the Reynolds number is based on the maximum velocity and half channel width. The values in the table show that double precision on the CDC 6000 series is necessary to achieve adequate precision of the higher eigenvalues. Indeed, it appears that for the larger Reynolds numbers even double precision will become inadequate for large N.

A remarkable feature of the calculations is that for high Reynolds numbers only the lowest modulus eigenvalue converges with any precision as the degrees of freedom are increased. This is demonstrated in Table III which gives the first (least) and second eigenvalues for $\alpha = 1$, $Re = 10^4$ and $N = 12$ to 36. The calculations are made with double precision (28 digits) and also given are the values of the least eigenvalue obtained by other authors.

The values of Thomas and of George and Hellums are obtained by extrapolating finite difference results to infinitely small division of the channel width. Comparing the modulus of the first eigenvalues we see that our results and

Table II

Infinite channel. Influence of computational precision on eigenvalues

Precision	Re	Maximum modulus		Minimum modulus	
		Real	Imaginary	Real	Imaginary
8 digits	10^2	- 121.06468	- 0.185488	- 0.244084	- 0.717723
	10^4	2.359512	0.688487	0.00521975	- 0.356596
14 digits	10^2	- 188.20099	0.149480	- 0.2444167473	- 0.7177408658
	10^4	- 1.865134	- 0.143499	0.00556314	- 0.3560377
28 digits	10^2	- 188.17340	0.149231	- 0.2444167471	- 0.7177408657
	10^4	- 1.864987	- 0.143300	0.00556275	- 0.3560376

Table III

Infinite channel; $Re = 10^4$. Convergence of first and second eigenvalues with number of terms in expansion

Degrees of Freedom N	First eigenvalue		Second eigenvalue	
	Real × 10³	Imaginary × 10	Real × 10	Imaginary × 10
12	5.562752	- 3.56037	- 2.130	- 2.847
13	5.500041	- 3.565258	- 2.543	- 2.479
14	5.557474	- 3.563398	- 2.976	- 2.440
15	5.654683	- 3.562153	- 2.321	- 3.189
16	5.569792	- 3.562860	- 2.496	- 2.812
17	5.597436	- 3.563203	- 2.598	- 2.695
18	5.552725	- 3.563293	- 2.428	- 2.666
19	5.628266	- 3.562799	- 2.915	- 2.681
20	5.609637	- 3.562786	- 2.853	- 3.135
30	5.610973	- 3.562893	- 2.748	- 2.815
36	5.6092650	- 3.562893	- 2.755	- 2.850
Thomas [12]	5.6106	- 3.562889		
Grosch and Salwen [14]	5.5215	- 3.56119		
George and Hellums [21]	5.52	- 3.561		

Table IV

Infinite channel; convergence of eigenvalues for $Re = 100$.

Eigenvalues in order of magnitude of moduli

Degrees of freedom $N = 12$		Degrees of freedom $N = 20$	
Real × 10	Imaginary × 10	Real × 10	Imaginary × 10
− 2.4441674708341 7	− 7.1774086570982	− 2.4441674708342 1	− 7.1774086570984
− 5.7386508599908	−10.3975308071 53	− 5.7386508599911	−10.3975308071 54
−13.0982522147 0	− 9.9249361266 57	−13.0982522145 7	− 9.9249361266 23
−23.4772434 87	− 9.9526659 256	−23.4772434 93	− 9.9526659 267
−36.8322 267	− 9.9736 833	−36.8322 226	− 9.9736 816
		32.8727	−29.1533
−53.14 05	− 9.98 307	−53.14 11	− 9.98 32
−72.4 30	− 9.9 11	−72.4 06	− 9.9 85
−96. 5	− 8. 93	−94. 7	− 9. 93

Thomas' agree to within six figures for $N \geqslant 30$, to five figures for $N \geqslant 19$, to four figures for $N \geqslant 14$ and to three figures for $N \geqslant 12$. No comparable agreement with other authors was found for the second and higher eigenvalues.

On the other hand, for low Reynolds numbers the first few eigenvalues listed in order of magnitude of their moduli do converge well as N is increased. Thus Table IV shows the results with 12 and 20 degrees of freedom and $Re = 100$, $\alpha = 1$, again calculated with double precision.

The most surprising feature of this table is the appearance of an unstable sixth eigenvalue for $N = 20$. The next root agrees to 1 part in 5000 with the sixth root of the $N = 12$ case. This makes it difficult to reject the unstable root as spurious. Similar results are obtained for lower Reynolds numbers. Between $Re = 1$ and 100 with $N = 20$ the first unstable root may be written in the form,

$$328/Re - 2.915i$$

and is the fifth root up to $Re = 10$ after which it is the sixth root. The frequency of the disturbance remains unchanged but the growth factor increases with the viscosity. Previous authors have not observed such instabilities and we are led to doubt the suitability of the polynomial functions for this application of the Galerkin method. On the other hand the high precision of the first root would then be difficult to explain, and the existence of only one "flutter" speed for the plate of the analogy does not seem probable. It is hoped to clear up these questions in a future paper.

Many of the numerical experiments necessary for the preparation of this paper were made by T. Angelopoulos with the help of G.A. Malejannakis and K. Straub. All the programming on the application of ASKA to Stokes flow and the calculations on the stability of Poiseuille flow were carried out by B. Bichat.

We also thank G. Grimm, H. Hager, K. Mai, K.A. Schneider and K. Strauss for their help in preparing the paper for the printer.

REFERENCES

1. Courant, R., "Variational methods for the solution of problems of equilibrium and vibrations," *Bull. Amer. Math. Soc.*, **49**, 1-23 (1943).
2. Ebner, H. and Koeller, H., "Zur Berechnung des Kraftverlaufes in versteiften Zylinderschalen," *Luftfahrtforschung*, **14**, No. 12 (1937).
3. Argyris, J.H., "Energy theorems and structural analysis," *Aircraft Engineering*, **26** (1954) and **27** (1955); reprinted by Butterworths (1960).
4. Strang, G. and Fix, G.F., "An analysis of the finite element method," Prentice-Hall, Englewood Cliffs (1973).
5. Zienkiewicz, O.C., "The finite element method in engineering science," McGraw-Hill, New York (1971).
6. Oden, J.T., "Finite elements in non-linear continua," McGraw-Hill, New York (1972).
7. Nemat-Nasser, S., "General variational principles in non-linear elasticity with applications," *Mechanics Today*, **1**, 214-261 (1972).
8. Argyris, J.H. and Mareczek, G., "Potential flow analysis by finite elements," *Ing.-Arch.*, **42**, 1-25 (1973).
9. Argyris, J.H. and Mareczek, G., "Finite-element analysis of slow incompressible viscous fluid motion," *Ing.-Arch.*, **43**, 92-109 (1974).
10. Taylor, C. and Hood, P., "A numerical solution of the Navier-Stokes equations using the finite element technique," *Computers and Fluids*, **1**, 73-100 (1973).

11. Heisenberg, W., "Über Stabilität und Turbulenz von Flüssigkeitsströmen," *Ann. Phys. Lpz.*, **74**, 577-627 (1924).

12. Thomas, L.H., "The stability of plane Poiseuille flow," *Phys. Rev.*, **91**, 780-783 (1953).

13. Dolph, C.L. and Lewis, D.C., "On the application of infinite systems of ordinary differential equations to perturbations of plane Poiseuille flow," *Q. Appl. Math.*, **16**, 97-110 (1958).

14. Grosch, C.E. and Salwen, H., "Stability of steady time-dependent plane Poiseuille flow," *J. Fluid Mech.*, **34**, 177-205 (1968).

15. Dowell, E.H., "Non-linear theory of unstable plane Poiseuille flow," *J. Fluid Mech.*, **38**, 401-414 (1969).

16. Grohne, D., "Über das Spektrum bei Eigenschwingungen ebener Laminarströmungen," *Z. angew. Math. Mech.*, **34**, 344-360 (1954).

17. Argyris, J.H., Dunne, P.C. and Bichat, B., "A Plate Analogy for Plane Incompressible Viscous Flow," Invited paper presented at International Conference on Computational Methods in Non-linear Mechanics, Texas Institute for Computational Mechanics, Austin (1974).

18. Gallagher, A.P., "On a proof by Petrov of the stability of plane Couette flow and plane Poiseuille flow," *Siam. J. Appl. Math.*, **17**, 765-768 (1969).

19. Finlayson, B.A., "The method of weighted residuals and variational principles," Academic Press, New York (1972).

20. Eberlein, P.J., "Solution of the complex eigenproblem by a norm reducing Jacobi type method," *Numer. Math.*, **14**, 232-245 (1970).

21. George, W.D. and Hellums, J.D., "Hydrodynamic stability in plane Poiseuille flow with finite amplitude disturbances," *J. Fluid Mech.*, **51**, 687-704 (1972).

FREE VORTEX SHEETS

K.W. Mangler

(University of Southampton)

INTRODUCTION

A wing of finite aspect ratio set at an angle of incidence in a steady uniform flow produces a lift and a drag force. Lift production is always associated with the shedding of a vortex sheet from the trailing edge and possibly also from the leading and side edges of the wing. In the absence of regions of break-away with thick wakes this flowfield can be modelled by an inviscid potential flow in which the vortex sheet consists of a stream surface, across which the velocity is discontinuous.

Such vortex sheets usually roll up at the edges. Experiments have shown that in many situations, e.g., in the case of a slender delta wing with leading edge separation, these vortex sheets assume rather stable shapes. The same is often true in the case of trailing sheets (wakes) behind a lifting wing. So it is meaningful to seek the solution of a steady flow problem in which the flow at a large distance from the wing is uniform, and the shape of the wing (which is a stream surface) is prescribed.

The shape and strength of the resulting vortex sheet follow from the conditions that it is a stream surface and the pressure difference across it must be zero (pressure condition). The shape and strength of the vortex sheet is such that the local vorticity vector is parallel to the local mean flow direction. At the origin of the sheet, namely at the side or trailing edges of the wing, a condition of smooth outflow applies. Obviously the strength depends on the lift distribution over the wing and the position of the lines of separation.

For incompressible flow we have to satisfy the three-dimensional Laplace equation, which is linear, so that the solution can be built up by singularities. The wing can be represented by a source distribution on its surface and by a doublet distribution (usually put in its interior). The free vortex sheet can be represented by a doublet distribution. Since the pressure condition is nonlinear, a solution can only be found by some kind of iteration. Because of its complexity the problem is unlikely to be solved by methods other than numerical computation. These computations can be considerably simplified by considering a slender delta wing. In this case slender wing theory applies and the problem becomes effectively two-dimensional (in a cross-sectional plane). This special case is discussed in the next section [1]. In the following section the wake behind a rectangular wing at incidence is considered and results are given as an illustration, which are based on work at NLR by Labrujere [2].

In the final section general comments are made on the computational difficulties which arise and suggestions are made on how to overcome them.

SLENDER WINGS

For a slender delta wing the flowfield can be assumed as conical and slender wing theory applies. Laplace's equation becomes (see [1]) (compare Fig. 1):

$$\phi_{yy} + \phi_{zz} = 0 \qquad (1)$$

where $Ux + \phi$ is the potential of the flow. The normal velocity condition valid on the wing and on the vortex sheet takes the form

$$\phi_n = \frac{\partial \phi}{\partial n} = - kU\frac{r}{s} \sin \phi \qquad (2)$$

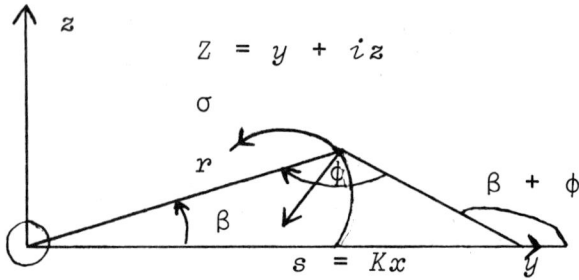

Fig. 1. Cross-sectional plane for slender delta wing at incidence α

The pressure condition (to be satisfied on the vortex sheet) is

$$D\phi = D\phi_\sigma (r \cos \phi - s\phi_{\sigma m}/(kU)) \qquad (3)$$

Here, D is the difference operator across the sheet, ϕ_σ is the tangential velocity, and $\phi_{\sigma m}$ is the mean value of the tangential velocity. In addition we demand smooth out-flow at the leading edge, i.e., the complex velocity,

$$qe^{-i\theta} = \frac{d}{dZ}(\phi + i\psi) \text{ finite at } Z = y + iz = \pm s \quad (4)$$

The complex velocity is then determined by singularities along the wing and the sheet, for instance for a wing without thickness,

$$\frac{d}{dZ}(\phi + i\psi) = v - iw$$

$$= iU\alpha + \frac{1}{2\pi i} \int_{\text{wing}} \frac{\gamma_w(\sigma)d\sigma}{Z - Z_w(\sigma)} + \frac{1}{2\pi i} \int_{\text{sheet}} \frac{\gamma_s(\sigma)d\sigma}{Z - Z_s(\sigma)} \qquad (5)$$

In many cases conformal mapping can be used which transforms the wing into part of the imaginary axis and the calculations can be performed in this new plane, where symmetry can be used to help to satisfy the normal velocity condition on the wing.

The projection of the sheet in the cross-flow plane is always infinitely long. Its behaviour near its core has been investigated by Mangler and Weber [3]. This shows that a single vortex of strength Γ, which is linked by a "cut" to the end of the finite sheet is a reasonable approximation to the inner part of the sheet. In the symmetric case, the last term in (5) can be written as:

$$\frac{1}{2\pi i} \int_{\text{sheet}} \frac{\gamma_s(\sigma) d\sigma}{Z - Z_s(\sigma)} = v^* - iw^*, \text{say}$$

where

$$v^* - iw^* = \frac{\Gamma}{2\pi i} \left(\frac{1}{Z - Z_v} - \frac{1}{Z + Z_v} \right)$$

$$+ \frac{1}{2\pi i} \int_0^{\sigma_m} \frac{\gamma_s(\sigma) d\sigma}{Z - Z_s(\sigma)} - \frac{1}{2\pi i} \int_0^{\sigma_m} \frac{\gamma_s(\sigma) d\sigma}{Z + Z_s(\sigma)} \quad (6)$$

An iteration procedure is set up in which the pressure condition (3) is used to determine the strength $\gamma_s(\sigma)$ of the finite part of the vortex sheet, the shape of which is approximately known. Since (3) is nonlinear the shape of the sheet and $\phi_{\sigma m}$ are assumed to be known from previous iterations and

$$\gamma(\sigma) = D\frac{\partial \phi}{\partial \sigma} = D\phi_\sigma$$

and

$$D\phi = \int_\infty^\sigma \gamma(\sigma) d\sigma = \Gamma + \int_{\sigma_m}^\sigma \gamma(\sigma) d\sigma \quad (7)$$

are calculated from a system of linear equations after (3) has been discretised. The position Z_v of the core is obtained from the condition that the total force on the vortex and its cut (linking it to the end $Z_E = Z(\sigma_m)$ of the finite sheet) is zero. Its strength Γ follows from the smooth outflow condition (4).

Then the shape of the finite sheet is changed in order to obtain a better approximation to the normal velocity condition (2). This is done by relaxation, since the difference between the calculated and the required normal velocities can only indicate the direction of the required change of slope and change of shape.

The discretisation of the integral in (6) can be done in different ways. Smith [1] selected a number of points $\sigma_0 \leqslant \sigma_j \leqslant \sigma_m$ and replaced the integral by a sum of integrals

$$\int_{\sigma_{2j-2}}^{\sigma_{2j}} \frac{\gamma(\sigma)d\sigma}{Z-Z^*(\sigma)} \approx \frac{\gamma(\sigma_{2j-1}) \times (\sigma_{2j}-\sigma_{2j-2})}{Z-Z^*(\sigma_{2j-1})} \quad (8)$$

where Z is never chosen closer to the pivotal point $Z^*(\sigma_{2j-1})$ than the intermediate points $Z = Z^*(\sigma_{2j-2})$ and $Z^*(\sigma_{2j})$ so that no singularities occur.

Fink and Soh [4] have pointed out that the accuracy of the integration can be improved by a careful choice of the stations σ_j. For values of Z not close to the segment considered above they write

$$\int_{\sigma_{2j-2}}^{\sigma_{2j}} \frac{\gamma(\sigma)d\sigma}{Z-Z^*(\sigma)} = \frac{\Gamma_{2j-1}}{Z-Z^*(\sigma_{2j-1})} \quad (9)$$

with

$$\Gamma_{2j-1} = \int_{\sigma_{2j-2}}^{\sigma_{2j}} \gamma(\sigma)d\sigma$$

and suggest that for the nearest pivotal point the integral should be ($d\sigma = dZe^{-i\theta}$)

$$\frac{\Gamma_{2j-1}}{\sigma_{2j}-\sigma_{2j-2}} e^{-i\theta j} \ln \left|\frac{Z-Z_{2j}}{Z-Z_{2j-2}}\right| \quad (10)$$

which vanishes if we choose $Z = \frac{1}{2}(Z_{2j} + Z_{2j-2})$.

Free Vortex Sheets

Perhaps one should pursue this argument further. Consider the integral, representing a linearly varying distribution of circulation along the real axis.

$$u - iv = \frac{1}{2\pi i} \int_{\sigma_1}^{\sigma_2} \frac{\gamma(\sigma) d\sigma}{Z - \sigma} \quad \text{with} \quad \gamma(\sigma) = \gamma_1 + \bar{\gamma}(\sigma - \sigma_1) \tag{11}$$

$$l = \sigma_2 - \sigma_1, \quad \gamma_2 - \gamma_1 = l\bar{\gamma}, \quad Z = x + iy$$

and σ is real.

Then with

$$x - \sigma_j = r_j \cos \theta_j, \quad y = r_j \sin \theta_j, \quad \delta = \theta_2 - \theta_1,$$

we have

$$u - iv = -\frac{\gamma_1 + \bar{\gamma}(x - \sigma_1)}{2\pi i} [\tfrac{1}{2} \ln \frac{r_2^2}{r_1^2} + i\delta]$$
$$- \frac{\gamma_2 - \gamma_1}{2\pi i} \left[1 - \tfrac{y}{l}\delta + i\tfrac{y}{l}\tfrac{1}{2} \ln \frac{r_2^2}{r_1^2} \right] \tag{12}$$

Choose

$$\sigma_0 = \tfrac{1}{2}(\sigma_1 + \sigma_2), \quad x - \sigma_0 = r_0 \cos \theta_0, \quad y = r_0 \sin \theta_0$$

then we find for $\frac{l^2}{r_0^2} \ll 1$

$$u - iv = \frac{\gamma_1 + \gamma_2}{4\pi i} \frac{l}{r_0} e^{-i\theta_0} [1 + \frac{l^2}{12 r_0^2} e^{-2i\theta_0}]$$
$$+ \frac{\gamma_2 - \gamma_1}{2\pi i} \frac{l^2}{12 r_0^2} e^{-2i\theta_0} \tag{13}$$

The first term corresponds to a single vortex placed at $\sigma = \tfrac{1}{2}(\sigma_1 + \sigma_2)$ with strength $\Gamma = \tfrac{1}{2} l (\gamma_1 + \gamma_2)$. This is the far-field term retained in the so called vortex lattice approximation. For the case $r_1 = r_2$, $x = \tfrac{1}{2}(\sigma_1 + \sigma_2)$ we find, from (12),

$$u - iv = -\frac{\gamma_1+\gamma_2}{4\pi}\delta - \frac{\gamma_2-\gamma_1}{2\pi i}\left(1 - \frac{y}{l}\delta\right) \quad (14)$$

here $\delta = \pm \pi$ for $y = \pm 0$. The first term gives the discontinuity in u. For $y \geqslant 0$ we find

$$\frac{y}{l}\delta = \left(\frac{\pi}{2} - \theta_1\right) \cot\left(\frac{\pi}{2} - \theta_1\right) \text{ and } \delta = \pi - 2\theta_1$$

These relations show the poor approximations obtained by simpler formulae. Perhaps this is one of the weaker aspects of the solutions obtained so far. These errors become more important in the rolled up part of the sheet, when two subsequent turns of the spiral come closer together. The error cannot be reduced by the simple device of using more vortices in the representation.

Fink and Soh [4] have suggested that it is advisable to choose the calculating (pivotal) point as mentioned above, (i.e., on the line of symmetry) in order to increase accuracy. If these points are not convenient for the further progress of the calculation, they find the values at the required points by interpolation. They recalculated various solutions, including time-dependent problems, and found more consistency and less apparent randomness in their behaviour, which had occurred in earlier work. Also they repeated one of the solutions given by Smith [1] for a slender delta wing. They found little change in the lift distribution over the wing, but calculated a more flattened shape for the vortex sheet. I am not sure whether this solution is better than Smith's (Fig. 2). Apparently their calculation was done without a strong vortex representing the core. This core was only represented by the free end of the sheet (consisting of 70 segments in contrast to Smith who used only about 20). Smith's strong vortex in the core carries about half the total vorticity on the sheet, so that the centre part is not properly represented by Fink and Soh, and no close agreement of the predicted shapes of the vortex sheet can be expected.

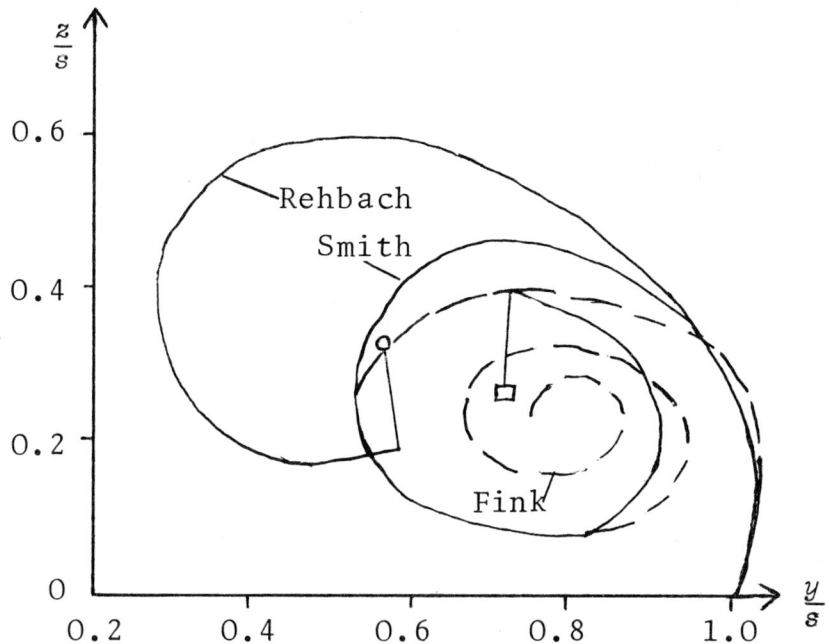

Fig. 2. Sheet shape for delta wing at incidence α (apex angle $\varepsilon = \alpha$). □——— Smith (ε small), ----- Fink and Soh (ε small), o——— Rehbach ($\varepsilon = 30°$), (o core from experiment)

In this context a recent paper by Moore [5] should be mentioned. He uses a point vortex representation to study numerically the evolution of an initially plane vortex sheet, which is moving in a direction normal to this plane. Previous solutions had shown spurious fluctuations, in particular near the ends of the rolled-up sheet. By introducing a tip vortex to represent the tightly rolled portion of the vortex sheet, the chaotic motion is eliminated and the details of the outer portion of the spiral are calculated. The rate of rolling-up is calculated and is shown to be governed by the similarity law predicted by Kaden [6], at least during the initial stages of the rolling-up process. The calculation is continued until 99 per cent. of the vorticity has gone into the core. Then the spiral sheet displays a marked ellipticity. This result confirms the importance of having a proper representation of the vortex core incorporated in the numerical model.

VORTEX SHEET BEHIND A LIFTING WING

Again Laplace's equation can be satisfied by the use of source distributions on the surface of the wing and a doublet distribution, preferably in the interior of the wing. The doublet distribution continues from the trailing edge (and possibly from the side edges) into the wake, which is also represented by a doublet distribution. A method developed by Labrujere [2] at NLR is followed. Similar work was done by Rehbach [7], [8], [9].

A doublet distribution $D\phi = \tau$ placed in a plane $z = 0$ represents a vector $\underset{\sim}{\gamma}$ giving the vortex strength with the components

$$\gamma_x = \frac{\partial \tau}{\partial y} \qquad \gamma_y = -\frac{\partial \tau}{\partial x} \qquad \underset{\sim}{\gamma} = \nabla \tau \times \underset{\sim}{n} \qquad (15)$$

The potential is given by ($r^2 = |PP'|^2$)

$$\phi(P) = -\frac{1}{4\pi} \iint \tau(P') \frac{\partial}{\partial n} \frac{1}{r} dS(P') \qquad (16)$$

and the velocity components are

$$\frac{\partial \phi}{\partial n} = -\frac{1}{4\pi} \iint \underset{\sim}{n}(P') \cdot \{\underset{\sim}{\gamma}(P') \times \nabla \frac{1}{r}\} dS(P')$$

$$\frac{\partial \phi^+}{\partial s} = \tfrac{1}{2} \underset{\sim}{s} \cdot (\underset{\sim}{\gamma} \times \underset{\sim}{n}) - \frac{1}{4\pi} \iint \underset{\sim}{s}(P') \cdot \{\underset{\sim}{\gamma}(P') \times \nabla \frac{1}{r}\} dS(P')$$

$$\frac{\partial \phi^+}{\partial s} - \frac{\partial \phi^-}{\partial s} = \underset{\sim}{s} \cdot (\underset{\sim}{\gamma} \times \underset{\sim}{n}) = |\underset{\sim}{\gamma}| \qquad (17)$$

Here $(\underset{\sim}{s}, \underset{\sim}{t}, \underset{\sim}{n})$ are unit vectors forming an orthogonal system on S, with $\underset{\sim}{n}$ normal to S, and $\underset{\sim}{s}$ parallel to $\underset{\sim}{\gamma}$.

If V_w denotes the velocity arising from the onset flow combined with the singularity distributions representing the wing, and V_v denotes the velocity induced by the trailing vortex sheet, then the normal velocity condition is

$$\underset{\sim}{V_m} \cdot \underset{\sim}{n} = \underset{\sim}{n} \cdot \{\underset{\sim}{V_w} + \tfrac{1}{2}(\underset{\sim}{V_v^+} + \underset{\sim}{V_v^-})\} = 0 \qquad (18)$$

The pressure condition follows from Bernoulli's equation

$$(\underset{\sim}{V_v^+} - \underset{\sim}{V_v^-}) \cdot \underset{\sim}{V_m} = |\gamma| \underset{\sim}{s} \cdot \underset{\sim}{V_m} = 0 \qquad (19)$$

$i.e.$, the vortices coincide with the mean streamlines.

Previous work at NLR on the calculation of pressure distributions over wings was based on a panel method. So far the trailing vortex sheet has been represented by a set of discrete vortices. On the sheet each vortex is represented by a set of straight line segments. Hence the integrals appearing in (17) are approximated by a summation over velocities induced by these line elements. These velocities are determined by means of the law of Biot-Savart. A line element $l = P_1 P_2$ (strength Γ) induces the velocity vector $\underset{\sim}{V}(Q)$ at point Q

$$\underset{\sim}{V}(Q) = \frac{\Gamma}{4\pi} \int \left(\nabla \frac{1}{r} \times d\underset{\sim}{s}\right) = \frac{\Gamma}{4\pi} \int_{P_1}^{P_2} \left(\nabla \frac{1}{r} d\xi\right) \times \underset{\sim}{l}$$

$$= \frac{\Gamma}{4\pi} |\underset{\sim}{l}| \frac{\underset{\sim}{l} \times \underset{\sim}{r}_1}{|\underset{\sim}{l} \times \underset{\sim}{r}_1|^2} (\cos \beta_1 + \cos \beta_2) \qquad (20)$$

Here $\underset{\sim}{r_i}$ is the vector QP_i and $r_i \cos \beta_i$ is the projection of $\underset{\sim}{r_i}$ on the vector $\underset{\sim}{l}$.

As stated before the velocities are always calculated at points half way between vortex lines. The values at other points are obtained by interpolation.

Since the discrete vortices coincide with streamlines, (18) and (19) can be interpreted as

$$\frac{dx}{V_{mx}} = \frac{dy}{V_{my}} = \frac{dz}{V_{mz}} \tag{21}$$

which have to be satisfied along these vortex lines. This can only be done by an iteration in which values of $\underline{V}_m$ obtained in part from previous iterations are used.

Mangler and Smith [9] have shown that along the trailing edge the vortex sheet is always parallel to either the upper or the lower surface of the wing, depending on the velocity distribution along the trailing edge. The method by Labrujere [2] proceeds as follows (see also [10]).

The trailing vortices are initially assumed to bisect the trailing edge angle. Then

1. by means of the NLR panel method the strength of the trailing vortices and the velocity along the trailing edge are obtained. This defines the correct position of the vortex sheet at the trailing edge [9].

2. With this new position the calculation is repeated (inner iteration) until the mean velocity and the vortex sheet strength are known.

3. With this distribution of Γ across the sheet the mean velocities are calculated and the shape of the sheet is found by integration of (21).

4. The NLR panel method is applied again, incorporating the effect of the new deformed vortex sheet on the wing.

Steps 3 and 4 are repeated until convergence occurs.

Labrujere applied the method in the case of a rectangular wing of aspect ratio 5 at an incidence of 15° ($C_L = 1.0$). The chordwise section of the wing was RAE101 ($t/c = 0.09$). According to this calculation the effect of the rolling-up of the vortex sheet on the lift distribution across the wing span is very small. The deformation of the trailing sheet is shown in Fig. 3.

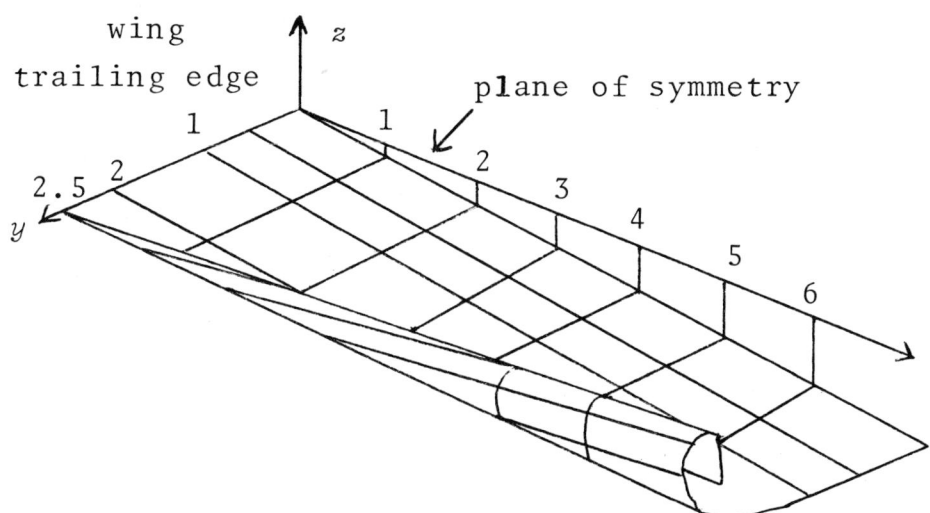

Fig. 3. Trailing sheet for rectangular wing at incidence $\alpha = 15°$ ($C_L = 1$)

Some features of the method, such as the effect of streamwise stepsize and spanwise discretisation of the vortex sheet, have been investigated numerically. The method and also Rehbach's approach [7] may be called an "engineering method" in so far as no definite rules can be given for the best discretisation and step sizes. Also the detailed description of the rolling up is not possible. According to my ideas the procedure must be modified by the introduction of a vortex representing the core of the rolled up sheet. The details depend on further work on the nature of the flow near the wing tips. The important result of this work is that the method appears to "converge" at least with the present set of vortices.

In a recent paper Rehbach [11] applied this method to a thin delta wing with leading edge separation (apex angle = 30°) (again applying a technique proposed by Butter and Hancock [10]). As a starting approximation he used a rectangular wing with the same span/chord ratio and changed it gradually (by sweeping the side edge) into a delta wing. Apart from some numerical problems near to the apex of the wing his solution converged to something close to a conical sheet. Agreement with results obtained by Smith [1] was fairly good, the differences partly arising from trailing edge effects and partly from the fact that the wing was perhaps not sufficiently slender for Smith's method to apply. Also it should be pointed out that the model of the vortex core used here was again not sufficient. Perhaps even for a rectangular wing a model of a slender vortex sheet as used by Smith should be incorporated in the first approximation.

CONCLUDING REMARKS

In conclusion it should be pointed out that (probably for reasons of computational economy) almost all numerical work published on this subject was based on modelling vortex sheets by sets of isolated vortices (sometimes known as vortex lattices). No attempt has been made in this paper to give a complete history of this work. A fairly complete review of the two-dimensional solutions (time dependent and slender) and the earlier work on trailing edge sheets was recently given by Fink and Soh [4], the background to the work on trailing edge sheets is to be found in the cited papers, for example by Rehbach [7] and Labrujere [2]. Work on slender wings has been extended to wings with thickness [12], with camber [13], wings with blowing from the leading edge [14] and yawed slender wings [15]. In all these papers similar techniques were used.

One important problem, which may affect the results based on discrete vortices, is that of stability of a set of isolated vortices. Rosenhead's work [16] seemed to indicate the rolling up of an infinite sheet near points where the vorticity is concentrated. Similar effects including apparent randomness were observed, for example, by Moore [17] when investigating the rolling up of a finite flat vortex sheet. As mentioned above the inclusion of a better model for the vortex core [5] cured this problem. This is an important lesson to be learnt.

A major difficulty is the inaccurate representation of the velocity field induced by a segment of a vortex sheet by single vortices (see pp. 199-206). Fink and Soh [4] and also Barsby [18] suggested certain choices for the position of the "calculating" points (pivotal points) which reduce the error. Perhaps some means will be found to improve the results by introducing suitable weighting coefficients into the formulae derived from vortex lattice theory or by replacing the vortex lattice by a continuous sheet, at least for the calculation of near-field effects. Similar remarks apply to the representation of the wing surface. There is hope that with further "computational" experience (and also by evaluation of suitable experiments leading to semi-empirical "engineering" methods) there will be methods available which are suitable for practical application in industry. The effects of compressibility will probably be allowed for by empirical correction factors, at least for the time being. A proper theoretical treatment of compressibility effects could only come from a solution of the three-dimensional field equations.

REFERENCES

1. Smith, J.H.B., "Improved calculations of leading-edge separation from slender delta wings," RAE TR.66070; *Proc. Roy. Soc. A.*, **306**, 67-90 (1968).

2. Labrujere, Th.E., "A numerical method for the determination of the vortex sheet location behind a wing in incompressible flow," NLR (Amsterdam), TR72091U (1972).

3. Mangler, K.W. and Weber, J., "The flow field near the centre of a rolled-up vortex sheet," RAE Tech Rep 66324: *J. Fluid Mech.*, **30**, 177-196 (1967).

4. Fink, P.T. and Soh, W.K., "Calculation of vortex sheets in unsteady flow and applications in ship hydrodynamics," University of New South Wales Report, Nav/Arch 74/1 (April 1974).

5. Moore, D.W., "A numerical study of the rollup of a finite vortex sheet," *J. Fluid Mech.*, **63**, 225-236 (1974).

6. Kaden, H., *"Aufwicklung einer unstabilen Unstetigkeits-fläche,"* Ing. Archiv, **2**, 140 (1931).

7. Rehbach, C., *"Etude numérique de l'influence de la forme de l'extremitie d'une aile sur l'enroulement de la nappe tourbillonaire,"* Rech. Aerospatiale No. 1971-6, 367-368 (1971).

8. Rehbach, C., *"Calcul d'ecoulements autour d'ailes sans épaisseur avec nappes tourbillonaires évolutives,"* Rech. Aerospatiale No. 1973-2, 53-61 (1973).

9. Mangler, K.W. and Smith, J.H.B., "Behaviour of the vortex sheet at the trailing edge of a lifting wing," *Aero. J. Roy. Aero. Soc.*, **74**, 906-908 (1970); RAETR69049.

10. Butter, D.J. and Hancock, G.J., "A numerical method for calculating the trailing vortex system behind a swept wing at low speed," *Aero. J. Roy. Aero. Soc.*, **75**, 564-568 (August 1971).

11. Rehbach, C., "Numerical investigation of vortex sheets issued along a separation line near the leading edge of a wing," Euromech Coll 41, Norwich GB (September 1973); Rech. Aerospatiale No. 1973-6, 325-330 (1973).

12. Smith, J.H.B., "Calculations of the flow over thick, conical, slender wings with leading edge separation," RAE TR 71057 (1971); ARC·R & M 3694.

13. Barsby, J.E., "Flow past conically-cambered slender delta wings with leading-edge separation," RAE TR 72179 (1972); ARC·R & M 3748.

14. Barsby, J.E., "Calculations of the effect of blowing from the leading edges of a slender delta wing," RAE TR 71077 (1971); ARC.R & M 3692.

15. Pullin, D.I., "Calculations of the steady conical flow past a yawed slender delta wing with leading edge separation," I.C. Aero Rep. 72-17, Imperial College London (July 1972).

16. Rosenhead, L., "The formation of vortices from a surface of discontinuity," *Proc. Roy. Soc. A.*, **134**, 170 (1931).

17. Moore, D.W., "The discrete vortex approximation of a vortex sheet," Cal. Tech. Rep AFOSR-1084-69 (1971).

18. Barsby, J.E., "Separated flow past a slender delta wing at incidence," *Aero. Q.*, 120-128 (1973).

SOME PROBLEMS OF UNSTEADY FLOW ABOUT AIRCRAFT

G.J. Hancock

*(Queen Mary College,
University of London)*

INTRODUCTION

As most passengers on commercial aircraft are aware, their flights from one destination to another are not always smooth or steady. Flight in slight to moderate atmospheric turbulence is not infrequent, whereas severe turbulence is encountered much less frequently. During excursions into atmospheric turbulence the intrepid passenger who dares to look out of the cabin window will have seen the wings flexing, sometimes in an alarming manner (at such times the passenger probably offers up a silent prayer that in building the wings the wing designer knew what he was doing, it is probably true to speculate that during the building of the wings the wing designer often offered up the same prayer). On other occasions, usually on a heavily loaded aircraft while either in cruise at high altitude or on approach to landing, the passenger may be aware of a continuous relatively high frequency judder; this phenomenon is associated with the breakdown of the smooth flow over the wings causing unsteadiness in the flow which excites the flexible wings and over-all airframe. Again on the approach to landing, a passenger may be aware of small variations in vertical acceleration and also of swaying from side to side as the pilot attempts to maintain a closely monitored descending flight path. Sometimes landings, at the moment of touchdown, can be heavy, inducing the wings to bounce up and down. All of these situations, which involve unsteadiness in the motion of the aircraft and structural vibrations in the wings, fuselage, etc., need to be fully understood and quantified so as to ensure efficient airframe design and aircraft operation while maintaining, first and foremost, complete safety but also providing as comfortable a ride as possible for the passengers and crew.

The aerodynamic aspects of the dynamic motion of flexible aircraft come within the over-all field of unsteady aerodynamics. There are however several distinctive classes of problem.

An important class of unsteady aerodynamic problems concerns the unsteady flow about aerofoils, wings and aircraft when the flow remains attached. In this type of flow the effect of viscosity is confined to relatively thin boundary layers around the aircraft surface area and to the downstream wakes which are formed from the shed boundary layers from the wing and tailplane trailing edges and from the separation lines on the fuselage.

When an aircraft is flying through slight to moderate atmospheric turbulence the flow about the aircraft usually remains attached; the unsteady atmospheric air flows induce unsteady loads on the aircraft which cause the aircraft to respond in a dynamic manner. However the dynamic response itself induces additional unsteady loads which in turn modify the actual dynamic motion. Two sets of aerodynamic information are therefore required, the aerodynamic loads associated with the unsteady gust inputs and the aerodynamic loads associated with dynamic response. When the flow about the aircraft remains attached during the response the superposition of all the various aerodynamic effects can be assumed. The basis of this assumption is that when the boundary layers are thin, linear theory gives good estimates for the lifting load distribution over the wings, tailplane and fuselage. The problem becomes nonlinear when the displacement effects of the boundary layers and wakes are taken into account but the degree of nonlinearity is small. Implicit in this superposition approach is the assumption that the loads arising from atmospheric gusts can be estimated from potential (irrotational) flow theory, even though atmospheric turbulence, by its nature and definition, is a non-potential (*i.e.*, rotational) flow. More research is required in this area to verify how far potential theory can be used to estimate gust loads; gusts of large wavelength in comparison to the wing chord can be regarded as potential flows at least in the neighbourhood of the aircraft, but the effects of gusts of smaller wavelength, which are responsible for exciting the higher structural frequencies, require further investigation.

As the incidence of an aircraft increases, at a critical incidence the flow no longer remains attached and flow separation occurs. On a swept wing aircraft this flow separation and breakdown into random unsteadiness first appears in the region of the wing tip. As incidence is further increased the region of flow breakdown grows and spreads inboard. When the region of flow breakdown extends sufficiently far across the wing span the aircraft goes out of control; this is known as aircraft stall. The random unsteadiness associated with flow breakdown is known as buffet, and the subsequent aircraft response, of both the over-all aircraft response and the structural flexibility modes, is known as buffeting. Little is known at present about the physical and even empirical characteristics of flow separation and breakdown on finite wings; attempts to model the flow mathematically have not really been made. One question in particular which needs resolution concerns the effect of the response back on the flow characteristics; flow breakdown implies a distribution of energy throughout a wide frequency spectrum, it is not known if the motion of the wing changes the magnitude or distribution of energy in the frequency bands.

Prediction of the onset of flow separation, especially in unsteady conditions, and the related problem of subsequent flow reattachment, is an active area of research, primarily by experiment.

One of the main advances in present day aeronautical technology is known as active control technology; with the appearance of powerful and fast acting hydraulic power systems and reliable electrical systems it is possible to utilise aerodynamic controls beyond their conventional rôles. It is argued that an aircraft need not be inherently stable if the elevator has a sufficiently fast response, through an internal sensor-feedback loop, to react to any destabilising motions. Similar fast acting controls on a wing could reduce the level of structural vibration and so alleviate vibration levels in gusty conditions. During the approach to landing, additional aerodynamic controls (known as direct lift controls) enable a pilot to keep closer to an indicated glide path. In all of these cases it is necessary to have a full understanding of the aerodynamic responses of controls, including spoilers, in general unsteady motion.

The problems outlined above represent the ultimate objectives in aircraft unsteady aerodynamics, remembering that answers are required for all types of aircraft configuration in speed ranges from subsonic, through transonic to supersonic. To build up experience to predict these complex flows it is desirable to follow a sequential programme of theoretical studies.

Up to the present most predictive methods are based on linearised wing theory which can be regarded as a first order theory for attached flows. In the UK the work of W.P. Jones at the NPL in the 1950's laid the foundation of unsteady linearised wing theory, this work was later extended and consolidated by H.C. Garner and his co-workers at the NPL and by D.L. Woodcock and E.D. Davies in the Structures Department, RAE. In the USA the numerical approach of the so called vortex doublet developed by Rodden, Giesing *et alia* is now a common tool. The status of unsteady linearised wing theory, utilising recent advances in computing power, can be said to cover many of the practical requirements for a wide range of wing configurations at subsonic and supersonic speeds although the unsteady aerodynamics of lateral, or antisymmetric, motions still require further work. Linearised wing theory is invalid at transonic speeds but work is in progress to fill this gap.

With the advent of the computer there has been tremendous activity and development in methods for prediction of the fields of steady aerodynamic flows about aerofoils, wings and complete aircraft. Most of these developments are described in the other papers in this volume. It is therefore important that a parallel effort be put into the complementary set of unsteady aerodynamics. From this standpoint several problems under consideration by myself and my co-workers are described here.

TWO-DIMENSIONAL AEROFOILS IN UNSTEADY MOTION

Steady "A.M.O. Smith" method

A standard technique to calculate the pressure distribution on a stationary aerofoil of arbitrary shape in a uniform stream, neglecting both viscous and Mach number effects, is the surface singularity distribution method associated with A.M.O. Smith and his co-workers [1]. In this section some of the extensions of the A.M.O. Smith approach to unsteady problems are described.

For the steady case, the solution to the problem can be obtained by the superposition of source and vortex flows so that the aerofoil profile remains a streamline. As shown in Fig. 1 the aerofoil surface is divided into N straight line elements, starting with element 1 on the bottom surface at the trailing edge and then proceeding clockwise such that element N is the element adjacent to the trailing edge on the upper surface. On each element is placed a uniform source distri-

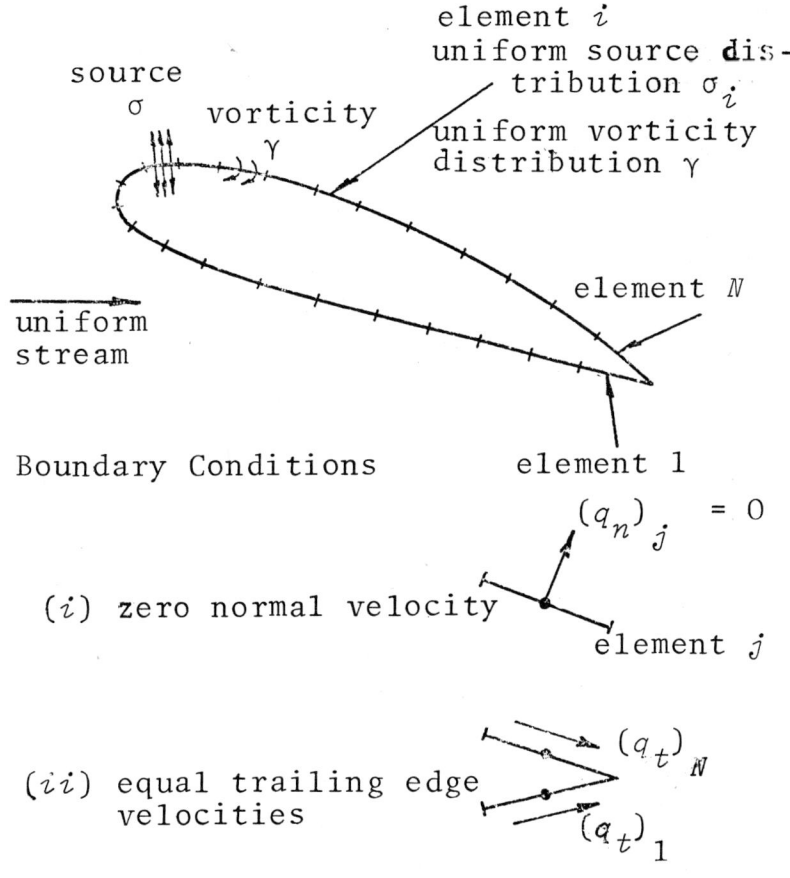

Fig. 1.

bution and a uniform vorticity distribution. The strength of the source distribution $\sigma_i (i=1,\ldots,N)$ varies from element to element, while the vorticity distribution γ is the same for all elements. Thus there are $N + 1$ unknowns; N boundary conditions are obtained by ensuring that the total velocity at the mid-point of each element is parallel to the element (total velocity means the velocities induced by all the source and vorticity distributions together with the mean flowfield); the final boundary condition is obtained from the Kutta condition of finiteness of flow in the neighbourhood of the trailing edge which is specified in terms of equal tangential velocities, both to be in the downstream direction, at the midpoints of the two trailing edge elements (elements 1 and N).

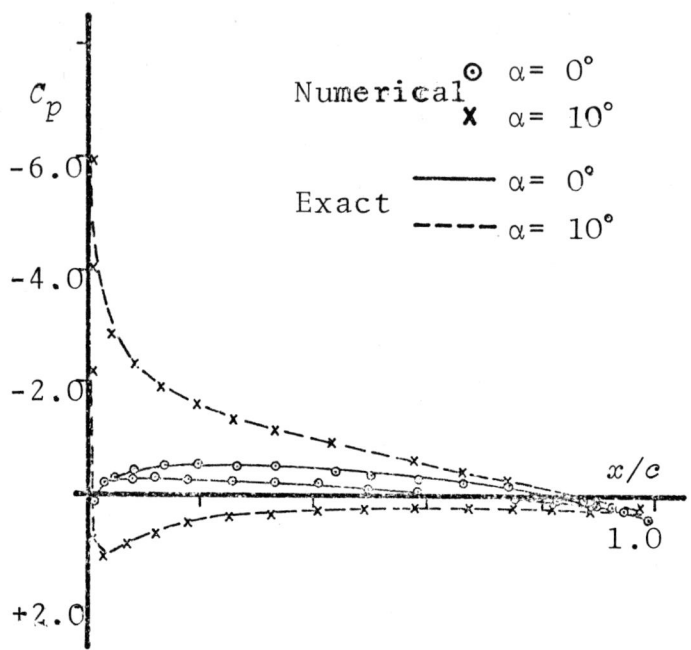

Fig. 2. Pressure distribution on a cambered Katman-Trefftz aerofoil

A typical result is shown in Fig. 2 for an aerofoil, which can be solved analytically, taking $N = 35$. Apart from the trailing edge region which has a different Kutta condition in the analytic solution, the comparison is good.

Boundary layer effects, when the flow remains attached, can be incorporated. Taking a pressure distribution as calculated from an A.M.O. Smith solution, assuming an inviscid flow, as described in the previous paragraph, a boundary layer calculation can then be performed along the upper and lower surfaces of the aerofoil. Usually the Thwaites method [2] is used for the initial laminar boundary layer and then either the Horton [3], Green [4] or the Bradshaw [5] method can be used for the turbulent boundary layers; the position of transition is difficult to predict especially on the lower surface. Once the displacement thickness of the boundary layers over the aerofoil are known, the displacement thickness of the downstream wake can either be simply assumed or calculated; on the whole the wake does not play an important rôle as far as the pressure distribution on the aerofoil is concerned. The mathematical model, allowing for the boundary layer and wake displacement effects, as developed in [6] is shown in Fig. 3.

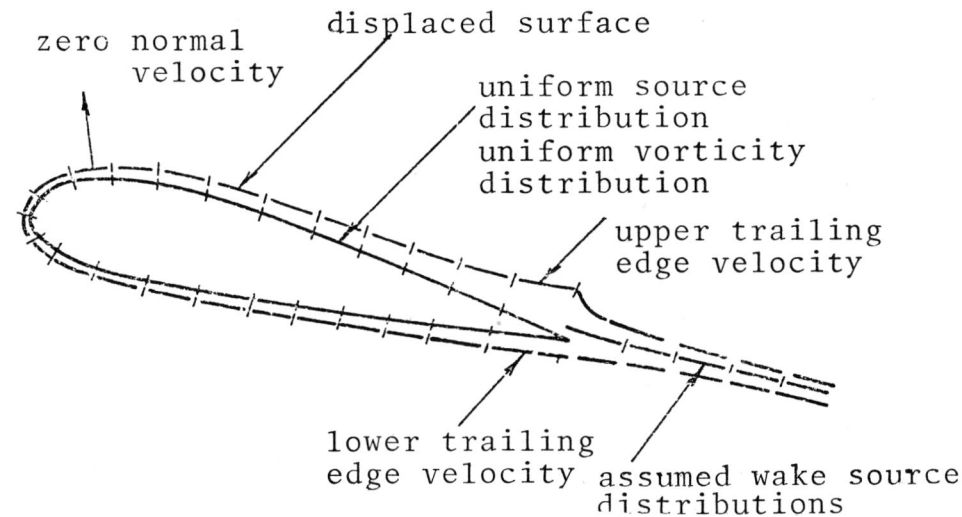

Fig. 3. Aerofoil with boundary layer

The singularity, source and vorticity, distributions are again distributed on the surface of the aerofoil, together with additional source distributions along the centre line of the wake. If the wake characteristics are assumed this is equivalent to assuming the wake source strengths. Again there are $(N + 1)$ unknowns; N conditions of adjacent flow are satisfied on the displaced surface, *i.e.*, on the edge of the boundary layer, and the Kutta condition is taken as equal velocities on the upper and lower surface elements on the edge of boundary layer at the trailing edge.

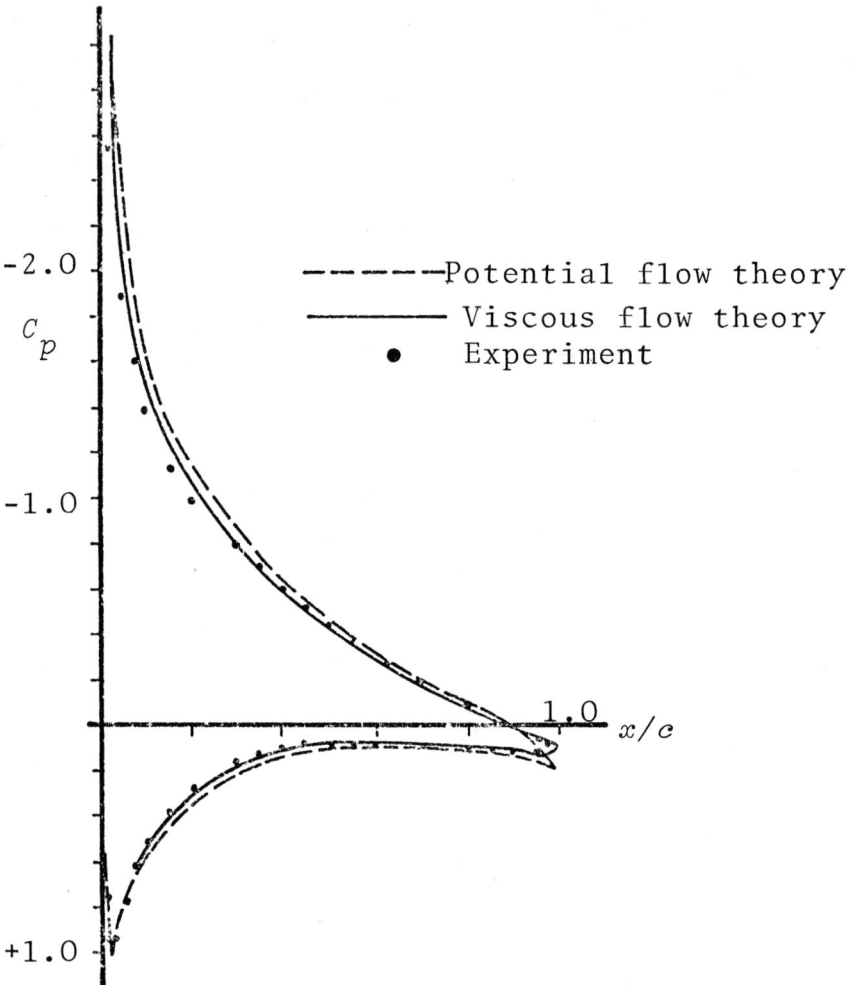

Fig. 4. Pressure distribution on RAE 101 aerofoil

A typical result [6] is shown in Fig. 4 where a theoretical pressure distribution is compared with experiment. Over-all the agreement is good, where differences do appear it is not clear whether these differences arise from deficiencies in the theory, or from experimental defects caused, for example, by minor variations in the shape.

Linearised oscillatory A.M.O. Smith method

An aerofoil oscillating in simple harmonic motion in some manner about a mean position, neglecting viscous effects, has been investigated [7] by extension of the A.M.O. Smith approach. At low speeds the differential equation for the velocity potential remains independent of time, the time dependence only comes into the boundary conditions, thus superposition of surface singularities can again be applied. Since displacements of the aerofoil from its mean position are assumed to be small it is assumed, as shown in Fig. 5, that singularities be placed on the mean steady profile and that the unsteady boundary conditions also be applied on the mean profile shape.

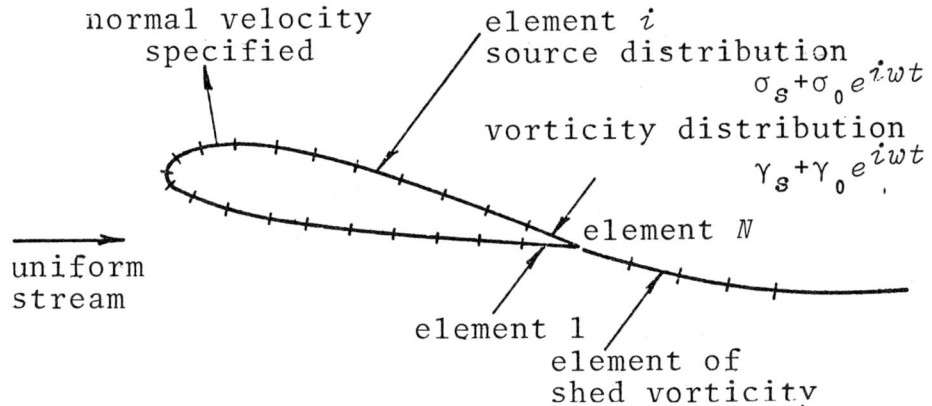

Fig. 5. Oscillatory linearised A.M.O. Smith model

Thus on each element there is a uniform source distribution $(\sigma_s + \sigma_0 e^{i\omega t})$, where σ_s refers to the mean steady condition and σ_0 is the amplitude of the oscillatory source distribution, and a uniform vorticity distribution $(\gamma_s + \gamma_0 e^{i\omega t})$; σ_s and γ_s are the solutions of the steady problem. Since the distribution $\gamma_0 e^{i\omega t}$ gives rise to an oscillatory circulation around the aerofoil, vorticity is shed to the wake; it is assumed that the shed vorticity is convected with the free stream velocity along the mean stream line emanating from the trailing edge. The vorticity distribution in the wake is known in terms of the time dependent variation of the circulation about the wing. In the near field up to one chord behind the wing the wake is divided into straight elements on which the vorticity distribution is assumed to be uniform; aft of one chord it is assumed that the remainder of the wake is parallel to the free stream direction and that this far field only induces downwash on the aerofoil.

For this linearised A.M.O. Smith approach there are again $(N + 1)$ unknowns which are now complex, σ_{qj} $(j = 1,...N)$ and γ_0; there are N conditions of (unsteady) adjacent flow at the mid-points of each element and one Kutta condition is required. In a similar exercise Giesing [8] took equal velocities at the mid-points of the two trailing edge elements (1 and N) but this assumption implies a difference in pressure between elements 1 and N. In [7] the condition of equal pressures on elements 1 and N has been taken on the argument that zero loading at the trailing edge would appear physically realistic and ensures consistency with the zero loading condition in the wake.

A result, taken from [7] with $N = 35$, is compared in Fig. 6 with an analytic solution by de Voorn and de Val. The agreement for both in and out of phase is good. Extensive results have been obtained for aerofoils in oscillatory pitch and heave, and for aerofoils in sinusoidal vertical gust conditions.

Research is currently in progress into the incorporation of unsteady boundary layer effects into the above unsteady A.M.O. Smith solution. A preliminary attempt to incorporate boundary layer effects, assuming that the boundary layer behaves in a quasi-steady manner, in a similar way to that shown in Fig. 3, gives the results shown in Fig. 7. There appears to be the correct trend as far as lift characteristics are concerned, but not for the moment characteristics.

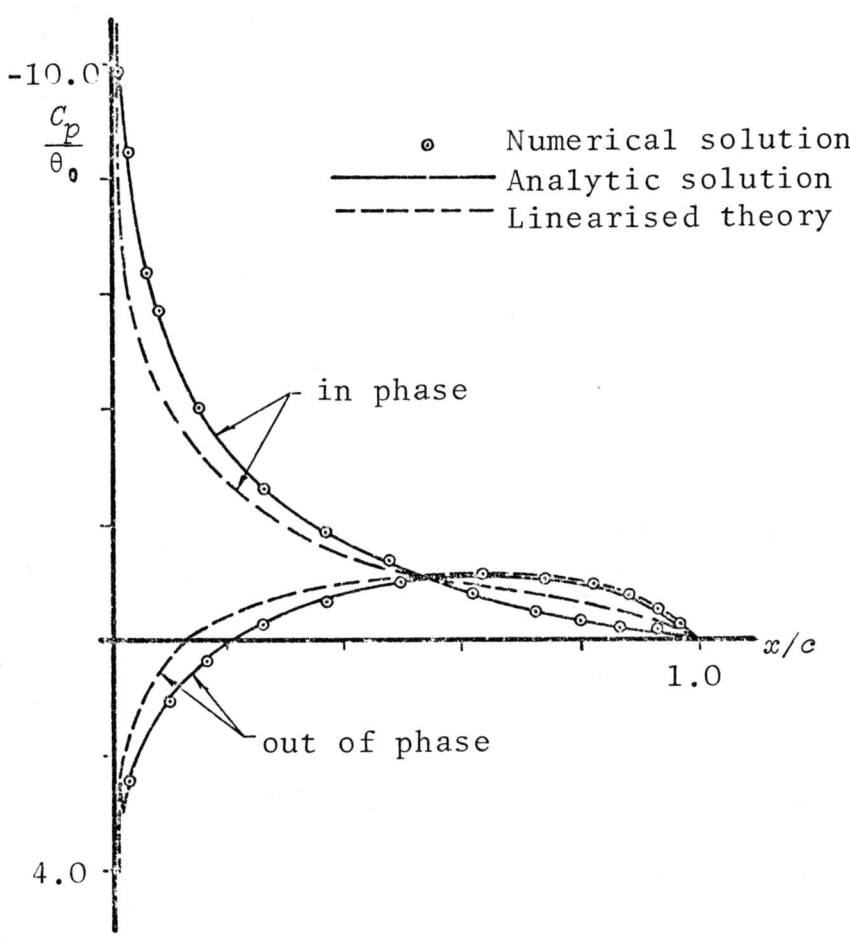

Fig. 6. Pressure distribution on oscillating aerofoil

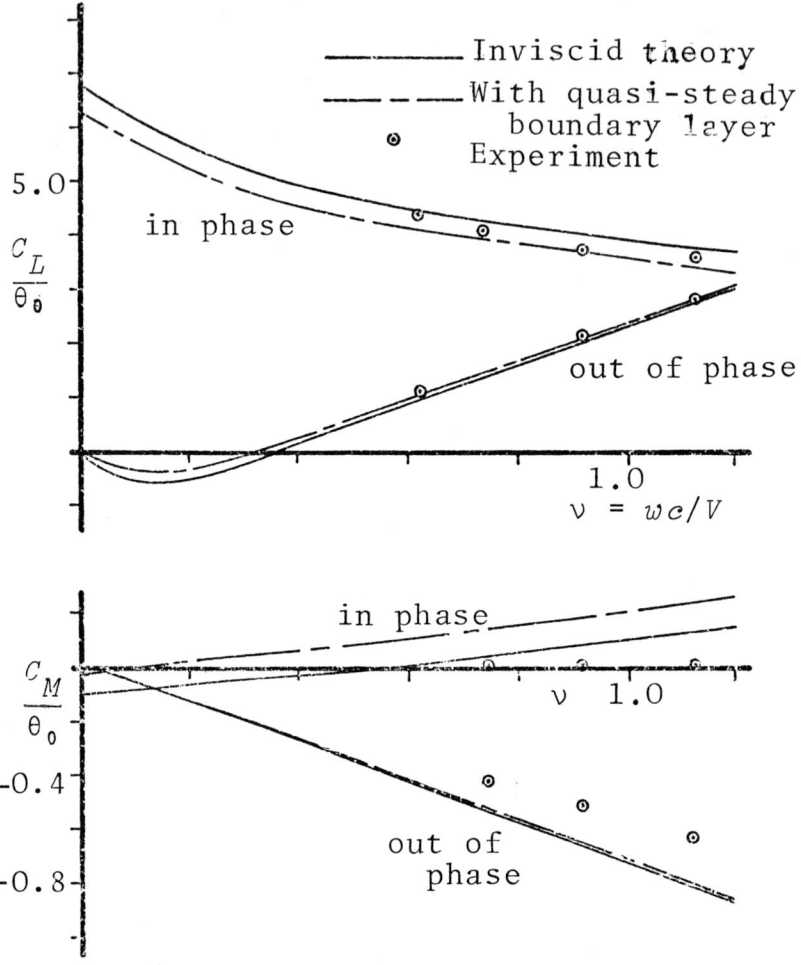

Fig. 7. Comparison of theory and experiment for oscillating aerofoil

Unsteady A.M.O. Smith method

A generalised unsteady inviscid solution has been obtained [9] for the solution of an aerofoil undergoing an arbitrary motion, assuming incompressible, inviscid flow. The solutions are determined at successive time intervals. The model for the solution at time t_k is shown in Fig. 8. Again N elements are distributed around the aerofoil surface at its location at time t_k. On each ele-

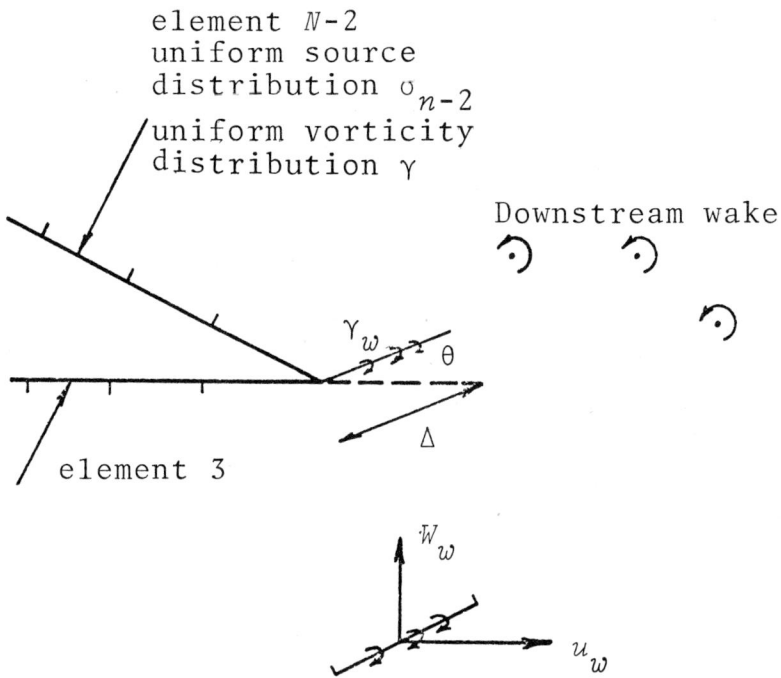

Fig. 8. Unsteady A.M.O. Smith model

ment is a uniform source distribution $\sigma_i (i=1,...N)$ and a uniform vorticity distribution γ, which is the same for all elements. A wake element from the trailing edge of length Δ and inclination θ has a uniform vorticity distribution γ_w; in this approach Δ and θ are unknowns to be determined; there is however the relationship that the circulation around this trailing edge element ($\gamma_w \Delta$) is equal to the difference in circulation $[\Gamma(t_k) - \Gamma(t_{k-1})]$ in the time interval $(t_k - t_{k-1})$. The downstream wake pattern is represented by discrete vortices whose strengths and positions at time t_k are known. There are $(N + 3)$ unknowns, $\sigma_i (i=1,...,N)$, γ, Δ and θ. There are N conditions of adjacent flow. The Kutta condition is taken as zero pressure difference between elements 1 and N. For zero loading across the wake trailing edge element, if u_w and w_w are the two components of total velocity at the mid-point of the wake element resulting from the entire singularity field and the mean flow but excluding the wake element itself,

$$\tan \theta = \frac{w_w}{u_w}$$

and

$$\Delta = (u_w^2 + w_w^2)^{\frac{1}{2}}(t_k - t_{k-1})$$

The calculation of Δ and θ involves a small iterative loop in the programme at time t_k.

Once the solution at t_k is evaluated, the velocities of the wake vortices can be calculated, leading to their new position at time t_{k+1}. The vorticity of the wake element on Δ at time t_k is assumed to take up the form of a discrete vortex, at time t_{k+1}, located along the direction θ at the appropriate distance.

The method has been applied to a number of problems.

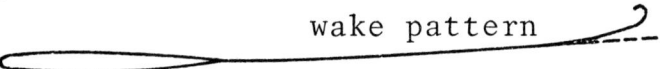

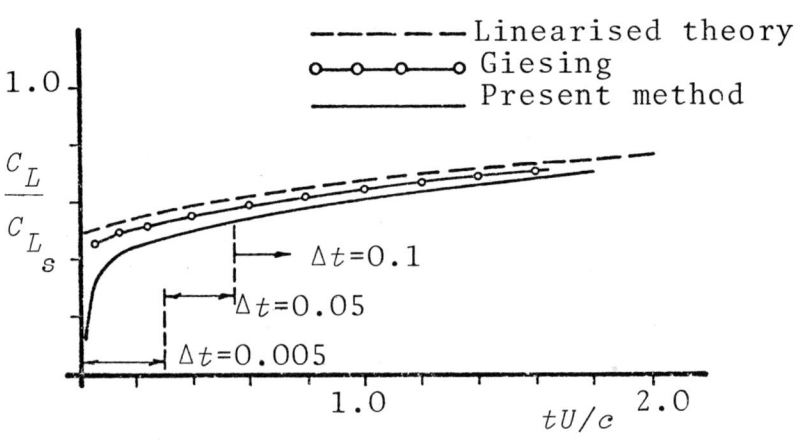

Fig. 9. Sudden change of incidence

(i) Results for a sudden change of incidence are shown in Fig. 9. Short time intervals have been taken in the initial response period and the time intervals are then lengthened as the rate of change of the response decreases. Comparisons are shown between the results obtained from the above model, results from a similar model developed by Giesing [10] but which does not satisfy the zero loading condition across the trailing edge, and linearised theory.

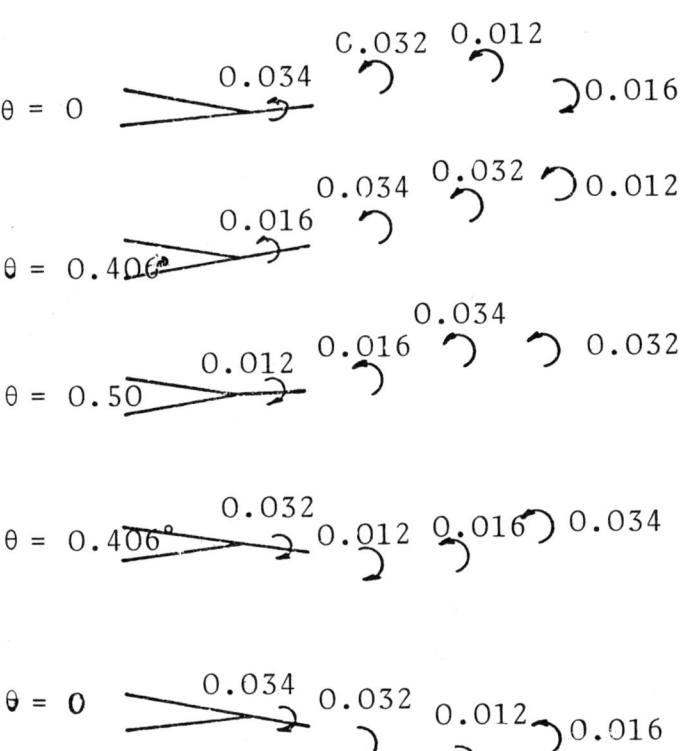

Fig. 10. Aerofoil pitching at high frequency

(ii) An aerofoil oscillating through an amplitude of about $\frac{1}{2}°$ at high frequency ($wc/V = 20$) is shown in Fig. 10. The rolled up wake in pairs of equal and opposite vortices is shown; this pattern closely resembles experimental observation. An enlargement of the theoretical flow in the region of the trailing edge during a half cycle shows the shedding and convection of vorticity aft of the trailing edge. The interesting point to note is that mostly the wake element from the trailing edge lies parallel to one surface or the other, depending on the direction of the shed vorticity, except when θ = 0, when the shed vorticity is changing sign. Maskell [11] has argued that, analytically, behaviour at the trailing edge indicates that vorticity leaves the trailing edge parallel to either the upper or lower surface. It is gratifying that the numerical model gives a close approximation to the local analytic solution in the region of the trailing edge.

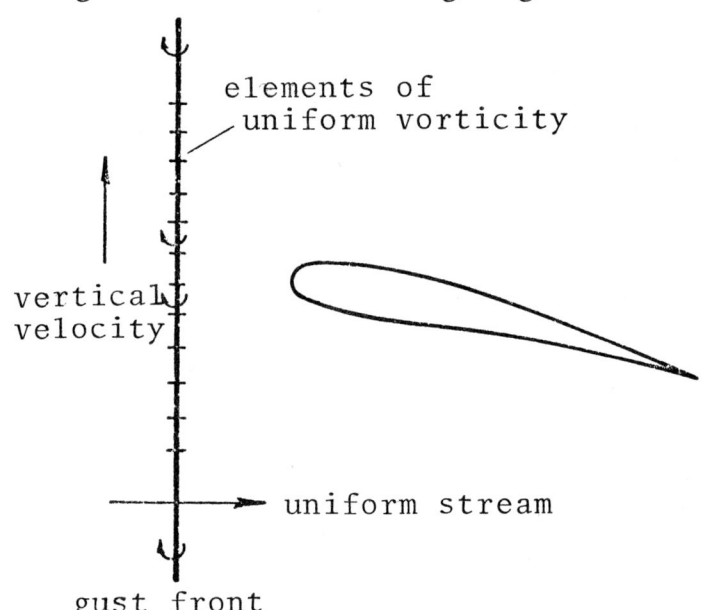

Fig. 11(*i*). Gust front approaching aerofoil

(*iii*) An attempt has been made to calculate the response of an aerofoil to a sharp edge vertical gust allowing for the deformation of the gust front as it passes over and behind the aerofoil. As shown in Fig. 11(*i*) when the gust front is ahead of the aerofoil, it is divided in its inner region into a finite number of elements of uniform vorticity, while in the outer region to infinity the gust front remains as a vortex sheet of uniform strength. At each time step the progress of each gust front element of vorticity can be calculated from the local velocity field; the outer gust front (to infinity) is assumed to connect with the uniform free stream. Suitable approximations are made in the region of the nose as the gust front wraps itself around the nose and then proceeds either side of the aerofoil.

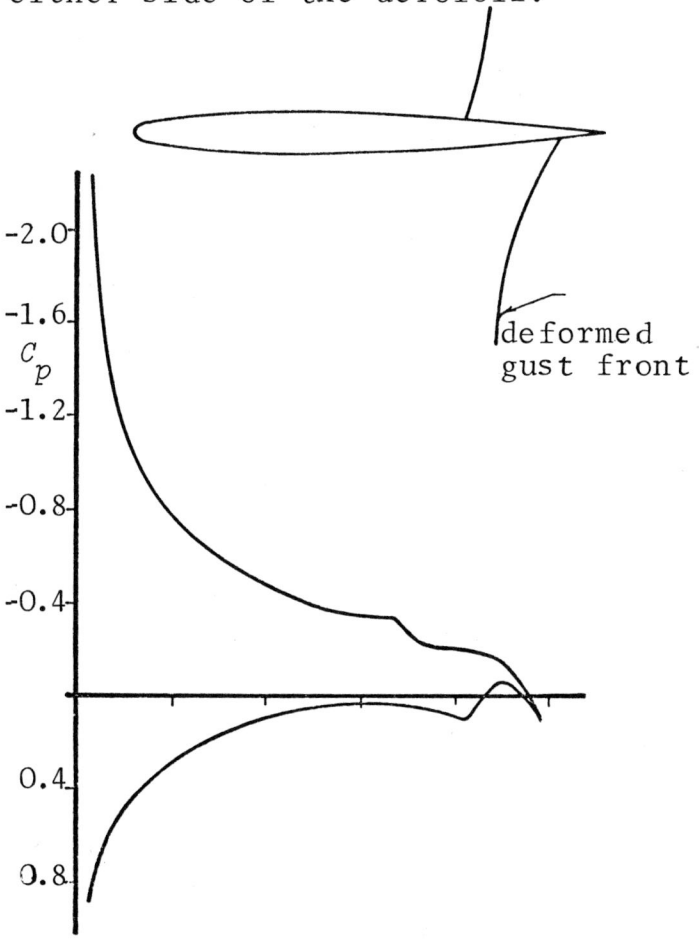

Fig. 11(ii). Gust front on the aerofoil

Fig. 11(*ii*) shows the gust front, and downstream shed vorticity when the gust front is passing over the aerofoil, together with the pressure distribution. Fig. 11(*iii*) shows the interaction pattern of shed vorticity and gust front vorticity when the gust front has passed aft of the aerofoil. It is interesting to speculate that if there were to be a tailplane downstream of the main aerofoil then the loading on the tailplane would be greatly influenced by the deformed gust front vorticity.

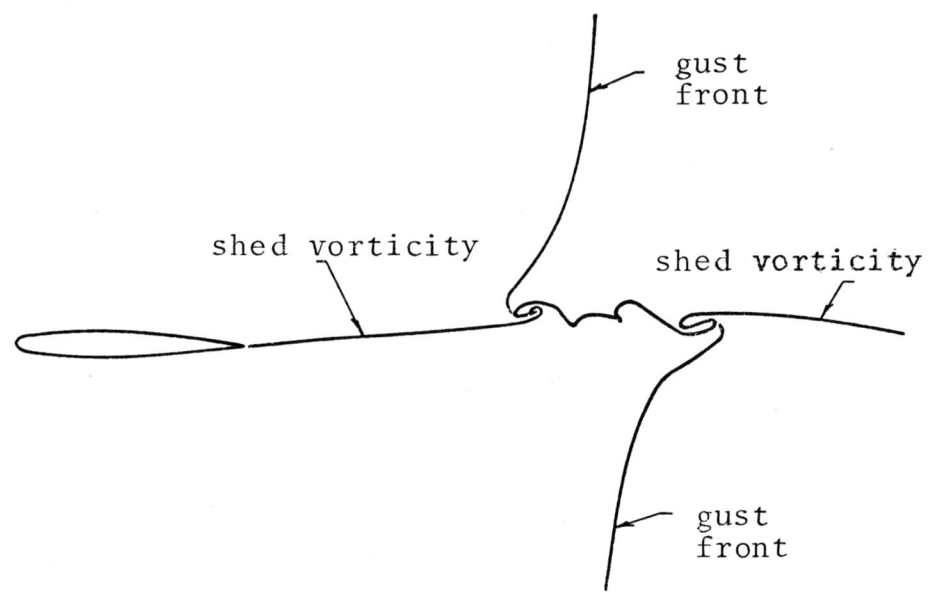

Fig. 11(*iii*). Gust front behind aerofoil

Two-dimensional control surfaces at low speeds

All of the steady and unsteady models described above can be applied to the determination of the characteristics of a two-dimensional aerofoil with a trailing edge control surface at low speeds. However a set of results for a NPL 1541 aerofoil at zero incidence with a control surface deflected through 4° is listed below.

	Lift coefficient C_L	Moment coefficient about 1/4 chord C_M	Control surface hinge moment C_H
Linearised theory	0.242	-0.044	-0.0025
A.M.O. Smith inviscid	0.275	-0.050	-0.0020
A.M.O. Smith and boundary layer	0.254	-0.046	-0.0018
Experiment	0.174	-0.034	-0.0015

It is seen that the best theory, including viscous effects, is totally inadequate for the prediction of the experimental results for C_L and C_M although C_H is moving in the right direction. These comparisons indicate a serious deficiency in the theory. It is suggested the critical factor is the small gap between the aerofoil and the leading edge of the control surface; there is little physical understanding of the effect and the formulation of a mathematical model to allow for the (viscous) flow through the gap poses a formidable task.

FINITE WINGS - VORTEX LATTICE METHODS

For the calculation of steady, and unsteady, load characteristics on finite wings linearised wing theory remains the basic approach. For the symmetric problem, *i.e.*, a finite symmetric wing in a uniform stream with the axis of symmetry of wing aligned with the uniform stream, the steady linearised wing theories based on Multhopp-Garner [12] or the vortex lattice theory of Hedman [13] and the oscillatory linearised wing theories of Garner [12] and the vortex doublet methods of

Rodden-Giesing [14] are well established and extensively used. However for asymmetric problems *i.e.*, when the axis of symmetry of the wing is not aligned with the uniform stream, especially unsteady ones, linearised theory has not yet been fully exploited. There are several problems, for example problems involving the aerodynamic interference between main wing and tailplane, where linearised (planar) theory is inadequate.

Some problems, involving the use of vortex lattice methods, are now described.

Steady vortex lattice method

For a steady finite wing in symmetric flight at low speeds the standard vortex lattice method involves dividing the wing plan form into a number of elements, placing a "lifting" or "bound" vortex on the quarter chord line of each element. Associated with each bound vortex are two semi-infinite trailing vortices in the downstream freestream direction to complete an elemental "horseshoe" vortex system, as shown in Fig. 12. Lift is experienced on the bound vortex only, not on the trailing vortices. The superposition of elemental horseshoe vortices gives the vortex lattice; the strength of each horseshoe vortex is determined from the satisfaction of the downwash condition at collocation points on the 3/4 chord line of each element. The induced downwash at a general point from a single horseshoe vortex can be expressed from the Biot-Savart law, so the superposition of the effects of all the vortices together with the solution of the resulting set of simultaneous equations is ideally suited to the computer. For those problems where the information required concerns over-all lift forces and moments, including some of the higher moments, steady vortex lattice methods have been shown to give reasonable results for a wide range of wing plan forms. Vortex lattice methods are inadequate for the prediction of detailed pressure distributions, more sophisticated panel methods are then required.

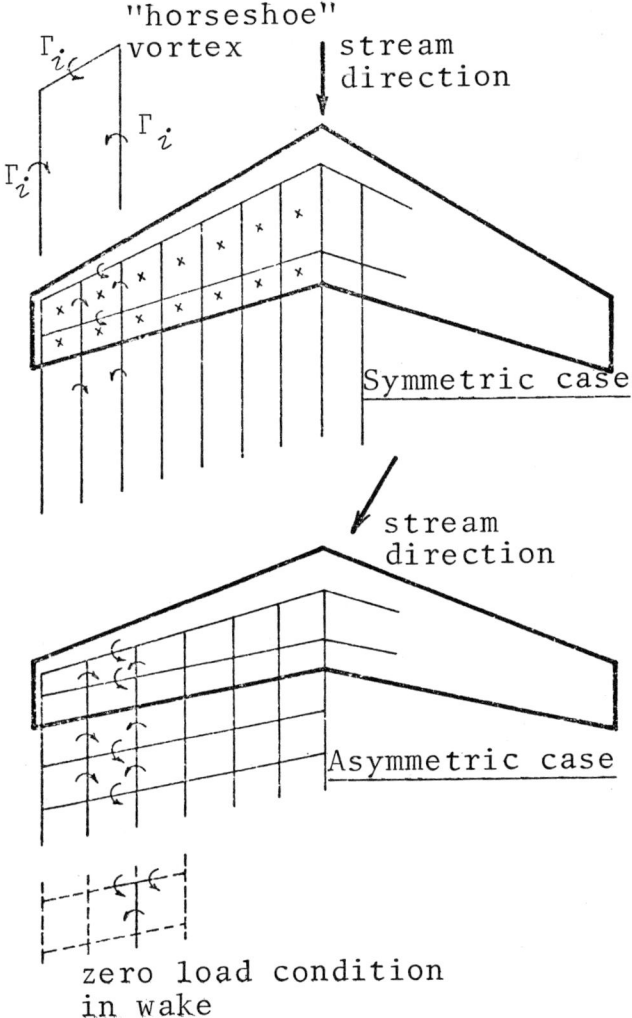

Fig. 12. Steady vortex lattice representation

For a steady wing in the asymmetric condition involving a steady sideslip as shown in Fig. 12, there are several variations of the vortex lattice pattern. For example, the trailing vortices aft of the trailing edge could be swept in the direction parallel to the uniform free stream direction. Alternatively, as shown in Fig. 12 trailing vortices on the wing and in the wake could be maintained in the direction parallel to the centre line of the wing but now bound vortices need to be incorporated in the wake. On the wing both "bound" and "trailing" vortices contribute to the lift distribution since neither is parallel to the free stream direction. In the wake the boundary condition that there shall be no lift is satisfied by making the total lift in the neighbourhood at each junction of "bound" and "trailing" vorticity equal to zero. Such mathematical models again lead to reasonable results for rolling moment, etc.

When the flowfield behind a steady wing needs to be calculated the planar vortex lattice model described above is unsatisfactory; it is necessary to "relax" the trailing vortices in the wake to attempt to satisfy the condition that vortex lines and streamlines coincide. Several numerical methods have been developed to do this; one approach is described in [15] and leads to the type of trailing vortex pattern shown in Fig. 13.

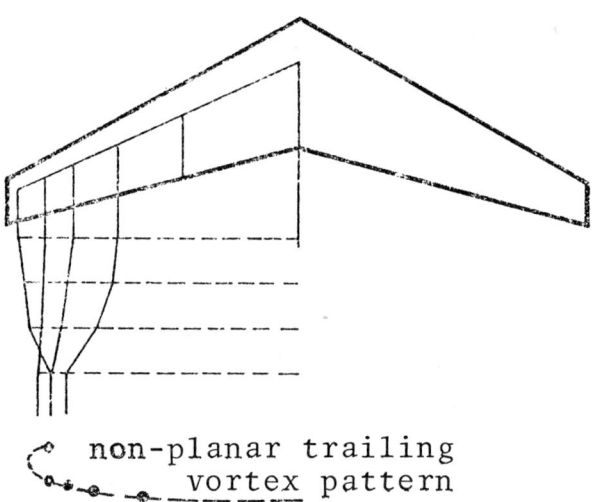

Fig. 13. Rolling up of trailing vorticity

Unsteady vortex lattice method

The above vortex lattice methods have been extended to a number of unsteady cases.

For a wing oscillating in simple harmonic motion in a uniform free stream, as the circulation around the wing at any spanwise station varies with time, vorticity is shed from the trailing edge and then convected downstream. Thus a vortex lattice model can be formulated involving, on the wing, oscillatory "bound" and "trailing" vortices; however the strengths of the wake oscillatory bound vortices are known in terms of the "bound" vortices on the wing. Again this type of model, as shown in Fig. 14, is ideal from the point of view of computational manipulation.

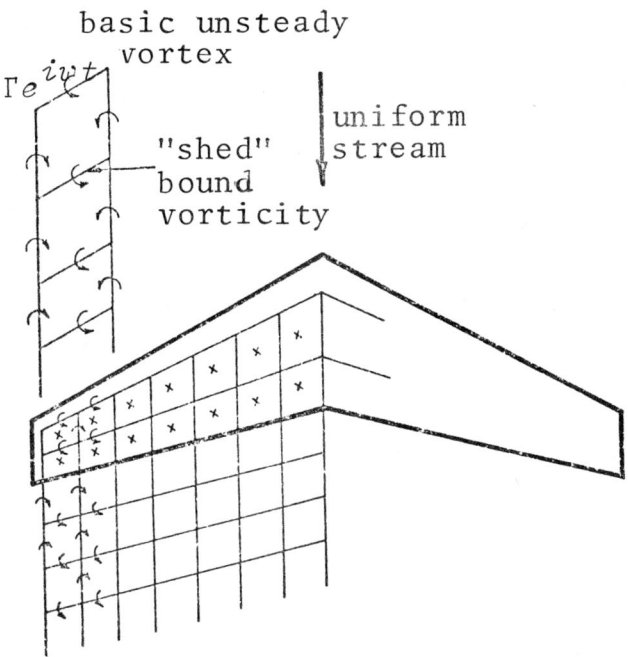

Fig. 14. Oscillatory vortex lattice model

Work is in progress, utilising this unsteady model, to calculate over-all unsteady longitudinal and lateral forces and moments on wings in symmetric and skew vertical two-dimensional sinusoidal gusts. Vertical two-dimensional sinusoidal gusts are symmetric relative to an aircraft when the motion of the aircraft is normal to the gust front, the effects become skewed when the motion of the aircraft is at an angle to the gust front. This theoretical work is complementary to an experimental programme of work in one of the Departmental gust tunnels.

The unsteady model is also being extended to calculate the unsteady forces and moments on a wing oscillating in ground effect by taking a complete image system. It was mentioned in the Introduction that in a heavy landing the wings vibrate through large amplitudes; it is necessary to calculate the over-all dynamic load distribution in the aircraft structure in this situation and the aerodynamic loads can play a significant rôle.

In problems which involve interference between a main wing and tailplane it is necessary to calculate the unsteady flowfield in the neighbourhood of the tailplane when the main wing is vibrating. Since the main wing has a mean lift acting on it there is a mean rolling up of the wake behind the wing about which the oscillatory field can be regarded as an oscillatory perturbation. Thus as a first approximation the oscillatory shed vorticity as determined from the planar model, as shown in Fig. 13, can be assumed to be convected down the steady rolled up wake as shown in Fig. 14, leading to the model shown in Fig. 15. From this model the oscillatory flowfield behind the wing can be determined.

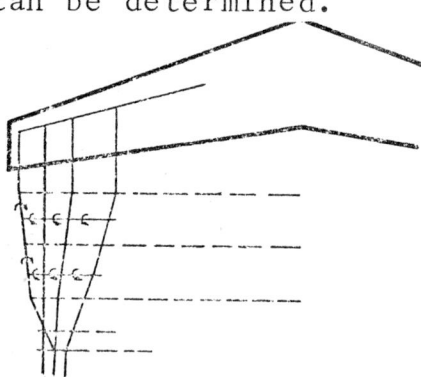

Fig. 15. Convection of unsteady vorticity along mean rolled up vortex wake

CONCLUDING REMARKS

In the previous sections a few problems which involve aspects of unsteady aerodynamics associated with the motion of aircraft have been outlined and numerical solutions described. However, all of the problems have been formulated in terms of the determination of the flow characteristics of an aerofoil or wing undergoing a specified motion. In practice the motion cannot be specified *a priori* because the motion which will comprise over-all longitudinal or lateral aircraft motion together with structural flexibility modal response depends on the unsteady loads. The practical problem combines a dynamic over-all aircraft plus structural response - aerodynamics interaction. Up to the present time structural response calculations and aerodynamic calculations have been calculated separately for simple harmonic motions and the two effects have been combined through an interface matching procedure; the general problem of damped response is usually resolved by Fourier transform techniques which are not altogether satisfactory. With the large computing resources now available it is feasible to combine the dynamic and aerodynamic inputs in a sequential successive time method, incorporating any nonlinear aerodynamics, and estimating directly all of the various responses to specified unsteady atmospheric and control inputs.

It is inferred from these remarks that unsteady aerodynamics and its applications to aircraft are important fields of research which offer considerable interest and challenge to all of those who become involved.

REFERENCES

1. Smith, A.M.O. and Hess, J.L., "Calculation of Potential Flow about Arbitrary Bodies," *Progress in Aeronautical Sciences*, **8** (1966).

2. Thwaites, B., "Approximate Calculations of the Laminar Boundary Layer," *Aero. Q.*, **1**, 245 (1949).

3. Horton, H.P., "Entrainment in Equilibrium and Non-Equilibrium Turbulent Boundary Layers," H.S.A. Hatfield/Research/1094/HPH (August 1969).

4. Green, J.E., "Application of Head's Entrainment Method to the Prediction of Turbulent Boundary Layers and Wakes in Compressible Flow," RAE Technical Report 72079 (1972).

5. Bradshaw, P., Ferris, D.H. and Atwell, N.P., "Calculation of Boundary Layer Development using the Turbulent Energy Equation," NPL Aero Report 1182 (1966); *J. Fluid Mech.*, **28**, 593 (1967).

6. Basu, B.C., "A Numerical Solution for a Steady Two-Dimensional Aerofoil in Incompressible Viscous Flow," QMC - EP 1008 (August 1973).

7. Basu, B.C., Hancock, G.J. and Padfield, G., "A Numerical Solution for Oscillating Two-Dimensional Aerofoils and Oscillating Control Surfaces in Inviscid Incompressible Flow," QMC - EP 1009 (1973).

8. Giesing, J.P., "Two-Dimensional Potential Flow Theory for Multiple Bodies in Small Amplitude Motion," Douglas Aircraft Company Report No. 67028 (April 1968).

9. Basu, B.C. and Hancock, G.J., "A Numerical Solution for a Two-Dimensional Aerofoil in General Unsteady Motion in an Incompressible Inviscid Flow," to be published as a QMC Report.

10. Giesing, J.P., "Non-Linear Two-Dimensional Unsteady Potential Flow with Lift," *J. Aircraft*, **5**, No. 2 (March - April 1968).

11. Maskell, E.C., "On the Kutta-Joukowski Condition in Two-Dimensional Unsteady Flow," ARC 33967, FM 4327; RAE Tech. Memo. 1451 (1973).

12. Garner, H.C., "Numerical Appraisal of Multhopp's Low Frequency Subsonic Lifting Surface Theory," ARC R & M No. 3634 (October 1968).

13. Hedman, S.G., "Vortex Lattice Method for Calculation of Quasi Steady State Loadings on Thin Elastic Wings in Subsonic Flow," FFA Report 105, Stockholm (1966).

14. Albano, E. and Rodden, W.P., "Doublet-Lattice Method for Calculating Lift Distributions on Oscillating Surfaces in Subsonic Flows," *AIAA J.*, **VII**, No. 2, 279-285 (1969).

15. Butter, D.J. and Hancock, G.J., "A Numerical Method for Calculating the Trailing Vortex System behind a Swept Wing at Low Speed," *Aero. J.* (August 1971).

TRANSONIC FLOWS*

M. G. Hall

(Royal Aircraft Establishment, Farnborough)

INTRODUCTION

Transonic flows are found in a wide variety of situations: around swept wing aircraft and helicopter rotors, around propellor, fan, compressor and turbine blades, around and through engine nacelles and around many forms of weapon. In all of these cases the behaviour of the transonic flow is a decisive factor in performance, which makes it important practically to gain an understanding of the flow and a capacity for predicting its behaviour. Wind tunnel testing is very expensive. For transonic flows such testing is subject to much greater uncertainties than either subsonic or supersonic testing. Here numerical methods have much to offer the designer. The aim, in most of the recent work, has been to help improve aerodynamic design by providing designers with numerical methods that are as advanced and as relevant to practice as possible.

The basic difficulty in the calculation of a transonic flow is that there is not just one type of flow but two - subsonic and supersonic - co-existing, with the boundary between them unknown. The physical difference is that a local disturbance in a subsonic flow is propagated in all directions while in a supersonic flow the effects are confined to the Mach cone downstream of the disturbance and shock waves can occur. Mathematically, the subsonic flow is described by elliptic equations while the supersonic flow is described by hyperbolic equations. For any but the simplest supersonic flows the representation

---

* © IMA; Controller HMSO London, 1975

of shock waves is a major problem. A set of measured pressure distributions for a range of free stream Mach numbers is shown in Fig.1 to demonstrate that transonic flows cannot be described by any simple extrapolation from known subsonic flows.

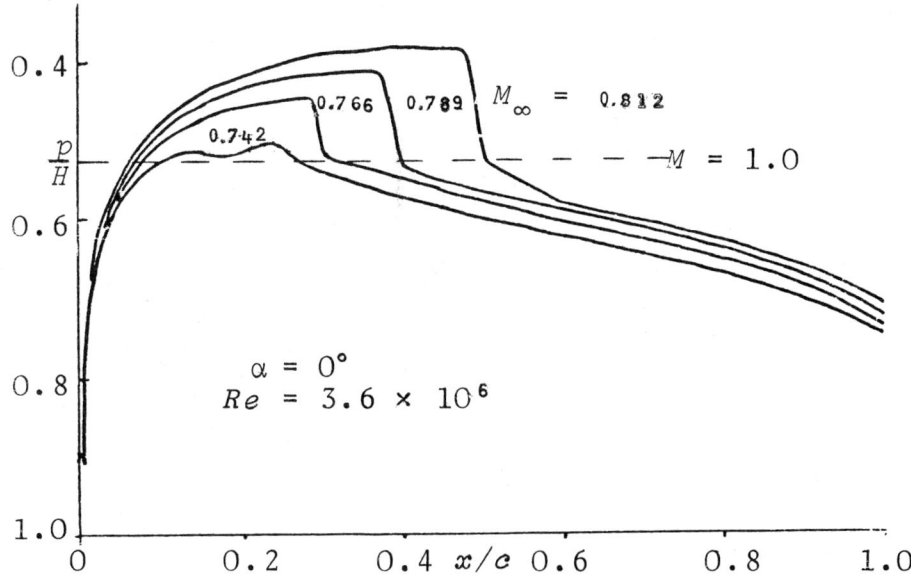

Fig. 1. Measured pressure distributions on NACA 0012 aerofoil for a range of Mach numbers

In spite of much effort, for over 20 years, no solutions could be obtained in general without such a drastic simplification in the governing equations, or in their solution, that important features of the real flow were lost. Only in the last 3 or 4 years has the situation changed. Methods have been developed, and are now in use, for both swept wings and axisymmetric bodies, although much more remains to be done. The first major advance was by Magnus and Yoshihara [1] who used the increased power of new computers to evade the mixed-flow problem! They obtained results by advancing in time to an asymptotic steady state, which reduces the mixed elliptic-hyperbolic problem to one that is entirely hyperbolic. The most important breakthrough, practically, was made by Murman and Cole [2], who boldly treated the mixed steady state problem by a novel adaptation of the cyclic relaxation method normally used for elliptic problems.

A brief comparison of the various types of method available is given in this review, to provide a guide to the possibilities for practical development and use. Hodograph methods, which are not suited to the treatment of flows with shock waves, are not considered here. The criteria applied are accuracy, adaptability for practical configurations and cost. An account follows of developments in methods of solution by relaxation, the type that has so far proved the most useful. Finally, possible advances by making modifications, developing alternative methods, or exploiting the capabilities of new, more powerful, computers are briefly discussed. For more detailed reviews reference can be made to surveys by Yoshihara [3] and Bailey [4].

OUTLINE OF CURRENT METHODS

Integral equation methods

These methods were developed by Oswatitsch [5] and others with the object of obtaining useful results for originally intractable problems. The problem was formulated in such a way that semi-empirical assumptions based on physical intuition could readily be made to render the problem tractable. The methods have proved moderately successful for subcritical flows but, for flows with shock waves, they have so far been found to be significantly less accurate than finite difference methods.

While the original formulation was with approximate semi-empirical solution in mind, the empirical element is not in fact necessary, and a first attempt at eliminating it has been made by Nixon (pp. 270 - 289). The essential step in the formulation is the transformation of the governing nonlinear differential equation for small perturbations in transonic flow into an integral equation that includes an integral over the entire flowfield. Oswatitsch reduced this field integral to a line integral by assuming a particular form for the variation of the velocity perturbation in the field in the direction normal to the plane of the wing. Nixon, however, treats the integral as an unknown to be determined by a process of iteration. In principle, any required degree of accuracy can now be obtained, but the cost in computation time is high and the present evidence is that the method will be slower than finite difference methods of equal accuracy. For a finite difference method of relaxation type the time for a given numerical solution is proportional to L^2MN, where L, M and N are the numbers of mesh points in the three coordinate directions. The corresponding time for the integral method is proportional to $L^2M^2N^2$. Thus if the integral method is only just as fast as the relaxation method for two-dimensional problems ($N = 1$) then it must be expected to be considerably slower for three-dimensional problems.

Method of integral relaxations

A few attempts [6], [7] have been made to apply the technique introduced by Dorodnitzyn [8] to transonic aerofoil problems. The flowfield is divided into a number of roughly streamwise strips and approximations are substituted for the variation of the flow variables across each strip, which reduces the problem of solving partial differential equations to one of solving a set of ordinary differential equations. The treatment of flows containing shock waves has proved to be difficult and so far no satisfactory results for such cases have been obtained. Moreover, computation times increase rapidly with increase in the number of strips, so much so that other methods would seem preferable even if a satisfactory treatment of flows with shock waves could be devised.

Finite difference methods

Most of the recent advances in the calculation of transonic flows have been through the use of finite difference methods, where difference approximations are introduced in place of the derivatives in the governing partial differential equations to reduce the problem to one of solving algebraic equations. The methods are usually divided into two main types, time dependent and relaxation, although the successive approximations in the relaxation process can be related to an advance in a time like sense. For time dependent methods the steady state solution required is regarded as an asymptotic condition to be obtained by advancing in time.

The essential idea behind the use of time dependent methods is that the introduction of derivatives with respect to time in the Eulerian equations of motion for the velocity components, density and so on, yields a first order hyperbolic system of partial differential equations that can be treated by the range of powerful methods available and being actively developed for calculating supersonic flows. The latter are discussed by MacCormack (pp.424-447), and also in [9]. The works of Magnus and Yoshihara [1], Laval [10] and Grossman and Moretti [11] are examples of this. In the first two methods shock waves are "captured" in the course of the computation, without special provision, and appear as a region of severe gradients, from the effect of an artificial viscosity. In the method of Grossman and Moretti shock waves are "fitted" in the sense that the equations of motion are solved for the flow only each side of a shock and the Rankine-Hugoniot conditions explicitly satisfied across the shock, which thus appears as a discontinuity. In the shock-capturing methods the governing equations must be written in conservation form for the shock waves to be calculated correctly; in most of these the simplifying assumption is made that the flow is isentropic so that only isentropic "shock" waves arise which differ from the physical Rankine-Hugoniot shock.

Nevertheless the capability for correct representation of shock waves is a major advantage of the time dependent methods.

The main disadvantage of time dependent methods is their relatively high cost in computer time. The reason lies in the condition for stability of the numerical solution to hyperbolic problems by explicit routines. According to this condition the size of the step taken in advancing in time must be smaller than a quantity fixed by the spatial mesh size. For practical problems in aeronautics, such as the calculation of the flow about an aerofoil, the requirements of accuracy demand a very fine spatial mesh, especially near a leading edge. The corresponding time steps have thus to be very small, so that thousands of time steps are needed, typically, to achieve a near-steady state in two space dimensions. The total computing times have so far been so large as to preclude the calculation of flows in three space dimensions.

Relaxation methods have turned out to be considerably faster than time dependent methods. The advantage is at present so marked that, in spite of shortcomings in the representation of shock waves, relaxation methods are almost the universal choice for the solution of practical problems. The idea, due to Murman and Cole [2], is to use backward rather than central difference approximations for the derivatives in the stream direction wherever the flow is locally supersonic, and then apply the successive line relaxation techniques originally developed for second order elliptic partial differential equations. If the flow is assumed to be adiabatic and irrotational a velocity potential exists and the governing equation for the velocity potential takes the required second order form. This equation has readily yielded solutions for three-dimensional as well as two-dimensional flows. The solutions include shock waves but special care is needed to ensure that the appropriate conditions are satisfied across the shocks. Even so, only non-physical isentropic shocks are obtained. The main mathematical features of the relaxation methods are explained and illustrated in the next section, with the aid of a simple example, and further developments are discussed in the following section.

Other methods

Three additional types of method are mentioned here. Each has attractive features and might, with development, prove useful in practice. The first type is the parametric or spectral method, including the use of **fast** elliptic solvers. These have been developed for solution of elliptic equations of the Laplace or Poisson type and their chief attraction is speed. They are much faster than relaxation methods, at least for linear or weakly nonlinear problems. In the method, the dependent variables are replaced by suitable finite series, such as a truncated Fourier series, involving a number of free parameters and satisfying the required boundary conditions. Substitution into the governing partial differential equation yields a finite set of ordinary differential equations which can be solved by standard techniques. So far only subcritical flows have been calculated successfully [12] and even in such cases there would be considerable difficulties if either the form of the boundary condition or the shape of the computational domain were complicated.

The second type is the finite element method. The method is highly developed in the field of structural analysis and applications to low speed flow calculations have been made, for example by Bratanow and Ecer [13]. The computational domain is divided into finite elements of suitable, say triangular, shape and the solution is represented by the values of the dependent variables at the nodal points. This enables flows past complex configurations to be treated relatively easily, which is the main asset of the method. On the other hand the equations to be solved are a variational form, or an equivalent, of the equations of motion, and it seems difficult to find such equations for supercritical flows. Moreover, even for subcritical flows, where results have been successfully obtained, the computing times have been relatively high.

A method with an asset similar to that of the finite element methods, and related to the integral equation methods outlined above, is the integral form approach described by MacCormack and Warming [9]. Complex configurations can readily be treated and results have been obtained for time dependent supersonic flows in two-and three spatial dimensions. Since shock waves can be represented accurately and embedded subsonic regions are admitted, the method is applicable also to the transonic flows considered here. A high price is paid, in programming logic and computer time, for such power. For steady two-dimensional flows relaxation methods are at present substantially faster. In the method, an integral form of the governing system of hyperbolic equations (in conservation form) is solved. The computational domain is divided into small volumes as required and the solution functions represented as averages over the corresponding small surfaces. For the most accurate calculation of shock waves a family of such surfaces is aligned with each shock surface.

THE BASIS OF THE RELAXATION METHOD

The essential features of the relaxation method are illustrated by an outline of the solution of the transonic small perturbation equation for the flow past a two-dimensional aerofoil. To formulate the problem it is assumed that any shock waves are weak enough for the flow to be considered isentropic and a velocity potential to exist. It is also assumed that the velocity perturbations are small and that the free stream velocity is near sonic. Thus the equation to be solved can be written

$$[K - (\gamma + 1)\phi_x]\phi_{xx} + \phi_{zz} = 0 \qquad (1)$$

where ϕ is the perturbation potential, x and z are rectangular Cartesian coordinates with x in the free stream direction, γ is the ratio of specific heats and K is a similarity parameter that is fixed by the aerofoil thickness and the free stream Mach number.

The boundary conditions are that the flow is tangential to the aerofoil surface and that at large distances from the aerofoil the flow approaches that of a single potential vortex. Hence these take the approximate form

$$\left. \begin{array}{l} \phi_z \text{ prescribed at } z = \pm 0 \text{ on the aerofoil chord,} \\ \phi = \dfrac{\Gamma}{2\pi} \tan^{-1}\left\{\dfrac{K^{\frac{1}{2}}z}{x}\right\} \text{ at infinity} \end{array} \right\} \qquad (2)$$

where $\Gamma \equiv [\phi(z = +o) - \phi(z = -o)]_{x = x_{te}}$ and $x = x_{te}$ at the trailing edge. Γ is the circulation around the aerofoil and is determined in the course of the calculation.

In addition there is a uniqueness or Kutta condition

$$[\phi(z = +o) - \phi(z = -o)]_{x > x_{te}} = \Gamma \qquad (3)$$

Equation (1) is a partial differential equation that is elliptic or hyperbolic in type depending on the sign of the coefficient of ϕ_{xx}. Where the coefficient is positive the equation is elliptic and the flow is locally subsonic. Where the coefficient is negative the equation is hyperbolic and the flow is locally supersonic. It would be expected that different techniques would be needed for different parts of the flow. The difficulty is that the boundary between the two is not known in advance.

The finite difference solution proposed by Murman and Cole has two essential features.

(*i*) Use an iterative relaxation technique for solving the finite difference equations, so that at any stage of the iteration previous results can be used to identify the local state of the flow.

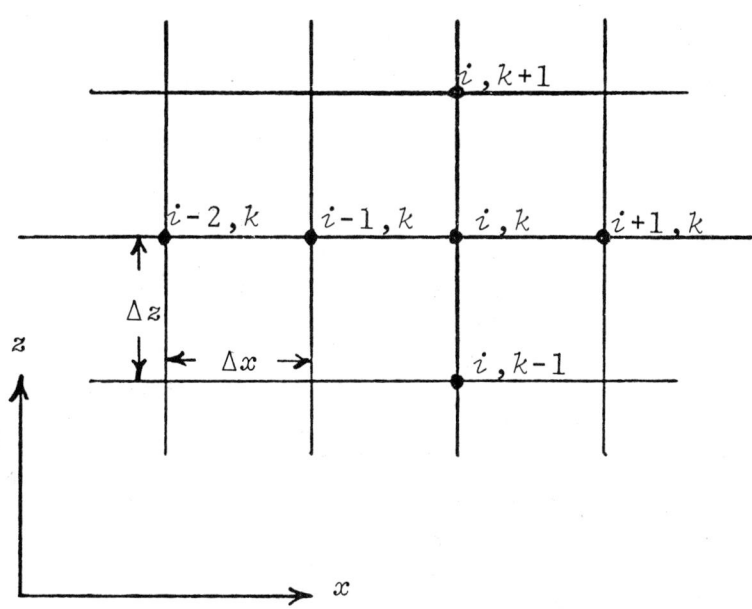

Fig. 2 Mesh points and coordinate system for basic relaxation method

(*ii*) If the flow at the mesh point (i,k), as shown in Fig. 2, is subsonic, use the central difference approximation

$$\phi_{xx}(i,k) = \frac{1}{(\Delta x)^2}(\phi_{i+1,k} - 2\phi_{i,k} + \phi_{i-1,k}) . \quad (4)$$

If the flow is supersonic use the backward difference approximation

$$\phi_{xx}(i,k) = \frac{1}{(\Delta x)^2}(\phi_{i,k} - 2\phi_{i-1,k} + \phi_{i-2,k}) \quad (5)$$

Whatever the state of the flow use

$$\phi_{zz}(i,k) = \frac{1}{(\Delta z)^2}(\phi_{i,k+1} - 2\phi_{i,k} + \phi_{i,k-1}) \quad (6)$$

Substitution in the differential equation (i) yields an algebraic equation for each point (i,k) which, for the nth iterative cycle, can be written

$$d_k \phi^{(n)}_{i,k-1} + c_k \phi^{(n)}_{i,k} + b_k \phi^{(n)}_{i,k+1} = a_k \quad (7)$$

where $a_k \ldots d_k$ depend on $\phi^{(n-1)}$ as well as x and z. The procedure in each cycle is to solve for ϕ on the lines i = constant in succession, in the direction of increasing i, making use of just calculated values on $i - 1$ and $i - 2$ and values from the previous cycle for $i + 1$. The simultaneous solution of the set of equations (7) for the points on i = constant is straightforward because the matrix of the coefficients d_k, c_k, b_k is tri-diagonal. The resulting values of ϕ on the line are then relaxed by setting

$$\phi^{(n)} = \omega\phi^{(n)} - (\omega-1)\phi^{(n-1)} \quad (8)$$

where $1.5 \leqslant \omega \leqslant 1.9$ for subsonic flow and $0.7 \leqslant \omega < 1.0$ for supersonic flow, the actual values being chosen from experience. The cycles are repeated until the change in ϕ is smaller than some prescribed tolerance.

The above method of successive line relaxation and adapted differences is the basis of a number of developments and extensions, and these are considered next.

DEVELOPMENTS AND EXTENSIONS OF THE BASIC RELAXATION METHOD

In the original Murman-Cole method the desired high density of mesh points near the leading edge of an aerofoil was obtained simply by using a non-uniform mesh. This complicated the programming logic and introduced numerical errors wherever the mesh size was changed. An obvious remedy is to stretch the coordinates so that while there is a high density of grid points where desired in the physical space, there is a uniform mesh in the computational space. At the same time the infinite physical domain can be transformed to a finite computational domain. An example of this is found in the method of Bauer, Garabedian and Korn [14] who solve the exact equation for the velocity potential, for the two-dimensional flow about an aerofoil. The coordinate transformation used is one introduced by Sells [15], in which the aerofoil surface is mapped on to a circle and the flowfield into the interior of the circle. Since the exact equation is essentially no more difficult to solve than the small perturbation equation, because the latter is non-linear and of mixed type, and the transformation cancels out the only advantage possessed by the small perturbation formulation (that of placing the points at which the flow tangency boundary condition is satisfied on a coordinate line), the exact equation can be solved as readily and rapidly as the small perturbation equation.* Thus, except for special applications, such as the calculation of the transonic flow past an aerofoil in a wind tunnel [17], where suitable mappings are difficult to find, the two-dimensional small perturbation formulation is little used. A similar method for the exact equation for the velocity potential has been developed by Jameson [18].

---

* It is worth noting that the transonic small perturbation equation can, with suitable adjustment of the available free parameters, yield results for a wide range of aerofoils that are bare-

ly indistinguishable for practical purposes from the corresponding solution of the exact equation.

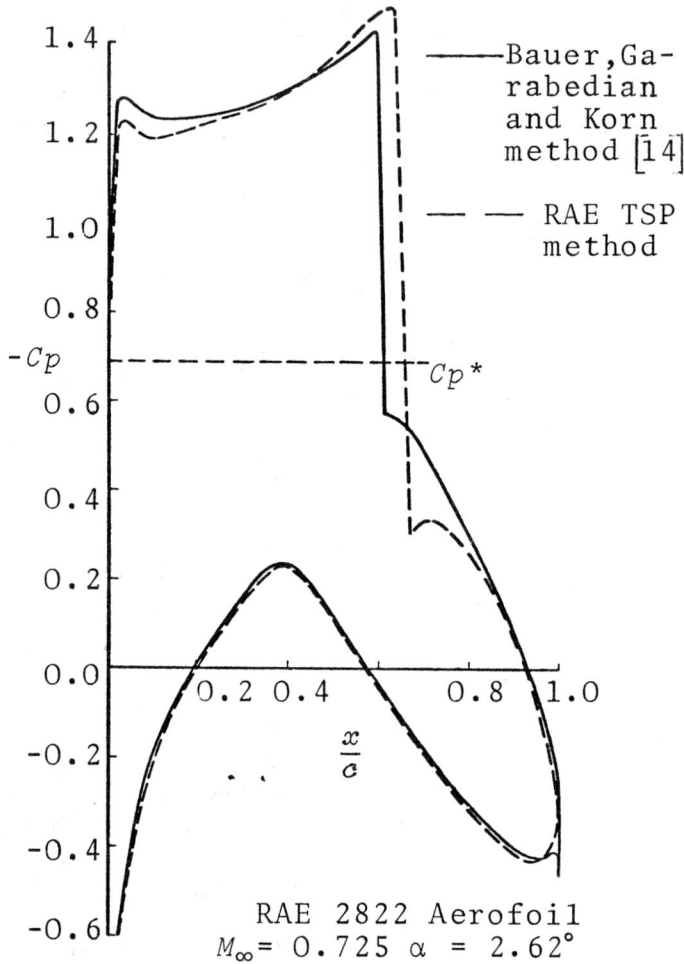

Fig. 3. Pressure distributions calculated from the transonic small perturbation and the exact equation

An example of this is given in Fig. 3, in which a pressure distribution obtained by Albone *et al.* [16] by solving the transonic small perturbation equation is compared with the corresponding solution of the exact equation by the method of Bauer,

Garabedian and Korn. The agreement, which is typical, is good, in spite of the fact that the perturbations are far from small and the free stream Mach number is not near one.

Two shortcomings soon became apparent when relaxation methods were applied to practical problems. The first is that the calculated pressure rise through a shock wave is appreciably smaller than that of the physically correct Rankine-Hugoniot shock and not larger, as the pressure rise through a mathematically consistent isentropic shock should be (cf [20]). The inconsistency effectively arises from the switching of difference approximations from backward(5) to central (4), in sweeping downstream through a shockwave in the calculation process. Examination shows that, as a consequence of the switch, the only quantity conserved across the shock is the potential. In particular mass flux is not conserved. For solutions of the transonic small perturbation equation, Murman [19] has recently shown how the mass flux can be conserved, and the correct isentropic shock pressure rise obtained, by a local modification to the numerical scheme. The same remedy is not obviously applicable to solutions of the exact equation for the velocity potential, but whatever remedy for inconsistency is found there would remain the shortcoming inherent in the use of the velocity potential: the isentropic shock pressure rise differs significantly from the pressure rise through a real shock when the Mach number of the component of flow normal to the shock, on its upstream side, exceeds about 1.2. Moreover, the flow behind a real shock can be rotational. A study of isentropic models of flows with embedded shocks has been presented by van der Vooren and Sloof [20].

The second shortcoming is a lack of convergence in the iterative process that occurs, for example in solutions by the method of Bauer, Garabedian and Korn [14], apparently when the flow is locally supersonic and the coordinate lines of the successive line relaxation process are not well aligned with the normals to the flow direction. The fault has been identified by Albone [21] and Jameson [22] as a failure to ensure that the domain of dependence of the difference scheme includes that of the differential equation, which is known to lead to numerical instability. This is illustrated in Fig. 4, where the velocity vector

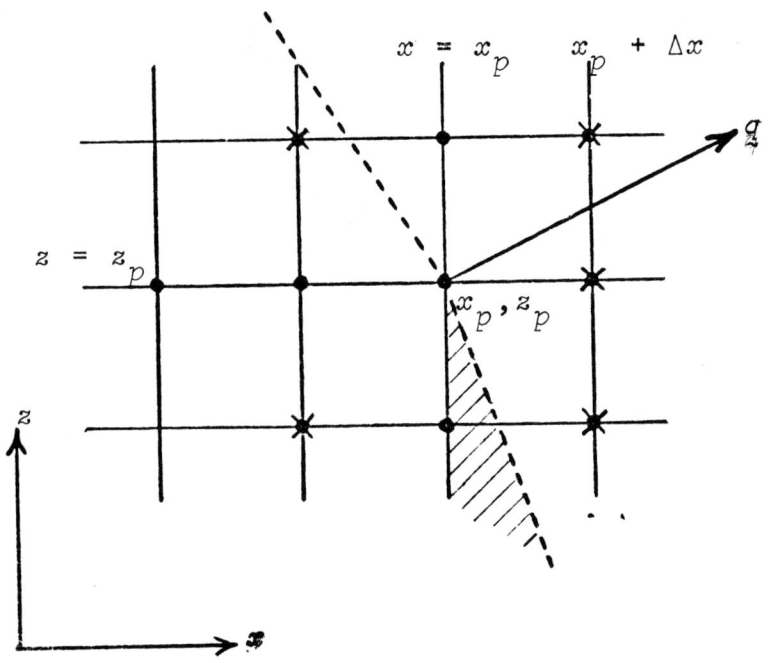

Key: ● active mesh points, Murman-Cole [2]
✗ additional points, Albone-Jameson [21], [22]

Fig. 4. Domains of dependence for supersonic flow

q at the point x_p, z_p is shown. When the flow is supersonic the domain of dependence of the difference scheme, for the point x_p, z_p is the region $x \leqslant x_p$, whereas the domain of dependence of the differential equation is the region enclosed by the characteristics shown as dotted lines. The hatched region is not included (as it should be) in the domain of dependence of the difference scheme. Essentially the same remedy, a split or rotated difference scheme, has been proposed by Albone and Jameson. Their aim is to express the differential equation in such a form that the domain of dependence of the corresponding difference equation includes that of the differential equation. They start with the canonical form of the equation for the full potential Φ

$$\underline{\left(1 - \frac{q^2}{a^2}\right)\Phi_{ss}} + \Phi_{nn} = 0 \tag{9}$$

where a and q are the local values of the speed of sound and the stream speed, respectively, and s and n are rectangular coordinates in the local stream direction and its normal, respectively. The underlined term in (9) is that which, following Murman and Cole, would for numerical stability be approximated by a backward difference approximation when the local flow is supersonic.

A transformation is then made from the s,n system to the required x,z system by rotating the coordinate axes. Thus

$$\Phi_{ss} = \frac{1}{q^2}[u^2\Phi_{xx} + 2uw\Phi_{xz} + w^2\Phi_{zz}]$$

$$\Phi_{nn} = \frac{1}{q^2}[w^2\Phi_{xz} - 2uw\Phi_{xz} + u^2\Phi_{zz}]$$

where

$$u = \Phi_x, \quad w = \Phi_z.$$

If the terms arising from the underlined term in (9) are themselves underlined and kept separate from the others, (9) becomes

$$\frac{(a^2 - q^2)[u^2\Phi_{xx} + w^2\Phi_{zz} + 2uw\underline{\Phi_{xz}}]}{+ a^2[w^2\Phi_{xx} - 2uw\Phi_{xz} + u^2\Phi_{zz}]} = 0 \quad (10)$$

This equation is no more than a rearrangement of the standard equation for the velocity potential, in which each of the second derivatives has been split into two parts. The Albone-Jameson procedure is to approximate only the underlined term by backward differences when the flow is locally supersonic. Otherwise central difference approximations are used. The consequence is that the domain of dependence of the difference scheme, for the point x_p, z_p in Fig. 4, now includes points on the line $x = x_p + \Delta x$ and thus includes the domain of dependence of the differential equation.

The split or rotated difference scheme provides an extra freedom in the choice of coordinate system, and has been successfully applied in a number of methods. It has greatly increased the range of capability of the methods for the solution for aerofoils of the exact equation for the velocity potential. In particular results right through the transonic range, for supersonic as well as subsonic free streams, can now be obtained. South and Jameson [23] have included the scheme in their method for solving the exact equation for the axisymmetric flow past arbitrary bodies. To ensure that the body surface is a coordinate surface a combination of stretched coordinate systems is used. Jameson [24] has also included the scheme in his method for solving the exact equation for the three-dimensional flow past a yawed or slewed wing, where there would otherwise be severe limitations. Probably the most important applications of the scheme have been in the methods developed at NASA, Ames [4] and at RAE, Farnborough [25], for the solution of the transonic small perturbation equation for the three-dimensional flow past swept wings.

Solutions of the transonic small perturbation equation have been sought because of the ease with which the approximate flow tangency condition can be satisfied for wings of arbitrary shape, where transformations of the type used for aerofoils are not directly applicable, and also because of the satisfactory degree of accuracy obtained in two dimensions (see Fig. 3). The equation to be solved for the perturbation potential takes the form

$$[1 - M_\infty^2 - (\gamma+1)M_\infty^2 \phi_x]\phi_{xx} + \phi_{yy} + \phi_{zz} - 2\phi_y \phi_{xy} - 2\phi_z \phi_{xz} = 0 \qquad (11)$$

where y is in the spanwise direction normal to the free stream.

Coordinate transformations have a major rôle. First, a shearing transformation of the wing planform is made so that the planform is near rectangular in the new coordinate space. The main purpose of this is to align the leading and trailing edges of the wing with coordinate lines, so that the mesh density needed for accuracy can be obtained economically. Then, in the RAE method, coordinate stretchings in all three directions are made, to provide a fairly uniform mesh density near the wing and a gradual reduction of density with increasing distance from the wing, and also to provide higher mesh densities near the leading edge, the centre section and the wing tip.

Finite difference approximations are then substituted in the transformed differential equation, as indicated in the previous section, and the resulting set of algebraic equations is again solved by successive line relaxation. In each cycle of the iteration we solve for ϕ along successive vertical lines on successive surfaces, beginning far upstream of the wing. In the RAE method it is usually sufficient to make 150 cycles on sweeps through a 60 x 25 x 40 mesh for a calculation of the flow about a given wing at a given Mach number and incidence, and this takes about 20 minutes to complete on an IBM 370/168 computer. Solutions for free stream Mach numbers ranging from 0.6 to 1.2 have been obtained, although flows with sonic or supersonic free streams take several hundred sweeps for convergence. Wings of curved or cranked planform and arbitrary spanwise variations of section and twist can be treated.   Some sample results are shown in Figs. 5, 6 and 7.   Fig. 5 shows a comparison of

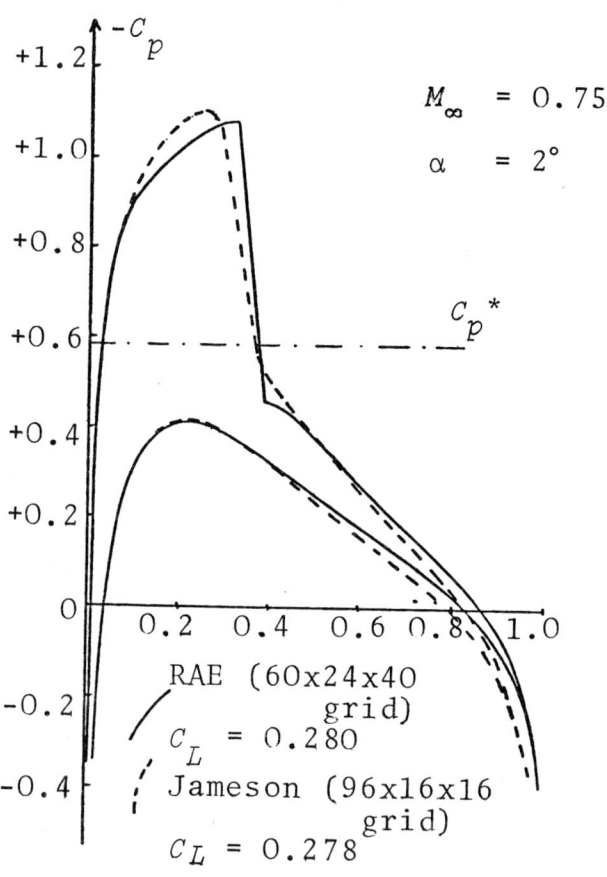

Fig. 5. Pressure distribution at the centre section of a rectangular wing, aspect ratio 6, NACA 0012 section

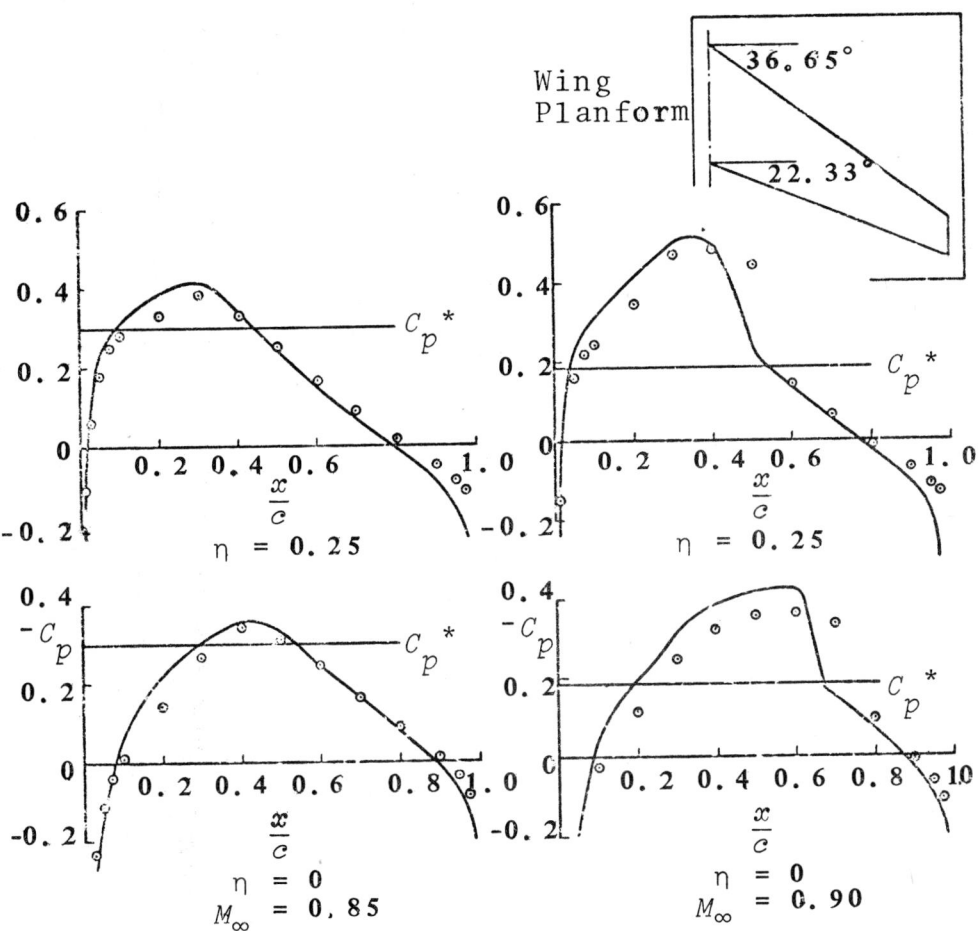

Fig. 6. Comparison with measured pressures on RAE wing A.
$\alpha = 0$, $Re = 1.0 \times 10^6$

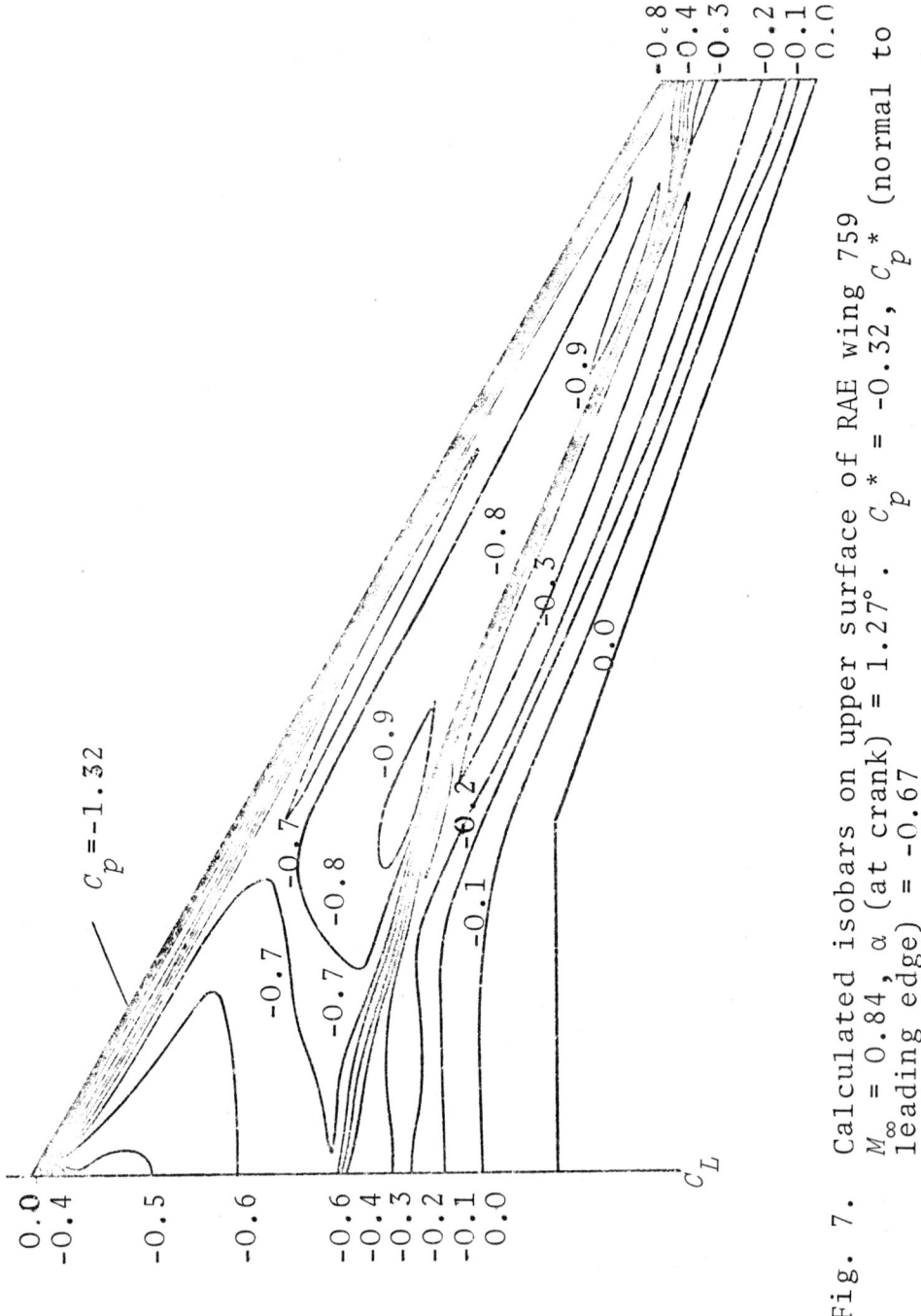

Fig. 7. Calculated isobars on upper surface of RAE wing 759 $M_\infty = 0.84$, α (at crank) = 1.27°, $C_p^* = -0.32$, C_p^* (normal to leading edge) = -0.67

calculated pressure distributions at the centre section of a rectangular wing of aspect ratio 6 and an NACA 0012 section. The comparison is with a solution of the exact equation for the velocity potential by the method of Jameson [24]. Fig. 6 shows comparisons with experimental measurements on a swept wing of symmetrical section, at zero incidence, conditions under which viscous and interference effects can be expected to be small. Fig. 7 is included to show the versatility of the method. It shows a calculated isobar pattern on the upper surface of a swept wing that is typical of modern transport design, with considerable twist and camber.

An extension for design purposes of the RAE method for swept wings has recently been made by Langley and Forsey at ARA [25]. Their design procedure is essentially a modification of the direct RAE method in which the flow tangency boundary condition is replaced, over part of the wing, by a surface pressure boundary condition. The wing planform and the leading edge geometry are fixed but otherwise there are several options in what is prescribed and what is to be calculated. For example, the over-all thickness distribution and the upper surface pressure distribution (aft of the prescribed part of the leading edge) can be prescribed and the camber and twist distributions calculated.

CONCLUDING REMARKS

Although the development of numerical methods for calculating transonic flows has reached a stage where they are used in aerodynamic design there is still a great deal to be done. Shortcomings in accuracy or penalties in computing time have to be accepted at present whenever shock waves of even moderate strength, or complexities in body geometry, are found. We are very far indeed from the ultimate aim of being able to compute accurately and at acceptable cost the transonic flowfield about a complete aircraft for all the relevant flight conditions.

For two-dimensional problems the ultimate aim is in sight, because of the possibility of mapping irregular physical domains on to regular computational domains and because the flow is simple enough for even the slow time dependent methods to be viable tools when coupled with the most powerful computer. Since the time dependent methods are capable of representing shock waves correctly, virtually any two-dimensional flow that is inviscid can be calculated as accurately as desired. The remaining problem is the treatment of the interactions between inviscid flows and viscous boundary layers, wakes or jets and, in particular, the treatment of the interaction between shock waves and boundary layers. For unsteady problems in two space dimensions the time dependent methods of course remain viable, but there is a formidable problem here in the interaction between an outer inviscid flow with a separated flow, which may be unsteady even when the body is moving with uniform velocity.

For problems in three spatial dimensions the time dependent methods are too demanding in computer time and the relaxation methods, although justifiably used, are deficient in accuracy. Since most practical problems are three dimensional there are high rewards for improvement. One obvious possibility is that computers will soon be so powerful that no further developments in numerical technique are needed. The impressiveness of some of the results obtained by use of the Illiac computer shows that even if computer growth does not remove the need for mathematical development it will strongly influence the latter. From among the several possible lines of development three can be singled out.

(i) Techniques may be sought for satisfying the flow tangency condition without making the body surface a coordinate surface, as is virtually the rule at present because otherwise there would be an unacceptable lack of smoothness in the boundary conditions. Coordinate transformations of the body to some regular surface appear to have reached a limit with swept wings. For wing-body or wing-body-nacelle combinations the integral form approach or finite element methods, with their freedom in choice of elemental volume, would enable the body surface to remain a coordinate surface, but the penalties in computer time and logical complexity are severe. If the coordinate system could be chosen without the constraint of having to fit the body surface and without the penalties of a lack of smoothness, there might be substantial improvements.

(ii) Techniques may be sought for increasing the rate at which time dependent solutions approach the asymptotic steady state. Advancing in a fictitious time is one possibility. Advancing in an implicit rather than explicit sense is another. One possiblity already being exploited is to advance in steps of different sizes in different parts of the field, using the largest practicable step everywhere.

(iii) Relaxation techniques may be applied to equations that enable shock waves and the flows downstream of shock waves to be more accurately represented. It may be necessary to give up, and not just modify, the use of the velocity potential. That the velocity potential is not essential has been demonstrated by Steger and Lomax [26], who took the velocity components (in an admittedly isentropic flow) as dependent variables.

REFERENCES

1. Magnus, R. and Yoshihara, H., "Inviscid transonic flow over aerofoils," AIAA Paper No. 70-47 (1970).

2. Murman, E.M. and Cole, J.D., "Calculation of plane steady transonic flows," *AIAA J.*, **9**, No. 1, 114-121 (1971).

3. Yoshihara, H., "A survey of computational methods for 2D and 3D transonic flows with shocks," *in* "Advances in Numerical Fluid Dynamics," AGARD-LS-64, Paper No. 6 (1973).

4. Bailey, F.R., "On the computation of two- and three-dimensional transonic flows by relaxation methods," *in* "Progress in Numerical Fluid Dynamics," AGARD Lecture Series No. 63 (1974).

5. Oswatitsch, K., *"Die Geschwindigkeitsverteilung an symmetrischen Profilen beim Auftreten lokaler Überschallgebiete,"* Acta Physica Austriaca, **4**, No. 2-3, 228-271 (1950).

6. Tai, T.C., "Application of the method of integral relations to transonic aerofoil problems," AIAA Paper 71-98 (1971).

7. Sato, J., "Inverse method of designing two dimensional transonic aerofoil sections," *AIAA J.*, **11**, No. 1, 58-63 (1973).

8. Dorodnitzyn, A.A., "A contribution to the solution of mixed problems of transonic aerodynamics," *Advances in Aero. Sci.*, **2**, 832-845 (1959).

9. MacCormack, R.W. and Warming, R.F., "Survey of computational methods for three-dimensional supersonic inviscid flows with shocks," *in* "Advances in Numerical Fluid Dynamics," AGARD-LS-64, Paper No. 5 (1973).

10. Laval, P., *"Méthodes instationnaires de calcul des effets d-interaction de paroi en écoulement bidimensionnel supercritique,"* Congrés Francais de Mécanique, Poitiers (September, 1973).

11. Grossman, R. and Moretti, G., "Time-dependent computation of transonic flows," AIAA Paper 70-1322 (October, 1970).

12. Martin, E.D. and Lomax, H., "Rapid finite-difference computation of subsonic and transonic aerodynamic flows," AIAA Paper No. 74-11 (May, 1974).

13. Bratanow, T. and Ecer, A., "Analysis of three dimensional unsteady viscous flow around oscillating wings," *AIAA J.*, **12**, No. 11, 1577-1584 (1974).

14. Bauer, F., Garabedian, P.R. and Korn, D.G., "A theory of supercritical wing sections, with computer programs and examples," *in* "Lecture Notes in Economics and Mathematical Systems," Springer-Verlag (1972).

15. Sells, C.C.L., "Plane subcritical flow past a lifting aerofoil," *Proc. Roy. Soc. A.*, **308**, 377-401 (1968).

16. Albone, C.M., Catherall, D., Hall, M.G. and Joyce, Gaynor, "An improved numerical method for solving the transonic small-perturbation equation for the flow past a lifting aerofoil," RAE Technical Report 74056 (1974).

17. Murman, E.M., "Computation of wall effects in ventilated transonic wind tunnels," AIAA Paper No. 72-1007 (September 1972).

18. Jameson, A., "Transonic flow calculations for airfoils and bodies of revolution," Grumman Aerospace Corp. Report 390-71-1 (1971).

19. Murman, E.M., "Analysis of embedded shock waves calculated by relaxation methods," Proceedings of AIAA Computational Fluid Dynamics Conference, Palm Springs, 27-40 (July 1973).

20. van der Vooren, J. and Sloof, J.W., "On inviscid isentropic flow models used for finite difference calculations of two-dimensional transonic flows with embedded shocks about aerofoils," NLR MP 73024 U: (Paper presented at Euromech 40: Transonic Aerodynamics, Saltsjöbaden, Sweden, September, 1973).

21. Albone, C.M., "A finite difference scheme for computing supercritical flows in arbitrary coordinate systems," RAE Technical Report TR 74090 (1974).

22. Jameson, A., "Iterative solution of transonic flows over airfoils and wings, including flows at Mach 1," *Comm. Pure Appl. Math.*, **27**, 283-309 (1974).

23. South, J.C. and Jameson, A., "Relaxation solutions for inviscid axisymmetric transonic flow over blunt or pointed bodies," Proceedings of AIAA Computational Fluid Dynamics Conference, Palm Springs, 8-17 (July 1973).

24. Jameson, A., "Numerical calculation of the three dimensional transonic flow over a yawed wing," Proceedings of AIAA Computational Fluid Dynamics Conference, Palm Springs, 18-26 (July 1973).

25. Hall, M.G. and Firmin, M.C.P., "Recent developments in methods for calculating transonic flows over wings," ICAS Paper No. 74-18 (1974): (Presented at 9th ICAS Congress Haifa, August 1974).

26. Steger, J.L. and Lomax, H., "Generalized relaxation methods applied to problems in transonic flow," Paper presented at Second International Conference on Numerical Methods in Fluid Dynamics (September, 1970).

NOTATION

a speed of sound

c wing chord

C_L lift coefficient

C_p pressure coefficient

C_p^* critical pressure coefficient (local $M = 1$)

K transonic similarity parameter

M	Mach number
q	local fluid speed
Re	Reynolds number
s	semi-span of wing
u, v, w	velocity components in (x,y,z) directions
x, y, z	Cartesian coordinates : x in free stream direction, y spanwise
α	angle of incidence
γ	ratio of specific heats
η	y/s
Φ	velocity potential
ϕ	perturbation velocity potential

AN EXTENDED INTEGRAL EQUATION METHOD FOR THE STEADY TRANSONIC FLOW PAST A TWO-DIMENSIONAL AEROFOIL

D. Nixon

(Queen Mary College, University of London)

INTRODUCTION

At present the most common method of calculating the inviscid distribution around aerofoils at transonic speeds is by using finite difference techniques. For subcritical flows one of the most accurate methods is that of Sells [1]. For supercritical flows with shock waves the initial work of Murman and Cole [2] has been followed by many developments, notably those by Murman and Krupp [3], Garabedian and Korn [4] and Jameson [5]. Recently, however, it was pointed out by Murman[6] that the difference scheme used in these methods is incorrect at the shock location, leading to incorrect shock jump relations; corrected results are presented in [6].

One of the earliest attempts to solve the transonic flow problem is the integral equation method first developed by Oswatitsch [7] with subsequent extensions by Spreiter and Alksne [8] and others [9],[10]. In the integral equation method the nonlinear small disturbance equation is inverted into integral form consisting of a line integral over the aerofoil chord and a surface integral, involving second order terms, over the flowfield. The line integral over the aerofoil chord is identical to one appearing in "linear" subsonic aerofoil theory and is easily evaluated using standard methods. In the standard integral equation method the surface integral is reduced to a line integral over the aerofoil chord by relating the variation of the perturbation velocity in the flowfield to its value on the aerofoil surface by a suitable approximation function. For subcritical flows the results obtained by the various formulations of the integral equation method [8],[9], [10] are fairly satisfactory when compared with the finite difference results. However, for supercritical flows with shock waves, the integral equation methods give results that do not agree satisfactorily with the finite difference results of Murman [6] calculated at the same value of the transonic similarity parameter, especially as regards the shock location. Since the fundamental equation used in all these calculations is the transonic small perturbation equation it can be assumed that the poor results of the integral equation methods are a result of an inaccurate evaluation of the field integral arising from inaccuracies in the approximation functions used in the reduction to a line integral.

In this paper an alternative means of evaluating the surface integral in the integral equation method is devised. The flowfield is divided into strips and the transverse variation of the perturbation velocity in each of these strips is approximated in terms of values on the strip edges. The surface integral is thus reduced to a line integral which is in turn evaluated by quadrature. The fundamental integral equation can thus be approximated by a set of nonlinear algebraic equations. The shock location is found by enforcing the condition of finite acceleration through the sonic line.

The pressure distribution around a biconvex aerofoil and a NACA0012 section is found for subcritical flows and also around a biconvex aerofoil in supercritical flow when shock waves are present. There is good agreement between these results and the results of the finite difference methods[1], [2],[6].

BASIC EQUATIONS

A two-dimensional Cartesian coordinate system is chosen with the origin at the leading edge; the x-axis is in the free stream direction and the z-axis normal to the free stream, x and z are non-dimensionalised with respect to the aerofoil chord.

It is assumed that the flow can be described by the second order equation

$$(1-M_\infty^2)\phi_{xx} + \phi_{zz} = k\,\phi_x\phi_{xx} + 2\,M_\infty^2\phi_z\phi_{zx} \qquad (1)$$

where ϕ is the perturbation velocity potential, defined as

$$\frac{\partial \phi}{\partial x} = u, \quad \frac{\partial \phi}{\partial z} = w \qquad (2)$$

where u and w are the perturbation velocities in the x and z directions, respectively, relative to the free stream velocity, U_∞; M_∞ is the free stream Mach number and k is a similarity parameter that is a function of M_∞ and the ratio of specific heats, γ.

For many practical applications the flow can be adequately described by the transonic small perturbation equation

$$(1-M_\infty^2)\,\phi_{xx} + \phi_{zz} = k\phi_x\phi_{xx} \qquad (3)$$

To simplify the subsequent analysis (3) is used and only the flow around symmetric aerofoils at zero incidence is considered.

For a non-lifting aerofoil the boundary conditions are that the resultant flow direction at the aerofoil surface is tangential to the aerofoil surface and that the derivatives of the perturbation potential vanish at infinity.

If $z = z_T(x)$ and $z = -z_T(x)$ denote the upper and lower surfaces of a symmetric aerofoil at zero incidence the tangency boundary condition can be written as

$$\frac{w(x, \pm z_T(x))}{[1 + u(x, \pm z_T(x))]} = \pm z_T'(x) \qquad (4)$$

where the prime denotes differentiation. If the usual thin aerofoil approximation is made then the boundary condition (4) can be written as

$$w(x, \pm 0) = \pm z_T'(x) \qquad (5)$$

The boundary condition is thus applied on the plane $z = 0$ rather than on the aerofoil surface.

The pressure coefficient, $C_p(x,z)$, can be found from Bernoulli's equation; thus

$$C_p(x,z) = \frac{2}{\gamma M_\infty^2} \{(1 + \frac{(\gamma-1)}{2} M_\infty^2 [1-q^2(x,z)])^{\frac{\gamma}{\gamma-1}} - 1\} \qquad (6)$$

where $q(x,z)$ is the resultant velocity and is given by

$$q(x,z) = \{(1 + u(x,z))^2 + w^2(x,z)\}^{\frac{1}{2}} \qquad (7)$$

If the parameter

$$\beta = (1-M_\infty^2)^{\frac{1}{2}}$$

and the variables

$$\bar{x} = x, \quad \bar{z} = \beta z, \quad \bar{\phi}(\bar{x},\bar{z}) = \frac{k}{\beta^2}\phi(x,z)$$

$$\bar{u}(\bar{x},\bar{z}) = \frac{ku(x,z)}{\beta^2}, \quad \bar{w} = \frac{k}{\beta^3}w(x,z) \qquad (8)$$

are introduced then (3) and (5) can be transformed to give

$$\bar{\phi}_{\bar{x}\bar{x}} + \bar{\phi}_{\bar{z}\bar{z}} = \bar{\phi}_{\bar{x}}\bar{\phi}_{\bar{x}\bar{x}} \qquad (9)$$

and
$$\bar{w}(\bar{x}, \pm 0) = \pm \bar{Z}'_T(\bar{x}) \qquad (10)$$
respectively, where
$$\bar{Z}_T(\bar{x}) = \frac{k}{\beta^3} z_T(x)$$

It can also be shown [8] that sonic conditions exist when
$$\bar{u}(\bar{x}, \bar{z}) = \bar{\phi}_{\bar{x}}(\bar{x}, \bar{z}) = 1$$

Equation (9) can be inverted into integral form by using Green's theorem. For a non-lifting aerofoil, assuming that any shock waves in the flow are normal to the free stream, (9) can be written [8] as
$$\bar{u}(\bar{x},\bar{z}) - \frac{\bar{u}^2(\bar{x},\bar{z})}{2} = \bar{u}_{TL}(\bar{x},\bar{z}) + I_T(\bar{x},\bar{z},\bar{x}_s) \qquad (11)$$
where $\bar{u}_{TL}(\bar{x},\bar{z})$ is the velocity found using linear potential theory, given by
$$\bar{u}_{TL}(\bar{x},\bar{z}) = \frac{1}{\pi} \int_0^1 \frac{\bar{Z}'_T(\bar{\xi})(\bar{x}-\bar{\xi})}{[(\bar{x}-\bar{\xi})^2 + \bar{z}^2]} d\bar{\xi} \qquad (12)$$
and where
$$I_T(\bar{x},\bar{z},\bar{x}_s) = -\frac{1}{4\pi} \iint_S \bar{u}^2(\bar{\xi},\bar{\zeta}) \psi_{\bar{\xi}\bar{x}}(\bar{x},\bar{\xi};\bar{z},\bar{\zeta}) ds, \qquad (13)$$
$$\psi(\bar{x},\bar{\xi};\bar{z},\bar{\zeta}) = \ln[(\bar{x}-\bar{\xi})^2 + (\bar{z}-\bar{\zeta})^2]^{\frac{1}{2}} \qquad (14)$$
and $\bar{x}_s$ is the shock location (if applicable).

It should be noted that the integral $I_T(\bar{x},\bar{z},\bar{x}_s)$ is a function of the unknown velocity $\bar{u}(\bar{\xi},\bar{\zeta})$. In the present formulation of the integral equation (11), both the singular point $(\bar{x},\bar{z})$ and any shock waves in the flow are excluded from the flow-field by infinitely thin strips in the $\bar{z}$ direction. The surface integral in (13) is then defined for $\bar{z} > 0$ as

$$\iint_S F ds = \lim_{\substack{\varepsilon \to 0 \\ \delta \to 0}} \Big\{ \int_{-\infty}^{\bar{x}-\varepsilon} \Big(\int_0^\infty F d\bar{\zeta}\Big) d\bar{\xi} + \int_{\bar{x}+\varepsilon}^{\bar{x}_s - \delta} \Big(\int_0^\infty F d\bar{\zeta}\Big) d\bar{\xi}$$

$$+ \int_{\bar{x}_s + \delta}^{\infty} \Big(\int_0^\infty F d\bar{\zeta}\Big) d\bar{\xi} + \int_{-\infty}^{\bar{x}_s - \delta} \Big(\int_{-\infty}^0 F d\bar{\zeta}\Big) d\bar{\xi}$$

$$+ \int_{\bar{x}_s + \delta}^{\infty} \Big(\int_{-\infty}^0 F d\bar{\zeta}\Big) d\bar{\xi} \Big\} \qquad (15)$$

The linearised velocity term $\bar{u}_{TL}(\bar{x},\bar{z})$, found using only first order terms in the boundary conditions and given by (12), is only adequate for sharp nosed aerofoils; for round nosed aerofoils a more accurate representation is required and can be obtained by retaining some higher order terms in the boundary conditions.

For subcritical flows, when $\bar{u}(\bar{x},\bar{z}) < 1$, (11) can be solved fairly easily since there are no discontinuities (or shocks) in the solution but in supercritical flows with shock waves additional restrictions have to be imposed to obtain a realistic solution.

For a shock wave normal to the free stream it can be shown that to the order of approximation of (9) the term $[\bar{u}(\bar{x},\bar{z}) - \bar{u}^2(\bar{x},\bar{z})/2]$ is continuous through the shock and that if $\bar{u}_a(\bar{z})$ and $\bar{u}_b(\bar{z})$ denote the values of $\bar{u}(\bar{x},\bar{z})$ ahead of and behind the shock, respectively, then

$$\bar{u}_a(\bar{z}) + \bar{u}_b(\bar{z}) = 2 \qquad (16)$$

Since in (11) $[\bar{u}(\bar{x},\bar{z}) - \bar{u}^2(\bar{x},\bar{z})/2]$ is continuous through a shock and since $\overline{u_{TL}(x,z)}$ is in general also continuous through a shock it follows that the double integral $I_T(\bar{x},\bar{z},\bar{x}_s)$ is continuous through a shock. A formal "solution" to (11) is given by

$$\bar{u}(\bar{x},\bar{z}) = 1 \pm \{1 - 2[\bar{u}_{TL}(\bar{x},\bar{z}) + I_T(\bar{x},\bar{z},\bar{x}_s)]\}^{\frac{1}{2}} \quad (17)$$

It can be easily seen that if there is a discontinuous jump in $\bar{u}(\bar{x},\bar{z})$ from the upper root of (17) to the lower root then the jump satisfies the shock jump conditions given by (16). A jump from the lower root to the upper root of (17) cannot be considered since this denotes an "expansion" shock which cannot occur in a real flow.

Since the free stream is assumed subsonic ($\bar{u} < 1$) the flow over at least some part of the aerofoil must be subsonic. When the flow velocity increases up to supersonic conditions ($\bar{u} > 1$) it can be seen from (17) that there must be a transfer in this accelerating flow from the lower (subsonic) root of (17) to the higher (supersonic) root. This transfer occurs at the sonic line ($\bar{u} = 1$). In order to avoid any kind of "expansion" shock the shock location, $\bar{x}_s$, must be such that the acceleration through the sonic line is finite.

On differentiating (11) with respect to $\bar{x}$

$$\bar{u}_{\bar{x}}(\bar{x},\bar{z})[1 - \bar{u}(\bar{x},\bar{z})] = [\bar{u}_{TL}(\bar{x},\bar{z}) + I_T(\bar{x},\bar{z},\bar{x}_s)]_{\bar{x}} \quad (18)$$

In order for $\bar{u}_{\bar{x}}(\bar{x},\bar{z})$ to be finite at the sonic line, where $\bar{u}(\bar{x},\bar{z})$ is unity,

$$\frac{\partial}{\partial \bar{x}}[\bar{u}_{TL}(\bar{x},\bar{z}) + I_T(\bar{x},\bar{z},\bar{x}_s)]\bigg|_{\bar{x}=\bar{x}_0(\bar{z})} = 0, \quad (19a)$$

where $\bar{x}_0(\bar{z})$ denotes the sonic line. And from (11)

$$[\bar{u}_{TL}(\bar{x}_0(\bar{z}),\bar{z}) + I_T(\bar{x}_0(\bar{z}),\bar{z},\bar{x}_s)] = \tfrac{1}{2} \quad (19b)$$

On differentiating (11) with respect to $\bar{z}$

$$\bar{u}_{\bar{z}}(\bar{x},\bar{z})[1-\bar{u}(\bar{x},\bar{z})] = \frac{\partial}{\partial \bar{z}}[\bar{u}_{TL}(\bar{x},\bar{z}) + I_T(\bar{x},\bar{z},\bar{x}_s)] \quad (20)$$

and in order to ensure a finite value of $\bar{u}_{\bar{z}}(\bar{x},\bar{z})$ at the sonic line

$$\frac{\partial}{\partial \bar{z}}[\bar{u}_{TL}(\bar{x},\bar{z}) + I_T(\bar{x},\bar{z},\bar{x}_s)]\bigg|_{\bar{x}=\bar{x}_0(\bar{z})} = 0 \quad (19c)$$

Thus to obtain a physically realistic solution (11) must be solved subject to the restrictions imposed by (19). It can be shown that (19a, b, c) are sufficient to give the sonic line $\bar{x}_0(\bar{z})$, the normal shock location, $\bar{x}_s$, and the slope of the sonic line, $\frac{d\bar{x}_0(\bar{z})}{d\bar{z}}$.

In the standard integral equation methods [7] to [10] the surface integral, $I_T(\bar{x},\bar{z},\bar{x}_s)$, is evaluated by approximating the velocity variation in the flowfield, $\bar{u}(\bar{\xi},\bar{\zeta})$ in terms of the ordinate $\bar{\zeta}$ the velocity on the aerofoil surface $\bar{u}(\bar{\xi},0)$ and the aerofoil geometry. It is assumed in this approximation of $I_T(\bar{x},\bar{z},\bar{x}_s)$ that errors in the approximation function for $\bar{u}(\bar{\xi},\bar{\zeta})$ away from the surface do not contribute to any great error in the evaluation of $I_T(\bar{x},\bar{z},\bar{x}_s)$ because of the effect of the rapidly vanishing kernel function $\psi_{\bar{\xi}_{\bar{x}}}(\bar{x},\bar{\xi};0,\bar{\zeta})$ given by (14). The system equations, (11), (19a, 19b), are applied only on the aerofoil surface, represented by $\bar{z} = 0$; the additional condition, (19c), is neglected. Several approximation functions have been used, [8], [9], [10], but the results are generally unsatisfactory especially the predicted shock location, $\bar{x}_s$.

EXTENDED INTEGRAL EQUATION METHOD

In order to evaluate the surface integral, $I_T(\bar{x},\bar{z},\bar{x}_s)$ in (11) a knowledge of the variation of $\bar{u}(\bar{\xi},\bar{\zeta})$ over the entire flowfield is required. An approximate, although still sufficiently accurate, evaluation of $I_T(\bar{x},\bar{z},\bar{x}_s)$ may be possible if $\bar{u}(\bar{\xi},\bar{\zeta})$ is known only at specified points in the flowfield and some interpolation function used to express $\bar{u}(\bar{\xi},\bar{\zeta})$ in the rest of the flowfield. In addition it can be seen that if $I_T(\bar{x},\bar{z},\bar{x}_s)$ is evaluated using estimated values of $\bar{u}(\bar{\xi}_i,\bar{\zeta}_j)$ at the specified points $(\bar{\xi}_i,\bar{\zeta}_j)$, (11) and (19) can then be used to calculate new values of $\bar{u}(\bar{\xi}_i,\bar{\zeta}_j)$ and an iterative scheme is evolved. It should be noted that the practical requirement is an accurate value of $\bar{u}(\bar{x},\bar{z})$ on the aerofoil surface, hence an acceptable solution to the problem of (11) and (19) need not necessarily be accurate away from the surface provided the surface velocities are calculated to the desired accuracy.

Let the flowfield be divided into $2N$ strips ($\bar{z}$ = constant) as shown in Fig. 1; as the present problem is symmetric only the upper half plane need be considered.

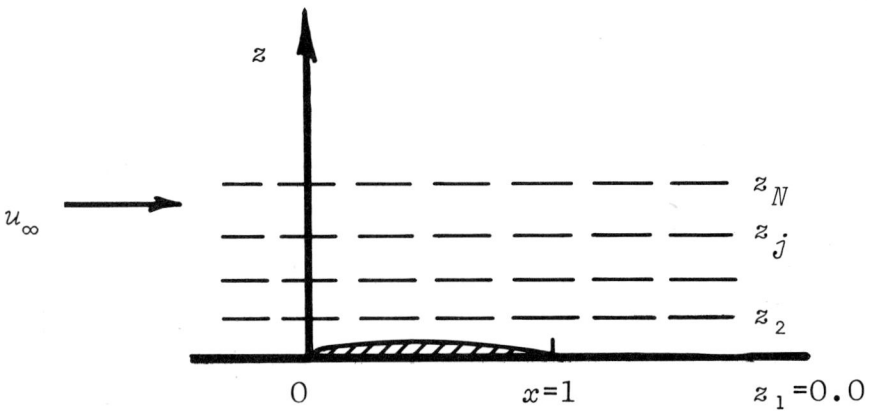

Fig. 1. Arrangement of strips for a half plane

The jth strip is defined by

$$\bar{\zeta}_j \leqslant \bar{\zeta} \leqslant \bar{\zeta}_{j+1} \quad (j = 1, N)$$

and

$$\bar{\zeta}_1 = 0 \tag{21}$$

The Nth strip in each half plane is assumed to stretch from $\bar{\zeta} = \bar{\zeta}_N$ to $\bar{\zeta} = \infty$, that is $\bar{\zeta}_{N+1} = \infty$. It is assumed that the velocity $\bar{u}(\bar{\xi},\bar{\zeta})$ is known on the strip edges

$$\bar{\zeta} = \bar{\zeta}_j \quad (j = 1, N+1)$$

The variation of $\bar{u}^2(\bar{\xi},\bar{\zeta})$ within each strip is represented by an interpolation function. In this paper straight line interpolation of $\bar{u}^2(\bar{\xi},\bar{\zeta})$ is used in the first $(N-1)$ strips in each half plane; thus

Integral Equations in Transonic Flow

$$\bar{u}^2(\bar{\xi},\bar{\zeta}) = \bar{u}^2(\bar{\xi},\bar{\zeta}_j) + \frac{[\bar{u}^2(\bar{\xi},\bar{\zeta}_{j+1}) - \bar{u}^2(\bar{\xi},\bar{\zeta}_j)]}{(\bar{\zeta}_{j+1} - \bar{\zeta}_j)} (\bar{\zeta} - \bar{\zeta}_j) \quad (22a)$$

for

$$\bar{\zeta}_j \leq \bar{\zeta} \leq \bar{\zeta}_{j+1}, \quad j < N$$

In the semi-infinite Nth strip in each half plane it is assumed that

$$\bar{u}^2(\bar{\xi},\bar{\zeta}) = \frac{\bar{u}^2(\bar{\xi},\bar{\zeta}_N)}{\left[1 + \frac{(\bar{\zeta} - \bar{\zeta}_N)}{b(\bar{\xi},\bar{\zeta}_N)}\right]^2}, \quad \bar{\zeta} \geq \bar{\zeta}_N \quad (22b)$$

where

$$b(\bar{\xi},\bar{\zeta}_N) = \frac{-2\bar{u}^2(\bar{\xi},\bar{\zeta}_N)}{\left\{\frac{\partial}{\partial \bar{\zeta}}[\bar{u}^2(\bar{\xi},\bar{\zeta})]\Big|_{\bar{\zeta}=\bar{\zeta}_N}\right\}}$$

$$\approx \frac{-2(\bar{\zeta}_N - \bar{\zeta}_{N-1})\bar{u}^2(\bar{\xi},\bar{\zeta}_N)}{[\bar{u}^2(\bar{\xi},\bar{\zeta}_N) - \bar{u}^2(\bar{\xi},\bar{\zeta}_{N-1})]} \quad (23)$$

The interpolation function given by (22b) and (23) is a generalisation of that used by Nixon and Hancock [11] in the "standard" integral equation method, when $\bar{\zeta}_N = 0$.

By using (22a and b) the integration with respect to $\bar{\zeta}$ in the surface integral $I_T(\bar{x},\bar{z},\bar{x}_S)$ can be performed and an approximate evaluation of $I_T(\bar{x},\bar{z},\bar{x}_S)$ can be obtained; the accuracy of this evaluation can be increased by simply increasing the number of strips.

In the standard integral equation method it has been found [8], [10] that for the purpose of evaluating the integral $I_T(\bar{x},\bar{z},\bar{x}_S)$ the perturbation velocity $\bar{u}(\bar{\xi},\bar{\zeta})$ can be assumed to be zero off the aerofoil. In the present analysis this assumption is again made; thus

$$\bar{u}(\bar{\xi},\bar{\zeta}) = 0 \quad \bar{\xi} < 0 \quad \bar{\xi} > 1$$

and

$$I_T(\bar{x},\bar{z},\bar{x}_s) = \frac{1}{4\pi}\int_0^1 \left\{ \sum_{j=1}^{N-1}\left\{ \bar{u}^2(\bar{\xi},\bar{\zeta}_j)g_0(X,Z_j,Z_{j+1})\right.\right.$$
$$+ \frac{[\bar{u}^2(\bar{\xi},\bar{\zeta}_{j+1})-\bar{u}^2(\bar{\xi},\bar{\zeta}_j)]}{(\bar{\zeta}_{j+1}-\bar{\zeta}_j)}g_1(X,Z_j,Z_{j+1})\right\}$$
$$\left. + \frac{\bar{u}^2(\bar{\xi},\bar{\zeta}_N)}{b(\bar{\xi},\bar{\zeta}_N)}g_2\left(\frac{X}{b},\frac{Z_N}{b}\right)\right\}d\bar{\xi} \qquad (24)$$

where

$$X = (\bar{\xi}-\bar{x}), \quad Z_j = (\bar{\zeta}_j-\bar{z})$$

and

$$g_0(X,Z_j,Z_{j+1}) = \left\{ \frac{Z_{j+1}}{(Z_{j+1}^2+X^2)} - \frac{Z_j}{(Z_j^2+X^2)}\right\} \qquad (25a)$$

$$g_1(X,Z_j,Z_{j+1}) = \left\{ \frac{Z_{j+1}(Z_{j+1}-Z_j)}{(X^2+Z_{j+1}^2)}\right.$$
$$\left. - \tfrac{1}{2}\ln\left[\frac{X^2+Z_{j+1}^2}{X^2+Z_j^2}\right]\right\} \qquad (25b)$$

$$g_2(X,Z) = \frac{1}{2[X^2+(1+Z)^2]^3}\left\{[3X^2-(1+Z)^2]\ln[X^2+Z^2]\right.$$
$$+ [(1+Z)^2+X^2][X^2-3(1+Z)^2]$$
$$\left. - 2[X^2-3(1+Z)^2]|X|\left[\frac{\pi}{2}+\arctan\left(\frac{Z}{|X|}\right)\right]\right\} \qquad (25c)$$

and $b(\bar{\xi},\bar{\zeta}_N)$ is given by (23).

The line integral, defined by (24), can be evaluated by approximating the range of integration by $(m-1)$ elements where m is a finite number, and assuming that $\bar{u}(\bar{\xi},\bar{\zeta}_j)$ and $b(\bar{\xi},\bar{\zeta}_N)$ can be represented by a mean value in each element. The influence functions g_0, g_1, g_2 are then integrated analytically over each element. The line integral $I_T(\bar{x},\bar{z},\bar{x}_s)$ can thus be represented by a finite sum. If the ith chordwise element is defined by

$$\bar{\xi}_i \leq \bar{\xi} \leq \bar{\xi}_{i+1}, \quad (i = 1, m-1) \quad (26)$$

and if $\bar{x}_i$, ($\bar{\xi}_i \leq \bar{x}_i \leq \bar{\xi}_{i+1}$) is the location of the mean values of $\bar{u}(\bar{\xi},\bar{\zeta}_j)$ and $b(\bar{\xi},\bar{\zeta}_N)$ in the ith element then

$$I_T(\bar{x},\bar{z},\bar{x}_s) = \sum_{i=1}^{m-1}\left\{\sum_{j=1}^{N-1}\left\{\bar{u}^2(\bar{x}_i,\bar{z}_j)f_0(X_i,X_{i+1},Z_j,Z_{j+1})\right.\right.$$

$$\left.+ \frac{[\bar{u}^2(\bar{x}_i,\bar{z}_{j+1}) - \bar{u}^2(\bar{x}_i,\bar{z}_j)]}{(\bar{z}_{j+1}-\bar{z}_j)}f_1(X_i,X_{i+1},Z_j,Z_{j+1})\right\}$$

$$\left.+ \bar{u}^2(\bar{x}_i,\bar{z}_N)f_2\left(\frac{X_i}{b},\frac{X_{i+1}}{b},\frac{Z_N}{b}\right)\right\} \quad (27)$$

where

$$X_i = (\bar{\xi}_i - \bar{x}), \quad Z_j = \bar{\zeta}_j$$

and

$$\left.\begin{array}{l} f_0(X_i,X_{i+1},Z_j,Z_{j+1}) = \dfrac{1}{4\pi}\displaystyle\int_{\xi_i}^{\bar{\xi}_{i+1}} g_0(X,Z_j,Z_{j+1})\,d\bar{\xi} \\[2mm] f_1(X_i,X_{i+1},Z_j,Z_{j+1}) = \dfrac{1}{4\pi}\displaystyle\int_{\xi_i}^{\bar{\xi}_{i+1}} g_1(X,Z_j,Z_{j+1})\,d\bar{\xi} \\[2mm] f_2\left(\dfrac{X_i}{b},\dfrac{X_{i+1}}{b},\dfrac{Z_N}{b}\right) = \dfrac{1}{4\pi}\displaystyle\int_{\xi_i}^{\bar{\xi}_{i+1}} g_2\left(\dfrac{X}{b},\dfrac{Z_N}{b}\right)\dfrac{d\bar{\xi}}{b} \end{array}\right\} \quad (28)$$

With $\bar{x} = \bar{x}_i$ and $\bar{z} = \bar{x}_j$ in (11) then

$$\bar{u}(\bar{x}_i,\bar{z}_j) - \frac{\bar{u}^2(\bar{x}_i,\bar{z}_j)}{2} = \bar{u}_{TL}(\bar{x}_i,\bar{z}_j) + I_T(\bar{x}_i,\bar{z}_j,\bar{x}_s) \quad (29)$$

where $I_T(\bar{x},\bar{z},\bar{x}_s)$ is given by (27). Thus (11) has been replaced by a set of nonlinear algebraic equations for the perturbation velocity $\bar{u}(\bar{x}_i, \bar{z}_j)$.

CALCULATION PROCEDURE

When there are no shock waves in the flow (29) can be solved directly. When shock waves are present (29) must be solved subject to the conditions given by (19). Hence two separate solution procedures for subcritical flows and for supercritical flows with shocks are required.

Subcritical flow

For subcritical flow (29) can be solved by evaluating $I_T(\bar{x}_i, \bar{z}_j)$ using estimated values of $\bar{u}(\bar{x}, \bar{z})$ and hence solving by iteration. In the standard integral equation theory for subcritical flows it is found [11] that if the integral equation for the surface velocity $\bar{u}(\bar{x}, 0)$ is written as

$$\bar{u}(\bar{x},0) - \frac{\bar{u}^2(\bar{x},0)}{4} = \bar{u}_{TL}(\bar{x},0) + \left\{ I_T(\bar{x},0) + \frac{\bar{u}^2(\bar{x},0)}{4} \right\} \quad (30)$$

and the nonlinear terms {} evaluated at each step of the iteration then convergence is very rapid. In addition it is also found [11] that a good first approximation for $\bar{u}(\bar{x},0)$ in subcritical flow is given by

$$\bar{u}(\bar{x},0) = 2\{1 - [1 - \bar{u}_{TL}(\bar{x},0)]^{\frac{1}{2}}\} \quad (31)$$

Thus, using the experience gained from applications of the standard integral equation method for subcritical flows, (29) is written in the form corresponding to (30), that is

$$\bar{u}(\bar{x}_i,\bar{z}_j) - \frac{\bar{u}^2(\bar{x}_i,\bar{z}_j)}{4} = \bar{u}_{TL}(\bar{x}_i,\bar{z}_j) + \left\{ \frac{\bar{u}^2(\bar{x}_i,\bar{z}_j)}{4} + I_T(\bar{x}_i,\bar{z}_j) \right\} \quad (32)$$

and the nonlinear terms {} evaluated at each step of the iteration. The initial estimate for $\bar{u}(\bar{x}_i, \bar{z}_j)$ is given by a generalisation of (31), that is

$$\bar{u}(\bar{x}_i,\bar{z}_j) = 2\{1 - [1 - \bar{u}_{TL}(\bar{x}_i,\bar{z}_j)]^{\frac{1}{2}}\} \quad (33)$$

Supercritical flows with shock waves

When shock waves are present in the flow (29) must be solved subject to the conditions given by (19). As noted earlier this system of equations gives $\bar{u}(\bar{x}_i,\bar{z}_j)$, the shock location, $\bar{x}_s$, and the location of the sonic line $\bar{x}_0(\bar{z})$. It is not practicable to satisfy (19) at each iterative step by altering the value of $\bar{x}_s$ since not only is the functional dependence of $I_T(\bar{x},\bar{z},\bar{x}_s)$ on $\bar{x}_s$ complicated but also the required $\bar{x}_s$ may be different on each strip edge, thus implying a curved shock wave which is not compatible with the mathematical model. Accordingly an alternative method of satisfying (19) during the iteration is used.

Equation (29) can be written

$$\bar{u}(\bar{x}_i,\bar{z}_j) - \frac{\bar{u}^2(\bar{x}_i,\bar{z}_j)}{2} = \bar{u}_{TL}(\bar{x}_i,\bar{z}_j) + \varepsilon(\bar{z}_j) I_T(\bar{x}_i,\bar{z}_j,\bar{x}_s) \quad (34)$$

where $\varepsilon(\bar{z}_j)$ is a parameter which is constant along each strip edge ($\bar{z} = \bar{z}_j$). A formal "solution" to (34) is given by

$$\bar{u}(\bar{x}_i,\bar{z}_j) = 1 \pm \{1 - 2[\bar{u}_{TL}(\bar{x}_i,\bar{z}_j) + \varepsilon(\bar{z}_j) I_T(\bar{x}_i,\bar{z}_j,\bar{x}_s)]\}^{\frac{1}{2}} \quad (35)$$

An initial estimate of the shock location $\bar{x}_s$ and the perturbation velocity $\bar{u}(\bar{x}_i,\bar{z}_j)$ incorporating the proper discontinuity at the shock is made and $I_T(\bar{x}_i,\bar{z}_j,\bar{x}_s)$ can be evaluated. The parameter $\varepsilon(\bar{z}_j)$ and the sonic point on the strip edge ($\bar{z} = \bar{z}_j$), $\bar{x}_0(\bar{z}_j)$ are evaluated by enforcing the conditions of finite streamwise acceleration ($\bar{u}_{\bar{x}}(\bar{x}_i,\bar{z}_j)$ finite), i.e., (19a, 19b) along each strip edge ($\bar{z} = \bar{z}_j$, $j = 1,N$). New values of $\bar{u}(\bar{x}_i,\bar{z}_j)$ are then computed and the process repeated until convergence of the $\varepsilon(\bar{z}_j)$. The shock location is then changed and the procedure outlined above repeated. The shock location is correct when the $\varepsilon(\bar{z}_j)$ are such that (19c) is satisfied in addition to (19a and b). It can be shown that if $\bar{x}_s$ is fixed and (34) used instead of (29), then (19c) implies that

$$\frac{d\varepsilon(\bar{z})}{d\bar{z}} = 0 \qquad (36)$$

The exact $I_T(\bar{x}_i, \bar{z}_j, \bar{x}_s)$ should lead to the values of all the $\varepsilon(\bar{z}_j)$ equal to unity but an adequate approximate solution to (34) can be obtained with less expenditure of computing time if the $\varepsilon(\bar{z}_j)$ are allowed to vary within a specified range on either side of unity.

Since high accuracy is required only on the aerofoil surface the accuracy of the solution in the flowfield can decrease as $\bar{z}_j$ increases provided the surface solution is not affected to any significant degree. In practice this is taken to mean that the difference between the converged value of $\varepsilon(\bar{z}_j)$ and unity is allowed to increase as $\bar{z}_j \to \infty$ and that (36) is satisfied only on the first few strip edges.

RESULTS

To test the method outlined in the previous sections calculations have been performed using the transonic small perturbation equation, (3), for the flow around a parabolic arc aerofoil of 6% thickness/chord ratio. Both subcritical and supercritical flows with shocks are considered. The pressure distribution around a NACA 0012 section in subcritical flow is also presented. This latter calculation was performed using the second order potential equation, (1), and the exact boundary conditions (4).

To compare results for the flow around the parabolic arc sections the similarity parameter K defined by Murman and Cole [2] is used, where

$$K = \frac{(1-M_\infty^2)(\gamma+1)^{\frac{2}{3}}}{(k\tau)^{\frac{2}{3}}}$$

where k is the parameter in (1) and (3) and τ is the thickness/chord ratio of the section. For the pressure distribution around the NACA 0012 section the parameter k is taken to be that derived by Hancock [12], namely

$$k = (3 + (\gamma-2)M_\infty^2)M_\infty^2$$

The pressure distribution around a 6% parabolic arc aerofoil at $M_\infty = 0.8$ ($K = 2.94$) is shown in Fig. 2, together with the results of Murman and Cole [2].

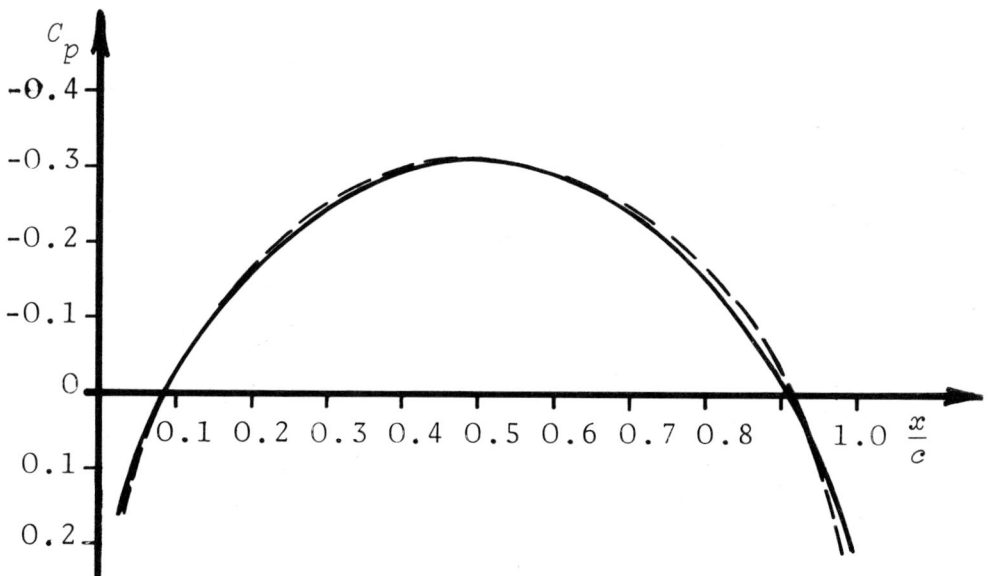

Fig. 2. Pressure distribution around a 6% biconvex aerofoil $M_\infty = 0.80$ ($K = 2.94$) ——— present method, ----- Murman and Cole [2]

It can be seen that there is good agreement between the results. This is to be expected as the nonlinear terms in this example are small. The pressure distribution around the 6% parabolic arc section at $M_\infty = 0.87$ ($K = 1.8$) is shown in Fig. 3 and is compared with the results of Murman [6] and the results of the standard integral equation method calculated by Norstrud [10]. It can be seen that the shock location predicted by the present method is in good agreement with the location predicted by Murman [6]. The pressure distributions agree satisfactorily except immediately ahead of the shock and in the neighbourhood of the leading edge. The calculation of Norstrud [10] does not give good agreement with the results of Murman [6].

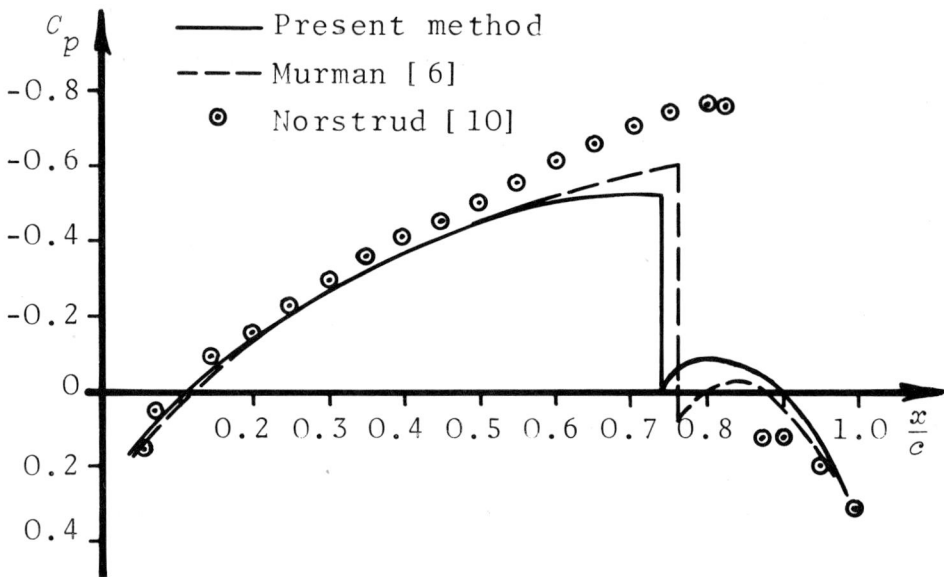

Fig. 3. Pressure distribution around a 6% biconvex aerofoil M_∞ = 0.87 (K = 1.8)

Finally the pressure distribution around the NACA 0012 section at M_∞ = 0.72 is shown in Fig. 4 and is compared with the finite difference results of Sells [1]; the result of the standard integral equation method [11] are also shown. It can be seen that the present calculation is in good agreement with that of Sells [1] and is an improvement over the standard integral equation result [11] especially in the neighbourhood of the leading edge.

In the above calculations 30 chordwise points and 3 to 5 strips were used in each half plane. In the subcritical examples convergence was obtained after 3 or 4 iterations and in the supercritical example convergence of the parameter $\varepsilon(\bar{z}_j)$ was obtained for each shock location after

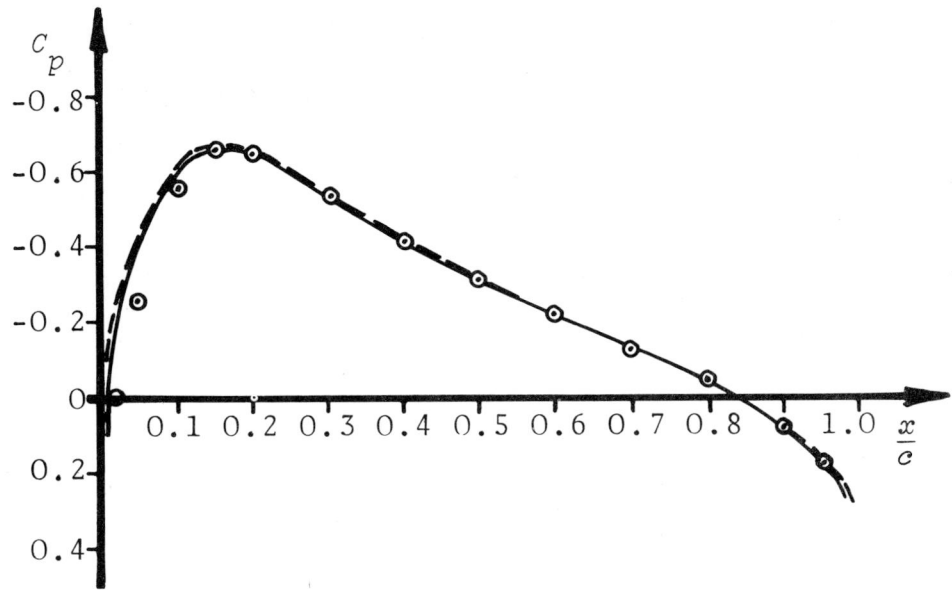

Fig. 4. Pressure distribution around a NACA 0012 aerofoil $M_\infty = 0.72$, —— present method, - - - - - Sells [1], ⊙ Nixon-Hancock [11]

about 8 iterations. The total computing time on an ICL1904S is about 50 seconds for the subcritical examples and 250 seconds for the supercritical example. These times are slightly greater than those required for the finite difference calculations but since the programs used in the present calculations were written mainly for experimental purposes it is probable that the over-all computing time could be considerably reduced with more efficient programming.

CONCLUSIONS

In this paper an extended integral equation method is developed which offers a considerable improvement in accuracy over the earlier integral equation methods [8], [9], [10], especially for supercritical flows with shock waves. The pressure distribution and, where applicable, the shock location have been calculated for the flow around a 6% parabolic arc aerofoil and a NACA0012 section at zero incidence. The results of this extended integral equation method compare favourably with the results of the finite difference calculations [1], [2], [6].

The number of iterative steps to a converged solution in the present method is small compared with the number required by the finite difference methods and it is suggested that with improved efficiency in the computer program the over-all computing time should be at least comparable with that of the recent finite difference methods.

REFERENCES

1. Sells, C.C.L., "Plane subcritical flow past a lifting aerofoil," RAE Tech. Rpt. 67146 (1967)
2. Murman, E. and Cole, J.D., "Calculation of plane steady transonic flows," *AIAA J.*, **9**, No. 1, 114-121 (1971).
3. Krupp, J.A. and Murman, E., "The numerical calculation of steady transonic flows past thin lifting aerofoils and slender bodies," AIAA Paper No. 71-566 (1971).
4. Garabedian, P. and Korn, D., "Analysis of transonic airfoils," *Comm. P.A. Math,*. **XXIV**, 841-851 (1971).
5. Jameson, A., "Transonic flow calculations for airfoils and bodies of revolution," Grumman Report 390-71-1 (1971).
6. Murman, E., "Analysis of embedded shock waves calculated by relaxation methods," *AIAA J.*, **12**, No. 5, 626-633 (1974).
7. Oswatitsch, K., "*Die Geschwindigkeitsverteilung au Symmetrischen Profilen beim Auftreten lokaler Uberschaltgebiete*," *Acta Physica Austriaca*, **4**, 228-271 (1950).
8. Spreiter, J.R. and Alksne, A.Y., "Theoretical prediction of pressure distributions on non-lifting airfoils at high subsonic speeds," NACA Rpt. 1217 (1955).

9. Crown, J.C., "Calculation of transonic flow over thick airfoils by integral methods," *AIAA J.*, **6**, 413-423 (1968).

10. Norstrud, H., "High speed flow past wings," NASA CR-2246 (1973).

11. Nixon, D. and Hancock, G.J., "High subsonic flow past a steady two-dimensional aerofoil," ARC C.P.-1280 (1974).

12. Hancock, G.J., "Some aspects of subsonic linearised wing theory with reference to second order forces and moments," ARC Paper ARC 34689 (1973).

RELAXATION NEAR A SONIC LINE

A. Roberts
(British Aircraft Corporation Limited)

INTRODUCTION

The complete transonic flow about a subsonic lifting aircraft can be regarded as a set of fields separated by interfaces. For each of these fields direct solution or rapid relaxation techniques are now available but there are still problems in achieving rapid relaxation over the interfaces. Consider a local supersonic field within a nonlinear subsonic far field. These fields are separated by the sonic line, the shock line and the boundary of a rectangular "working section" (Fig. 1).

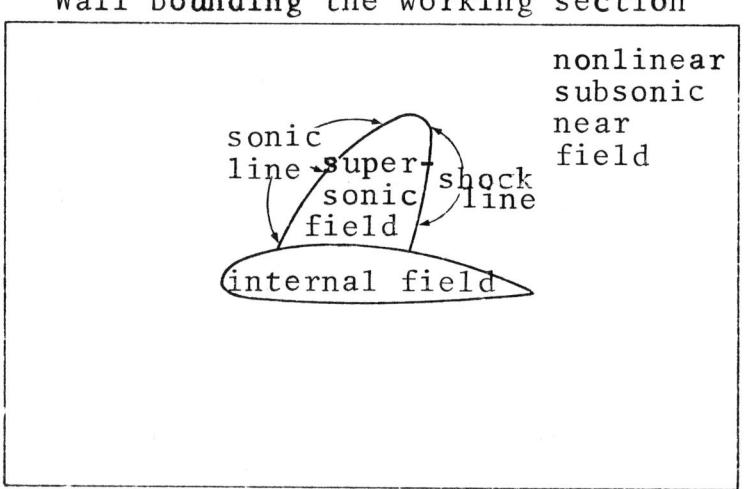

Fig. 1. The complete transonic field is regarded as a set of component fields separated by interfaces

The supersonic field away from the sonic line can be treated effectively using a marching sequence. An "elliptic solver" provides rapid local relaxation in the subsonic field away from the sonic line. The far field can be represented as the potential field induced by a relatively small number of point sources.

When relatively slow relaxation techniques are used the interfaces do not appear to be a limiting factor. However, when fast relaxation techniques are used it is necessary to develop comparable efficiency near the interfaces. It will be shown that for low Mach numbers a simple treatment of the walls of the "working section" is effective. Extension to higher Mach numbers should not be unduly difficult.

The sonic line region presents special problems. The geometry of the sonic line is unknown. There is a need for a relevant mathematical model which takes into account the change in geometry of the sonic line when the fields on either side are disturbed.

THE FIELD EQUATIONS

We assume that the steady isentropic field equation is to be satisfied over the physical plane. This equation for the super-velocity potential ϕ can be expressed in the two-dimensional form

$$\underset{\sim}{L}\phi = 0$$

where

$$\underset{\sim}{L}\phi = (c^2-u^2)\frac{\partial^2 \phi}{\partial x^2} + (c^2-v^2)\frac{\partial^2 \phi}{\partial y^2} - 2uv\frac{\partial^2 \phi}{\partial x \partial y} \quad (1)$$

where

x,y are the coordinates of the physical plane
u,v are the local velocity components
u_0, v_0 are the free stream velocity components
c is the local speed of sound
c_0 is the free stream speed of sound
γ is the ratio of specific heats

$$u = \frac{\partial \phi}{\partial x} + u_0 \qquad v = \frac{\partial \phi}{\partial y} \qquad (2)$$

$$\left(\frac{c}{c_0}\right)^2 = 1 - \left(\frac{\gamma-1}{2}\right)(u^2 + v^2 - u_0^2) \qquad (3)$$

Let X, Y be the coordinates in the Glauert transformation

$$X = x; \qquad Y = \beta y \quad \text{where} \quad \beta = \sqrt{1 - (u_0/c_0)^2}$$

For small perturbations from the free stream velocity we have

$$\underset{\sim}{L}\phi \to c_0^2 \, \beta^2 \left(\frac{\partial^2 \phi}{\partial X^2} + \frac{\partial^2 \phi}{\partial Y^2}\right) \qquad (4)$$

The limiting form of the field equation is then the Prandtl-Glauert equation

$$\frac{\partial^2 \phi}{\partial X^2} + \frac{\partial^2 \phi}{\partial Y^2} = 0$$

THE WALL PROBLEM

The computing planes of the wall problem

The physical plane is regarded as the near field inside a rectangular "working section," together with the far field outside (see Fig. 2). The computing sequences treat two complete planes - the near field plane and the far field plane. The ϕ distribution over the near field plane is defined below in terms of reflections of the ϕ distribution *within* the working section where the ϕ distribution is the same as that in the physical plane. The field equation over the complete far field plane is assumed to be the Prandtl-Glauert equation. The ϕ distribution *outside* the working section is the same as that in the physical plane.

The field equation in the near field plane is systematically modified so that $\underset{\sim}{L}\phi$ has the same reflection properties as the ϕ distribution. Typically ϕ is reflected without change of sign in the downstream wall and any wall representing a line of physical symmetry and is reflected with a

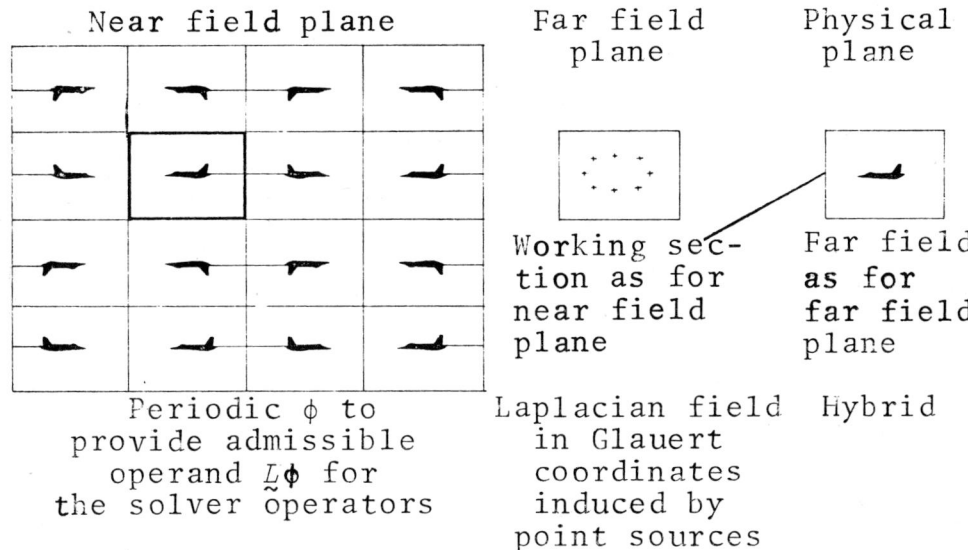

Fig. 2. The problem is posed as coupling between two complete planes

reversal of sign in all other walls. The special treatment of the downstream wall is to facilitate extension to include trailing doublet sheets penetrating this wall. These reflection laws ensure that ϕ is periodic with respect to both coordinates with a cycle length of four times the length of the working section. It is then possible to use numerical operators which require periodic operands.

For each step in the iteration these operators are used to obtain an approximate solution of the Poisson equation and so provide a semi-direct method of solving the nonlinear equation $\underset{\sim}{L}\phi = 0$. The equation is:

$$\left(\frac{\partial^2}{\partial X^2} + \frac{\partial^2}{\partial Y^2}\right)\left(\phi_{t+1} - \phi_t\right) = \sigma \underset{\sim}{L} \phi_t \qquad (5)$$

or the three-dimensional generalisation.

This inversion is expressed formally by

$$\phi_{t+1} = \phi_t - \mathcal{H} \sigma \underset{\sim}{L} \phi_t \qquad (6)$$

(In these equations σ is a local over-relaxation factor defined as a function of local velocity. This factor is introduced to improve the local rate of convergence of the error distribution $\underset{\sim}{L}\phi_t$.)

This inversion can be performed directly using a computing sequence known as an elliptic solver. An exact inversion of the finite difference form of the differential equation can be obtained by use of the discrete Fourier transformation as described in Appendix I. The computing effort can be reduced by using a fast Fourier algorithm. Alternatively an even faster method, based on a set of progressively sparser arrays, can be used to obtain an approximate solution of the Poisson equation at each step.

As we are solving the nonlinear equation $\underset{\sim}{L}\phi = 0$, the approximate solution of the Poisson equation requires a slightly larger number of steps to converge but it does not affect the accuracy of the final solution.

Here we are primarily interested in the treatment of interfaces between fields. There are two types of rigid interfaces - one represents the aerofoil and the other represents the walls of the "working section."

At present a simplified model of the aerofoil is used. The difference in $\partial\phi/\partial z$ across the aerofoil is represented by the discontinuity of $\partial\phi/\partial n$ across a mean surface and the magnitude remains constant throughout the calculation. (Ultimately this will be replaced by discontinuities across the three-dimensional wetted surfaces with distributions adjusted during the calculation to satisfy the exact boundary condition $\partial\phi/\partial n = 0$ over the wetted surfaces.)

In the physical plane ϕ and grad ϕ must be continuous across the walls of the working section. The reflection laws in the near field plane (required by the elliptic solver) combined with the continuity condition in the physical plane can only be satisfied if ϕ or grad ϕ is discontinuous across the walls in the near field plane. These discontinuities are adjusted at each stage of the iteration.

The effect of both constant and variable discontinuity functions is introduced by modifying the formulae for numerical first and second derivatives near the discontinuity lines as described in Appendix II. This modification is defined in such a way that the correct values of derivatives are obtained near discontinuities of ϕ or grad ϕ when the magnitude and location of the discontinuities are known.

Analytic "wall" conditions

The continuity and reflection effects are shown in Fig. 3.

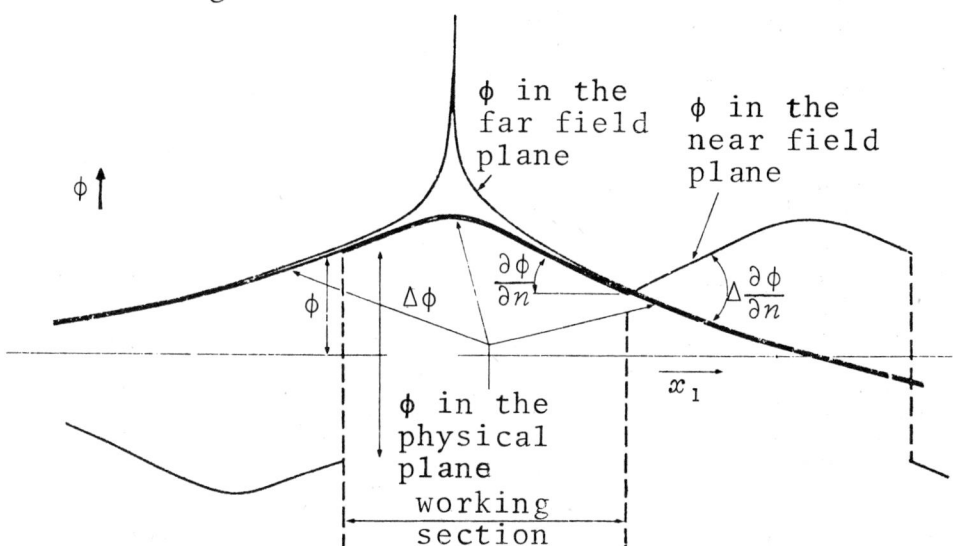

Fig. 3. Physical plane continuity and near field plane reflections link near field plane discontinuities to far field plane wall values

The φ distribution in the physical plane is equal to that in the near field plane within the working section and to that in the far field plane outside of the working section. When φ is reflected with a sign reversal such as at the upstream face there is no discontinuity in $\partial\phi/\partial n$. The antisymmetry condition in the near field plane combined with the continuity condition in the physical plane implies that the discontinuity in φ in the near field plane is precisely twice the value of φ at the "wall" in the far field plane. At the upstream wall the value of $\partial\phi/\partial n$ in the far field plane should be the same as that in the near field plane.

Similarly when φ is reflected without change of sign, such as at the downstream wall, there is continuity of φ and discontinuity of $\partial\phi/\partial n$. From the symmetry condition in the near field plane and the continuity condition in the physical plane the discontinuity in $\partial\phi/\partial n$ in the near field plane must be twice the value of $\partial\phi/\partial n$ in the far field plane. The value of φ in the far field plane should be the same as that in the near field plane.

The continuous function is φ at symmetric walls and $\partial\phi/\partial n$ at antisymmetric walls. The discontinuous function is $\partial\phi/\partial n$ at symmetric walls and φ at antisymmetric walls. The condition at infinity and distribution of the continuous function over the external side of all walls is sufficient to define the φ distribution over the far field. When the near field equation is nearly Laplacian the symmetry conditions and the distribution of the discontinuous function over all walls is sufficient to determine φ over the near field plane. Thus we appear to have a correctly posed compound problem. We now convert these principles into literal computing techniques.

Wall relaxation

Because the far field is Laplacian in Glauert coordinates and the higher derivatives of φ are small in the far field we can regard the far field as that produced by a relatively small number m of point sources in the far field plane (Fig. 4).

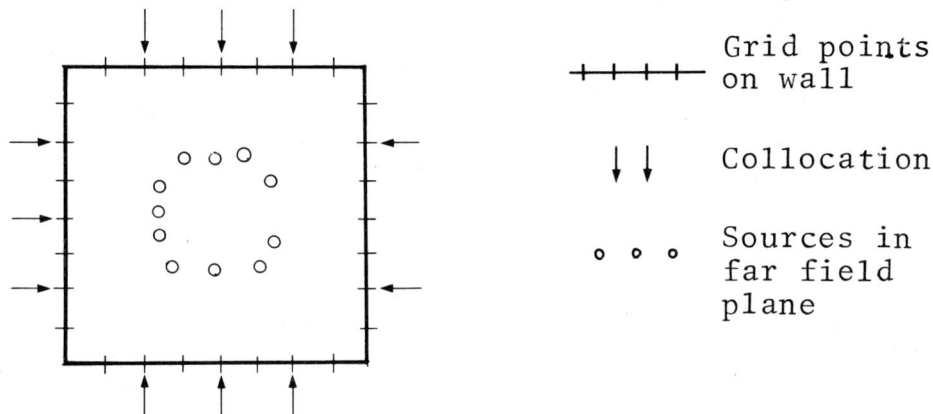

Fig. 4.   The boundary problem is reduced to matrix form

$S(j)$ = scaling factor of the jth point source ($j = 1, \ldots, m$)

$C(i)$ = spot value of the continuous function at the ith collocation point on the wall

$D(k)$ = spot value of the discontinuous function at the kth grid point on the wall

$A(i,j)$ = value of $C(i)$ induced by unit $S(j)$ in the far field plane

$B(k,j)$ = value of $D(k)$ induced by unit $S(j)$ in the far field plane

S_t, C_t, D_t = values of the S C D arrays at the tth step

Then in the final steady state the S array S_∞ must satisfy

$$A \times S_\infty = C_\infty \qquad (7)$$

and

$$B \times S_\infty = D_\infty \qquad (8)$$

During the relaxation, C_t is obtained by numerical sampling or differentiation in the near field plane. The array D_{t+1} is used implicitly in the $\underset{\sim}{L}$ operator in the evaluation of $\underset{\sim}{L}\phi_t$. This mixture of generations (D_{t+1} and ϕ_t) results in a numerical equivalent of source and doublet distributions near the walls in the near field plane as discussed in Appendix II.

In the near field plane $\phi_{t+1} - \phi_t$ is derived from $\underset{\sim}{L}\phi_t$. We start with ϕ_0 zero and $\underset{\sim}{L}\phi_0$ zero except near the aerofoil, where the implicit source and doublet distributions are obtained from the given constant discontinuity functions.

In the far field plane we need to derive $(S_{t+1} - S_t)$ from $(C_t - A \times S_t)$ so that in the steady state $S_{t+1} = S_t$ and $C_\infty = A \times S_\infty$. We start with $S_0 = 0$. Since ϕ_0 is zero except near the aerofoil we have C_0 zero so S_1 must also be zero. The upper limit on the rate of initial convergence at the wall would give $S_2 = S_\infty$. This would have to be computed from C_1.

Now the field ϕ_1 is that induced by the given discontinuity functions plus that induced by *all* reflections of those discontinuity functions at the aerofoil. The far field induced by the given discontinuity functions in a Laplacian field *without* reflections is equal to that induced by some equivalent set of sources S_e. Then near the walls ϕ_1 (from which C_1 is computed) is equal to the field induced by S_e and *all* reflections of S_e. Since C_1 is known we can compute S_e for any new aerofoil by solving

$$R \times S_e = C_1 \qquad (9)$$

where $R(i,j)$ = value of $C(i)$ at the ith collocation point induced by $S(j) = 1$ with *all* images of $S(j)$ with all other elements of S zero in a Laplacian field.

The values of the elements of $R(i,j)$ can be computed numerically for all values of i from one solution for ϕ using the elliptic solver with the aerofoil replaced by the single unit source corresponding to $S(j)$; when the nonlinear compressibility effects are negligible the set of source scaling factors S_e is equal to the required set S_∞. The field ϕ_2 is computed by implicitly adding to ϕ the Laplacian field induced by the discontinuity function D_2 over the walls and the explicit solution of the Poisson equation. The resulting error distribution $\underset{\sim}{L}\phi_2$ will also have an equivalent source field $\tilde{S}_e$.

The computing sequence is

START $\Delta\phi$ and $\Delta(\partial\phi/\partial y)$ at $y = 0$ given.

$\qquad S_0, S_1 = 0$

$\qquad \phi_0 \quad\quad = 0$

$\qquad D_0, D_1 = 0$

LOOP INVERT $\quad R \times (S_{t+1} - S_t) = C_t - A \times S_t$
EVALUATE $D_{t+1} \qquad\qquad\qquad\quad = B \times S_{t+1}$

SOLVE $\quad \left(\dfrac{\partial^2}{\partial X^2}+\dfrac{\partial^2}{\partial Y^2}\right)\left(\phi_{t+1}-\phi_t\right) = -\sigma \underset{\sim}{L}\phi_t$

$\hfill (10)$

RESULT $\quad C_\infty = A \times S_\infty$
$\qquad\qquad D_\infty = B \times S_\infty \qquad\qquad (11)$
$\qquad\qquad \underset{\sim}{L}\phi_\infty = 0$

where $\underset{\sim}{L}\phi = (c^2-u^2)\dfrac{\partial^2\phi}{\partial x^2} + (c^2-v^2)\dfrac{\partial^2\phi}{\partial y^2} - 2uv\dfrac{\partial^2\phi}{\partial x \partial y}$

$\qquad\qquad$ + (discontinuity corrections)

x, y are the coordinates in the physical plane

X, Y are the coordinates in the Glauert plane

$\sigma \quad$ is a local relaxation factor defined as a function of local velocity components

The matrix R has no effect on the final solution. It was designed to achieve rapid initial convergence when compressibility effects were small.

The first test case was a point source at a Mach number of 0.0001 and near zero free stream velocity. The grid used was 16 × 32. The contours are shown for ϕ_1, ϕ_2 and ϕ_5 (see Fig. 5).

ϕ_1 - zero wall correction
ϕ_2 - first wall correction
ϕ_5 - steady solution

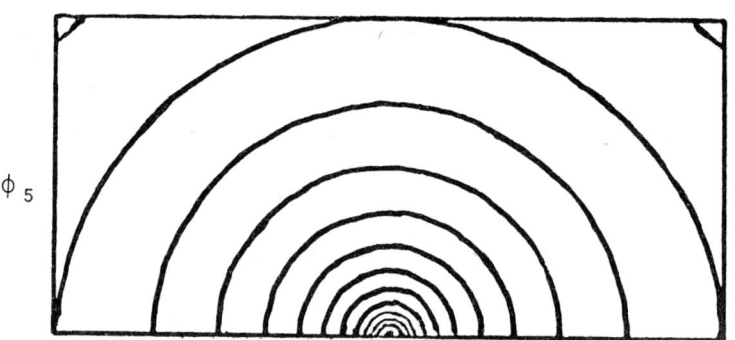

Fig. 5. ϕ contours for a point source $M \simeq 0$

The ϕ_1 computation has zero wall correction because ϕ_0 is zero at the wall. This distribution is nearly the field induced by the point source with images. The ϕ_2 contours are a fair approximation to those for a point source without images. Detailed examination of the tabulated results showed a reduction in error of about 90% per step. This is slightly lower than that with no wall correction.

The second test case used a discontinuity in $\partial\phi/\partial y$ derived from a solution of the Neumann problem [3] for the aerofoil NACA 0012. The pressures at the geometric surface were obtained by interpolation. Rapid convergence was obtained with the wall correction and the final pressure distribution was close to the Neumann solution (see Fig. 6).

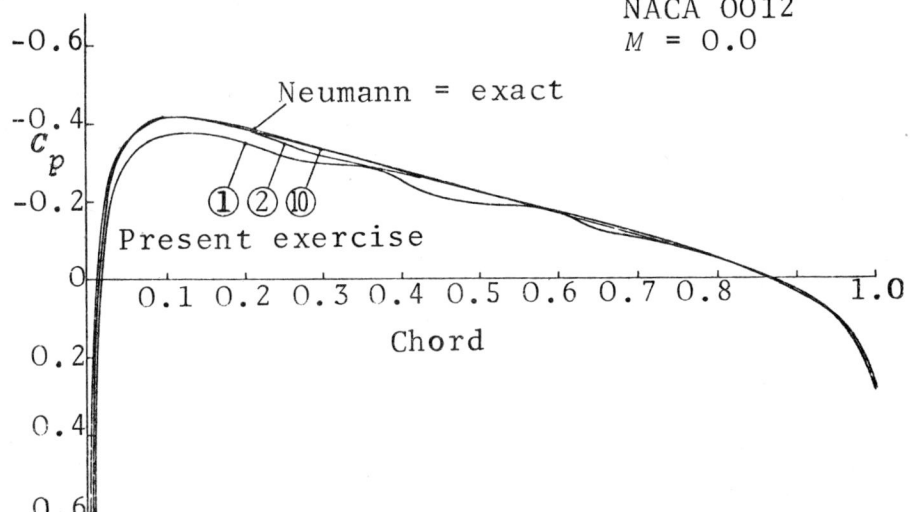

Fig. 6. Initial convergence with wall corrections without compressibility

The third test case was the same section at $M = 0.72$ without the wall correction. Moderately rapid convergence was achieved (see Fig. 7) and the final result was by chance close to the Neumann solution. The error resulting from the uncorrected wall together with violation of the aerofoil surface condition was approximately the same as the error in the Neumann solution.

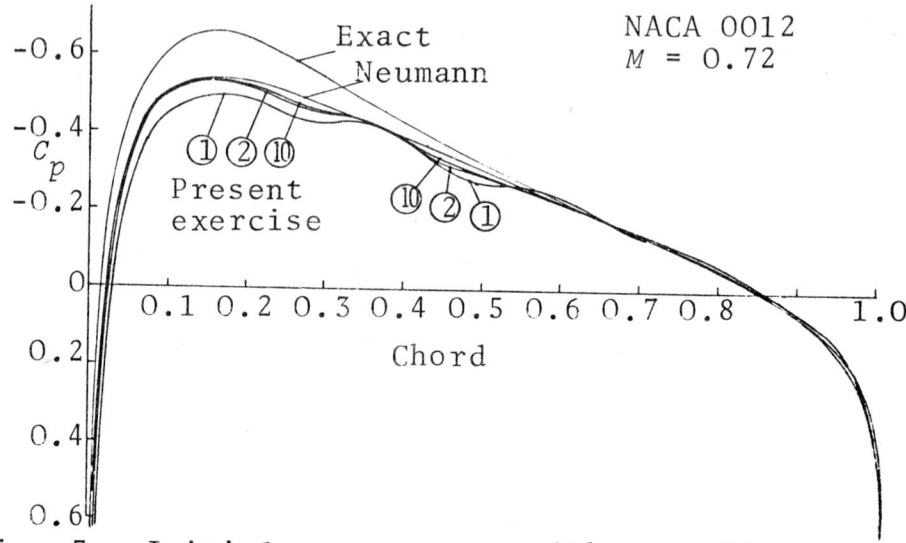

Fig. 7. Initial convergence without wall corrections with compressibility

The fourth test case was the same aerofoil and the same Mach number but with the wall correction included. This case failed to converge (see Fig. 8).

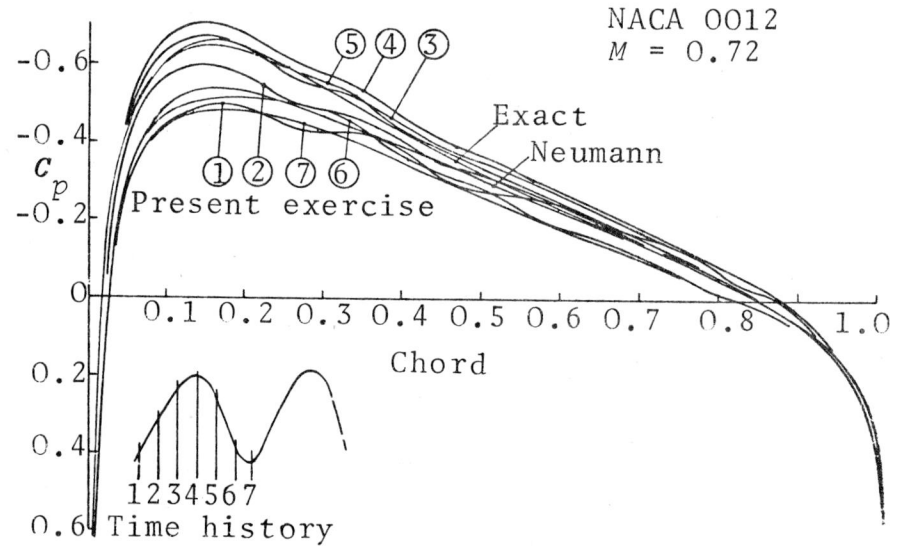

Fig. 8. Initial convergence with wall corrections with compressibility: original σ formula

The computed normal velocity at the wall was derived from the difference between the value of ϕ at the wall (defined as half the *required* discontinuity in ϕ) and the value at the adjacent point in the field representing half the *achieved* discontinuity. This difference was sensitive to the value of σ defined by a convergence criterion away from the wall. The fourth test case was repeated with σ modified at one interval from the wall

$$\sigma = \beta^2 c_0^2/(c^2 - u^2) \text{ along the upstream wall}$$
$$\sigma = c_0^2/(c^2 - v^2) \text{ along the } Y_{max} \text{ wall and the corner point.}$$

This time the initial rate of convergence was moderately rapid (see Fig. 9). The final result did not agree with the exact result because the discontinuity distribution was taken from the Neumann solution and ignored errors in normal velocity at the aerofoil surface in the compressible field.

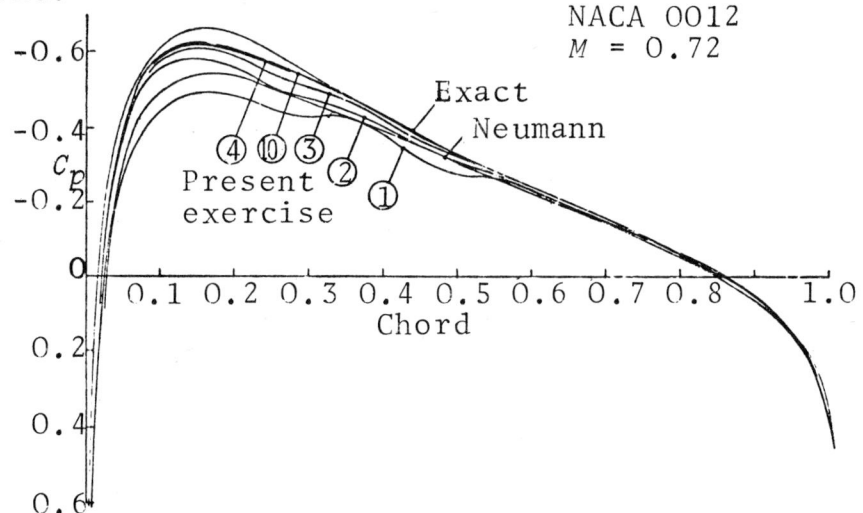

Fig. 9. Initial convergence with wall corrections with compressibility: modified σ formula

A previous exercise without wall correction had shown that moderately rapid convergence could be achieved with a given ϕ discontinuity for both two-dimensional and three-dimensional problems.

Ultimately it will be necessary to include an outer loop which corrects the normal velocity in the presence of a boundary layer. However this is beyond the scope of the present investigation of interface stability.

The conclusion from the first phase of this exercise is that the combination of the boundary matrices A, B and R provides a quick, rapidly convergent and well conditioned treatment of the interface with the far field. (The rate of convergence is limited by the action of the elliptic solver in regions of near sonic velocity.) The test case used of a 20" chord aerofoil inside a 32" "working section" is a more severe test of wall correction than would be met with a wing with a working section significantly larger than a wing span.

Detailed examination of intermediate results and subsequent numerical analysis suggests that the computing sequences used were significantly off optimum but further development was deferred until the viability of treatment of the sonic and shock surfaces had been demonstrated.

THE SONIC LINE PROBLEM

The type of inner solution required

We are developing a hybrid iteration sequence which applies an elliptic solver to the complete near field plane and then marches downstream through a local supersonic region. We require a rate of convergence near the sonic line comparable to that which can be achieved elsewhere in the plane. The local rate of convergence on the subsonic side approaches zero as the sonic line is approached. The downstream step length of the marching sequence approaches zero as the sonic line is approached. Techniques which are stable on both sides of the sonic line, such as line relaxation and time stepping, have slow convergence away from the sonic line (or over the complete plane). The fundamental difficulty is that disturbances of the solution on either side of the sonic line cause a change in geometry of the sonic line. We need inner solutions which are valid in the close neighbourhood of the sonic line. Consider Fig. 10.

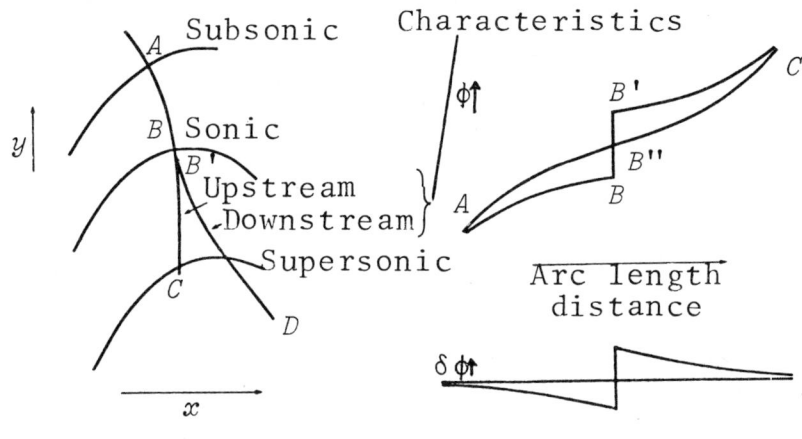

Fig. 10. The required continuous inner solution is obtained by adding a discontinuous small incremental solution to a discontinuous approximate solution

Suppose that both ϕ and grad ϕ are given along the velocity contours through A and C. In principle it is then possible to march towards the sonic line from both directions. The result will be discontinuity of both ϕ and grad ϕ along the assumed geometry of the sonic line. The required inner solutions cannot match both ϕ and grad ϕ on both sides. We need the less stringent constraint of matching the asymptotes on the subsonic line and along the upstream characteristics. Subject to these constraints we wish to satisfy the nonlinear field equation for ϕ in the range A, B, C. We assume that the given discontinuous ϕ distribution is a fair approximation to the required solution which has the required asymptotic form. Thus we need to find the small incremental solutions, with zero asymptotes on both sides, which neutralise the discontinuities.

Reduction of the field equation in the hodograph plane

The coordinates of the physical and the hodograph planes are related to the velocity potential ϕ and the space potential (or transformed potential of the Legendre transformation)

$$u = \frac{\partial \phi}{\partial x}; \quad v = \frac{\partial \phi}{\partial y}; \quad x = \frac{\partial \xi}{\partial u}; \quad y = \frac{\partial \xi}{\partial v} \qquad (12)$$

shown elsewhere [1].

Thus the coordinates of the hodograph plane are the derivatives of the velocity potential ϕ in the physical plane and the coordinates of the physical plane are derivatives of the space potential ξ in the hodograph plane with ξ and ϕ related by

$$\phi + \xi = ux + vy \qquad (13)$$

It is convenient to express location in the hodograph plane in polar coordinates w, θ

$$w^2 = u^2 + v^2 \qquad \tan\theta = u/v$$

where w is the modulus of the velocity vector.

Along the sonic line $w = w^* = c^*$ where w^* and c^* represent the velocity and speed of sound along the sonic line.

The local speed of sound c can be expressed as

$$(c^2 - c^{*2}) = -\left(\frac{\gamma-1}{2}\right)(w^2 - w^{*2}) \qquad (14)$$

The isentropic field equation $\mathcal{L}\phi = 0$ transforms into

$$\xi_{ww} + p\xi_w + q\xi_{\theta\theta} = 0 \qquad (15)$$

where $p = p(w) = (1 - (w/c)^2)/w$ and $q = p/w$.

For the treatment of harmonic solutions it is convenient to regard the physical space potential as the real part of the complex variable satisfying the same equation. In order to separate variables we assume the trial form of solution

$$\xi = Qg(w)e^{im\theta} \qquad (16)$$

where Q is a complex constant. Substituting for ξ in the field equation leads to the real ordinary differential equation

$$g_{ww} + pg_w - m^2qg = 0 \qquad (17)$$

To reduce the equation to standard form near the sonic line we can regard g as a function of a transformed variable η.

$$g = g(\eta(w))$$

This function is defined by

$$\eta = 0 \quad \text{at} \quad w = w^*$$

$$\frac{d\eta}{dw} = \left(\frac{d\eta}{dw}\right)^* e^{-\int_{w^*}^{w} p(w)\,dw} \qquad (18)$$

$$= \left(\frac{d\eta}{dw}\right)^* (1 + \text{order}\,(w - w^*)^2)$$

Then near $\eta = 0$ the differential equation becomes

$$\frac{d^2g}{d\eta^2} - m^2 \left\{\frac{q}{(d\eta/dw)^2}\right\} g = 0$$

and for

$$\frac{d\eta}{dw} = 1/\sqrt{dq/d\eta} \quad \text{at} \quad \eta = 0$$

it becomes

$$\frac{d^2g}{d\eta^2} - m^2\eta g = \text{order}(m^2 g\eta^2).$$

Thus near the sonic line we consider the equations

$$\frac{d^2g}{d\eta^2} - m^2\eta g = 0 \qquad (19)$$

The solution to this equation is expressed in terms of Bessel functions.

<u>Approximate particular harmonic solutions</u>

Let R and $(\gamma - \pi/6)$ be the modulus and argument of the complex value of the Bessel function of the third kind, i.e.,

$$H^{(1)}_{-\frac{1}{3}}(\lambda) = Re^{i(\gamma-\pi/6)} \text{ with } \gamma = 0 \text{ at } \lambda = 0 \quad (20)$$

Then from the standard Bessel identities [2] the Bessel functions of the first kind may be expressed as

$$\left.\begin{array}{rl} J_{-\frac{1}{3}}(\lambda) &= R(e^{+i(\gamma-\pi/6)} + e^{-i(\gamma-\pi/6)}) \\ J_{+\frac{1}{3}}(\lambda) &= R(e^{+i(\gamma-\pi/2)} + e^{-i(\gamma-\pi/2)}) \end{array}\right\} \quad (21)$$

in which $R = R(\lambda)$ and $\gamma = \gamma(\lambda)$.

The modified Bessel functions of the first kind have the property

$$I_{-\frac{1}{3}}(\lambda) - I_{+\frac{1}{3}}(\lambda) \to 0 \text{ as } \lambda \to \infty \quad (22)$$

The solution to the near sonic equation can then be expressed in the usual form for

$$\lambda = \sqrt{m^2|\eta|^3/9} \quad (23)$$

That is

$$\left.\begin{array}{rl} \xi &= \left(\frac{\lambda}{2}\right)^{\frac{1}{3}} \{AI_{-\frac{1}{3}} + BI_{+\frac{1}{3}}\}e^{im\theta} \text{ subsonic side} \\ \xi &= \left(\frac{\lambda}{2}\right)^{\frac{1}{3}} \{CJ_{-\frac{1}{3}} - DJ_{+\frac{1}{3}}\}e^{im\theta} \text{ supersonic side} \end{array}\right\} \quad (24)$$

The associated discontinuities across the sonic line are:

$$\Delta\xi = Me^{im\theta}/\Gamma(2/3) \text{ where } M = A - C$$

and $\Delta\left(\frac{\partial\xi}{\partial\eta}\right) = \left(\frac{m}{3}\right)^{\frac{2}{3}} Ne^{im\theta}/\Gamma(4/3)$ where $N = B - D$ \quad (25)

However by using the Bessel function identities these equations for ξ can be mixed to produce the more relevant forms

$$\xi = \left(\frac{\lambda}{2}\right)^{\frac{1}{3}} \{E(I_{-\frac{1}{3}} + I_{+\frac{1}{3}}) + F(I_{-\frac{1}{3}} - I_{+\frac{1}{3}})\} e^{im\theta} \quad \text{subsonic side}$$

$$\xi = \left(\frac{\lambda}{2}\right)^{\frac{1}{3}} R\{Ge^{i(m\theta+\gamma)} + He^{i(m\theta-\gamma)}\} e^{im\theta} \quad \text{supersonic side} \tag{26}$$

The function associated with the E coefficient increases with distance from the sonic line and represents the asymptote on the subsonic side. The function associated with the F coefficient decreases with distance from the sonic line and represents the required incremental solution.

The lines $m\theta \pm \gamma = $ constant are known as the pseudo-characteristic lines. The function associated with the G coefficient is a harmonic function parallel to the upstream pseudo-characteristic lines and represents the asymptotic form on the supersonic side. The function associated with the H coefficient is a harmonic function parallel to the downstream characteristic lines and represents the required incremental solution.

The coefficients are related by:-

$$\left. \begin{array}{ll} A = E + F & 2Gc = C - e^{-i\pi/3}D \\ B = E - F & 2H/c = C - e^{+i\pi/3}D \\ E = (A+B)/2 & \mu C = e^{i\pi/2}G - e^{-i\pi/2}H \\ F = (A-B)/2 & \mu D = e^{i\pi/6}G - e^{-i\pi/6}H \end{array} \right\} \tag{27}$$

where $\mu = e^{i\pi/3} - e^{-i\pi/3}$.

From these relations we can obtain

Coefficient matrices   Associations

$$\begin{bmatrix} A \\ B \\ C \\ D \end{bmatrix} = \begin{bmatrix} +2b & +a & +a & -b \\ +2a & -a & -a & +b \\ +2b & +a & -b & -b \\ +2a & -a & -a & -a \end{bmatrix} \times \begin{bmatrix} E \\ 2Gc \\ M \\ N \end{bmatrix} \begin{array}{l} \leftarrow \text{Subsonic} \\ \leftarrow \text{Upstream supersonic} \\ \leftarrow \Delta\xi \\ \leftarrow \Delta\left(\frac{\partial\xi}{\partial n}\right) \end{array} \tag{28a}$$

$$\begin{bmatrix} E \\ \sqrt{3}F \\ G \\ H \end{bmatrix} = \begin{bmatrix} 1 & 0 & 0 & 0 \\ -e^{i\pi/2} & e^{i\pi/3} & e^{i\pi/6} & -e^{-i\pi/6} \\ 0 & 1 & 0 & 0 \\ 2e^{-i\pi/6} & e^{2i\pi/3} & -e^{-i\pi/6} & -e^{-i\pi/6} \end{bmatrix} \times \begin{bmatrix} E \\ G \\ M \\ N \end{bmatrix} \quad (28\,b)$$

where
$$a = e^{i\pi/6}/\sqrt{3} = \tfrac{1}{2} + \frac{im}{2\sqrt{3}},$$

$$b = e^{-i\pi/6}/\sqrt{3} = \tfrac{1}{2} - \frac{im}{2\sqrt{3}}, \quad c = e^{i\pi/6},$$

$$M = A-C, \quad N = B-D$$

Most applications of the hodograph transformation to aerofoil problems have used particular solutions for the space potential, or the stream function, which are expressed as harmonic functions of ϕ only, multiplied by real functions of w. With such particular solutions there is no distinction between upstream and downstream. While this symmetric form is appropriate to reversible shock free flow it is not necessarily appropriate to relaxation near a sonic line upstream of a shock.

Contours of the particular solutions

For the approximate form of the field equation an asymmetric form of solution of the type

$$\xi = R \cos(m\theta + \gamma + \alpha)$$

with $R = R(w)$ and $\gamma = \gamma(w)$ is derived from the Bessel function of the third kind. This asymmetric form can be regarded as the result of adding two symmetric particular solutions. Conversely the asymmetric form above may be added to its mirror image $R \cos(m\theta - \gamma + \alpha)$ to form a symmetric solution. By adjusting the value of α the complete cycle of symmetric solutions can be obtained. The combinations with α selected for $\partial \xi/\partial w = 0$ and for $\xi = 0$ at $\eta = 0$ are shown in Figs. 11a and 11b as contours of ξ in the θ, η plane.

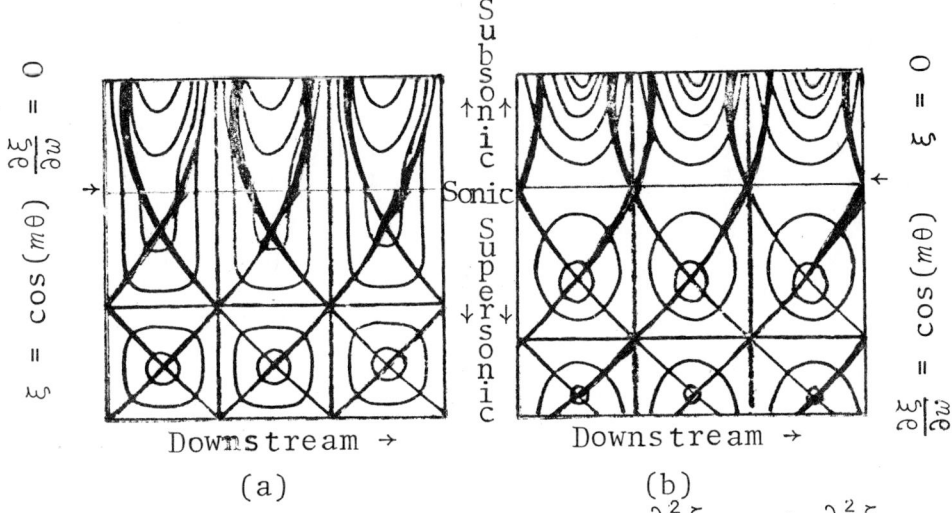

Fig. 11. Particular solutions of $\frac{\partial^2 \xi}{\partial \eta^2} + m^2 \eta \frac{\partial^2 \xi}{\partial \theta^2} = 0$ of the form $\xi = g(w)\cos(m\theta)$

ξ_1 represents the continuation through the sonic line of the asymmetric form of solution on the supersonic side. The contours of this solution are shown in (Fig. 12(a)). The amplitude of the harmonic function of θ and the phase $(\gamma + \infty)/m$ both increase monotonically with η. In graphical terms the ridge lines start to turn downstream before the sonic line is reached.

ξ_2 is the result of adding this asymmetric solution to its mirror image. The contours are shown in Fig. 12(b) together with the ridge lines of the component solutions.

The contours of the two discontinuous solutions ξ_3 and ξ_4 are shown in Figs. 12(c) and (d). These may be regarded as derived from the supersonic side of the asymmetric solution and the subsonic side of the symmetric solution. By adjusting the relative phases and amplitudes we can obtain zero discontinuity of $\partial \xi/\partial w$ or zero discontinuity of ξ. For both the discontinuous solutions the subsonic asymptote and the upstream supersonic asymptotes are zero. Thus the continuous solutions permit matching of the asymptotes

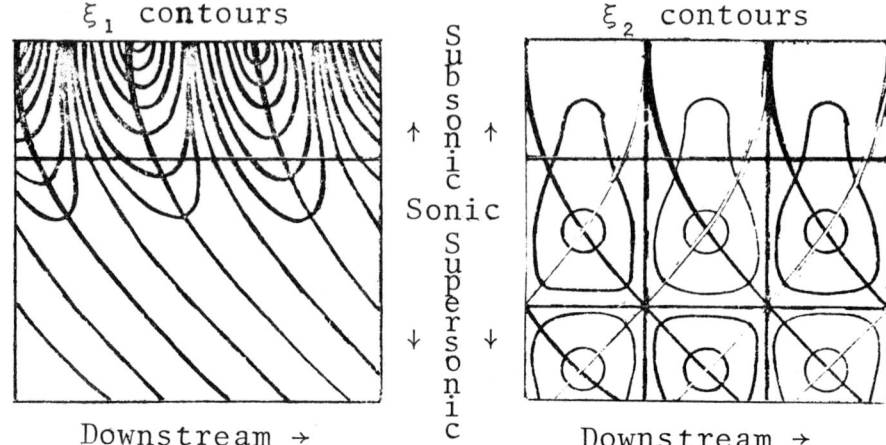

(a) Subsonic disturbance (b) Upstream supersonic disturbance

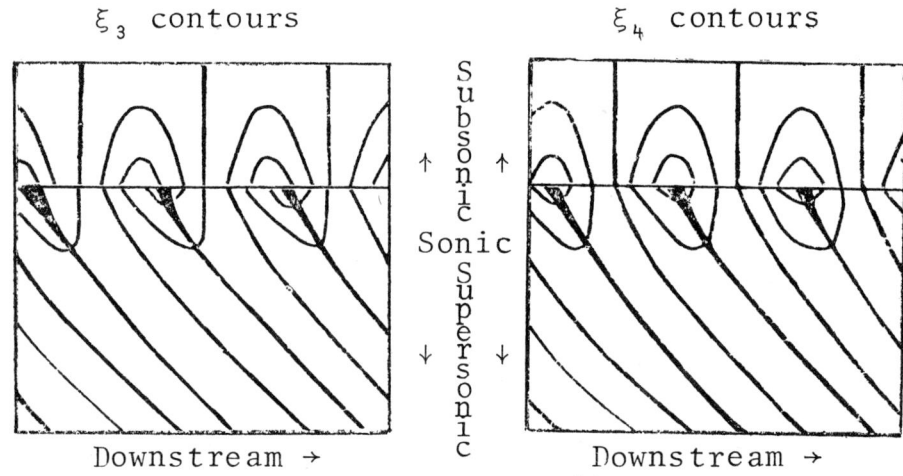

(c) ξ discontinuity (d) $\frac{\partial \xi}{\partial \eta}$ discontinuity

Fig. 12. The solution of $\frac{\partial^2 \xi}{\partial \eta^2} + m^2 \eta \frac{\partial^2 \xi}{\partial \theta^2} = 0$ for each value of m is
$\xi = E\xi_1 + G\xi_2 + M\xi_3 + N\xi_4$

and the discontinuous solutions can provide incremental solutions to neutralise discontinuities without disturbing the asymptotes.

Arbitrary discontinuity functions

Ideally we wish to obtain inner solutions of the nonlinear field equations which match the asymptotic form on both sides of the sonic line. To linearise the problem we assume that the required solution can be expressed as a small perturbation on a known ϕ distribution. By marching towards the sonic line from both sides we can obtain a ϕ distribution which satisfies the field equation on both sides but with an arbitrary discontinuity of ϕ and $\nabla\phi$ along the sonic line. We want to derive the incremental ϕ and grad ϕ along the sonic line which must be added to the known discontinuous ϕ distribution to obtain ϕ and grad ϕ of the required inner solution.

In the hodograph plane we consider an arbitrary discontinuity in ξ and in $\partial\xi/\partial\eta$

$$\Delta\xi = f_1(\theta)$$
$$\Delta\frac{\partial\xi}{\partial\eta} = f_2(\theta) \qquad (29)$$

The incremental solution must have the given discontinuities without change to the required asymptotic form. We consider first the case of $f_2(\theta)$ zero.

The function $f_1(\theta)$ can be expressed as the sum of its harmonic components

$$f_1(\theta) = \frac{1}{2\pi}\int_{-\infty}^{+\infty} e^{im\theta} g_1(m)\, dm$$

where $\qquad (30)$

$$g_1(m) = \int_{-\infty}^{+\infty} e^{im\theta} f_1(\theta)\, d\theta$$

For each harmonic component of f_1 there is harmonic component of ξ at the sonic line approached from the subsonic side. With m tacitly assumed to be positive these are related for E, G,

N all zero by the third column of the coefficient matrix

i.e., with
$$\Delta\xi = \frac{M}{\Gamma(\frac{2}{3})} e^{im\theta}$$

we have the associated harmonic component for ξ at $\eta = 0$ approached from the subsonic side

$$\xi = \frac{A}{\Gamma(\frac{2}{3})} e^{im\theta} = \left(\frac{A}{M}\right)\left(\frac{Me^{im\theta}}{\Gamma(\frac{2}{3})}\right) = \frac{e^{i\pi/6}}{\sqrt{3}} \Delta\xi \qquad (31)$$

By repeating the derivation of the coefficient matrix *ab initio* with m assumed to be negative and still fitting along the upstream pseudocharacteristic and comparing results we obtain a form valid for m positive or negative

$$\xi = \frac{e^{i\,\text{sign}(m)\,\pi/6}}{\sqrt{3}} \Delta\xi \qquad (32)$$

i.e., for all m the real part of ξ is a cosine function with a phase shift of 30° in the upstream direction relative to $\Delta\xi$.

Now

$$e^{i\,\text{sign}(m)\,\pi/6} = \frac{\sqrt{3}}{2} + \frac{i\,\text{sign}(m)}{2} \qquad (33)$$

Since the formula for the ratio of the harmonic component of ξ to the harmonic component of $\Delta\xi$ is valid for all harmonic components we can sum the harmonic components of ξ

$$\xi(\theta) = \frac{1}{2\pi} \int_{-\infty}^{+\infty} e^{im\theta} (\tfrac{1}{2} + g_0(m)) g_1(m) \, dm \qquad (34)$$

where $g_0(m) = \dfrac{i\,\text{sign}(m)}{2\sqrt{3}}$

From the Fourier transformation identities the transform of the resultant of two convolution factors is the product of the transforms of the convolution factors. From this identity we obtain the inverse transform of the products $\tfrac{1}{2} + g_0(m)$ and $g_1(m)$ as a convolution integral along the sonic line

i.e.,
$$\xi(\theta) = \frac{\Delta\xi}{2} + \int_{-\infty}^{+\infty} k_0(\theta_1 - \theta) f_1(\theta_1) d\theta_1 \qquad (35)$$

where $k_0(\theta) = \frac{1}{2\pi} \int_{-\infty}^{+\infty} e^{im\theta} g_0(m) dm$

Again from a harmonic component of ξ we can obtain the corresponding harmonic component of $\partial\xi/\partial\eta$

$$\frac{d\xi}{d\eta} = \frac{1}{2} \frac{\Gamma(\frac{2}{3})}{\Gamma(\frac{4}{3})} \left(\frac{|m|}{3}\right)^{\frac{2}{3}} \xi \qquad (36)$$

Summing such harmonic components we obtain

$$\frac{\partial\xi}{\partial\eta} = \frac{1}{2}\frac{\Gamma(\frac{2}{3})}{\Gamma(\frac{4}{3})} \fint_{-\infty}^{+\infty} k_1(\theta - \theta_1) \xi(\theta_1) d\theta_1 \qquad (37)$$

where

$$k_1(\theta) = \frac{1}{2\pi} \int_{-\infty}^{+\infty} e^{im\theta} \left(\frac{|m|}{3}\right)^{\frac{2}{3}} dm$$

where $\fint$ denotes the generalised principal part integral.

Similarly for $f_1(\theta) = 0$ and a general $f_2(\theta)$ discontinuity in $\partial\xi/\partial\eta$ we have

$$f_2(\theta) = \frac{1}{2\pi} \int_{-\infty}^{+\infty} e^{im\theta} g_2(m) dm \qquad (38)$$

where

$$g_2(m) = \int_{-\infty}^{+\infty} e^{im\theta} f_2(\theta) d\theta$$

and with E, G, M all zero we obtain from the fourth column of the coefficient matrix the value of $\partial\xi/\partial\eta$ as the sonic line is approached from the subsonic side

$$\frac{\partial\xi}{\partial\eta} = \frac{e^{-i\,\text{sign}(m)\pi/6}}{\sqrt{3}} \Delta\left(\frac{\partial\xi}{\partial\eta}\right) \qquad (39)$$

and
$$\xi = \frac{2\Gamma(\frac{4}{3})}{\Gamma(\frac{2}{3})} \left(\frac{|m|}{3}\right)^{-\frac{2}{3}} \frac{\partial \xi}{\partial \eta} \qquad (40)$$

By summing the harmonic components of $\partial\xi/\partial\eta$ and ξ we obtain

$$\frac{\partial \xi}{\partial \eta} = \tfrac{1}{2}\Delta\left(\frac{\partial \xi}{\partial \eta}\right) - \int_{-\infty}^{+\infty} k_0(\theta - \theta_1) f_2(\theta_1) d\theta_1 \qquad (41)$$

$$\xi = \frac{2\Gamma(\frac{4}{3})}{\Gamma(\frac{2}{3})} \int_{-\infty}^{+\infty} k_2(\theta - \theta_1) \frac{\partial \xi}{\partial \eta} d\theta_1 \qquad (42)$$

where k_0 is as before

and
$$k_2(\theta) = \int_{-\infty}^{+\infty} e^{im\theta} \left(\frac{|m|}{3}\right)^{-\frac{2}{3}} dm \qquad (43)$$

CONCLUSIONS

Rapid relaxation of the transonic flow about a lifting aircraft is still a long way away. However interesting techniques have been examined for solving parts of the complete problem. Previous exercises showed that elliptic solvers are effective in subsonic fields when the discontinuity distributions are constant. This applied to two-dimensional and three-dimensional non-lifting and lifting cases. We already have computing sequences based on spline modes which provide velocity component influence matrices associated with the source-doublet discontinuity functions which may be used to neutralise any given distribution of error in normal velocity over the wetted surfaces of an aircraft. Marching sequences can be devised for supersonic regions. The far field can be represented by the field induced by point sources. We still have to develop linkages between these computing sequences.

A semi-direct method in which an approximate elliptic solver is used has been applied in Glauert space with a local over-relaxation factor. This gives rapid local convergence where the velocity is near the free stream value. By adjusting the elliptic solver computing sequence to take account of local velocity it may be possible to achieve the same rapid rate of local convergence over a wider velocity range.

Based on Laplacian field concepts the wall relaxation sequence also gives rapid convergence when the compressibility effects are small. However this simple formula gives a one step delay in the coupling with the far field, $i.e.$, ϕ_1 gives a fair approximation to the solution with images and ϕ_2 gives a fair approximation to the solution without images. This delay destabilises the relaxation with strong nonlinear compressibility effects.

The sonic line analysis breaks away from Chaplygin's particular solutions for which there is no distinction between forward and reversed flow solutions. Thus a physical distinction between upstream and downstream has been reflected in the mathematical model. However the effect of the near sonic approximation, the finite length of the sonic line, the presence of an adjacent solid surface, the downstream shock, the numerical implementation and the three-dimensional effects have not been taken into account. The sonic line problem applies to any matching technique and is not special to the realisation of the hybrid method used at Weybridge. There are many engineers who would be interested in a more rigorous analysis of this problem by mathematicians. Extension to three-dimensional sonic surfaces must be empirical as the field equation in hodograph space does not appear to be amenable to the present approach.

REFERENCES

1. Guderley, K.G., "The Theory of Transonic Flow," The International Series of Monographs in Aeronautics and Astronautics, **3**, Pergamon Press (1962).
2. Watson, G.N., "Theory of Bessel Functions," Cambridge (1952).

3. Roberts, A. and Ruddle, K., "Computation of incompressible flow about bodies and thick wings using the spline mode system," BAC Aerodynamics Report MA 19 (1972).

APPENDIX I

ELLIPTIC SOLVERS

An elliptic solver is an algorithm for computing the direct solution of a Poisson equation. Such solvers may be used within an iterative loop to obtain solutions of nonlinear elliptic differential equations. Used in this way they may be regarded as completing the progression of point relaxation, line relaxation, area relaxation and volume relaxation. Alternatively the action of an elliptic solver may be regarded as integration with Green's function in two or three dimensions.

A primitive form of elliptic solver is provided by literal expansion of the discrete Fourier transform identities as follows.

An array A is periodic (or polycyclic) if

$$A(x_1 + N, x_2) = A(x_1, x_2 + N) = A(x_1, x_2)$$

The discrete Fourier transform of such an array is also a periodic array with the same cycle size N and is denoted by $\underline{J}A$ where

$$\widehat{\underline{J}A}(w_1, w_2) = \frac{1}{N} \sum_{x_1=1}^{N} \sum_{x_2=1}^{N} \left\{ e^{-\lambda(w_1 x_1 + w_2 x_2)} A(x_1, x_2) \right\}$$

where $\lambda = 2\pi i/N$.

The discrete inverse Fourier transformation is provided by

$$A(x_1, x_2) = \frac{1}{N} \sum_{w_1=1}^{N} \sum_{w_2=1}^{N} \left\{ e^{+\lambda(w_1 x_1 + w_2 x_2)} \widehat{\underline{J}A}(w_1, w_2) \right\}$$

The "product" of two periodic arrays A and B is also a periodic array denoted by $A \times B$ as is

defined by
$$\widetilde{A \times B}(x_1, x_2) = A(x_1, x_2) B(x_1, x_2)$$
i.e., for each x_1, x_2 combination the values are multiplied. The resultant of two periodic arrays A and B is also a periodic array denoted by $A \otimes B$ and is defined by

$$\widetilde{A \otimes B}(x_1, x_2) = \frac{1}{N} \sum_{y_1=1}^{N} \sum_{y_2=1}^{N} A(y_1, y_2) B(x_1 - y_1, x_2 - y_2)$$

Then from the discrete Fourier transform identities
$$\underline{J}(A \otimes B) = (\underline{J}A) \times (\underline{J}B)$$
$$\underline{J}(A \times B) = (\underline{J}A) \otimes (\underline{J}B)$$

Let P, F, G, be periodic arrays with all elements of P zero except for

$$P(0, 0) = -4N$$
$$P(0, 1) = N$$
$$P(0, -1) = N$$
$$P(1, 0) = N$$
$$P(-1, 0) = N$$

and all periodic images of these non-zero elements near the origin.

Then
$$\widetilde{P \otimes F}(x_1, x_2) = +F(x_1, x_2 - 1) + F(x_1, x_2 + 1)$$
$$+ F(x_1 - 1, x_2) + F(x_1 + 1, x_2)$$
$$- 4F(x_1, x_2)$$

Thus if F is the array of spot values of the function f then $P \otimes F$ may be recognised as the first order numerical approximation to $\nabla^2 f$.

Let $P \otimes F = G$

Then $\widetilde{\underline{J}P}(w_1, w_2) \widetilde{\underline{J}F}(w_1, w_2) = \widetilde{\underline{J}G}(w_1, w_2)$

We then have $F = \underline{J}^{-1}(\underline{J}F)$

where $\vec{JF}(w_1,w_2) = \dfrac{\vec{JG}(w_1,w_2)}{\vec{JP}(w_1,w_2)}$

Thus the solution of $P \otimes F = G$ to obtain F for a given G is the numerical approximation to the solution of the Poisson equation

$$\nabla^2 f = g$$

to obtain f for a given g. The numerical result is then obtained by finding $\underline{J}G$ from G, finding $\underline{J}F$ from

$$\vec{JG}(w_1,w_2)/\vec{JP}(w_1,w_2)$$

and finally finding F from $\underline{J}F$. This compound operation is expressed formally by $F = \underline{H}G$.

Algebraically equivalent computing sequences for determining $\underline{J}G$ and $\underline{J}^{-1}F$ can be obtained using one of the fast Fourier algorithms. A typical algorithm of this type is based on the structure of the B array defined by

$$B(r + mp, c) = \sum_{q=0}^{2^c-1} e^{\lambda qpm} B(r + mq, 0)$$

where

$$m = 2^{d-c}; \quad \lambda = 2\pi i/n; \quad n = 2^d$$

Let $B(r,0)$ represent an array before transformation then $B(p,d)$ represents an array after transformation,

i.e., $$B(p,d) = \sum_{q=0}^{n-1} e^{\lambda pq} B(q,0)$$

The algorithm is derived by separating the odd and even values of q in the summation to obtain $B(r + mp, c)$.

The algorithm is
for $c = 1 \to d$

for $r = 0 \to m-1$
for $p = 0 \to \frac{n}{2m}-1$

$$B(r + mp) = B(r + 2mp, c - 1) + e^{\lambda mp} B(r + m + 2mp, c - 1)$$

$$B(r + mp + \frac{n}{2}) = B(r + 2mp, c - 1) - e^{\lambda mp} B(r + m + 2mp, c - 1)$$

The number of complex products is proportional to nd rather than n^2.

Even faster computing sequences can be used if we only require an approximate solution of the Poisson equation within an iterative loop. The reduction in computing time per step is then partially offset by a slight increase in the number of steps required to reach convergence. The approximate inversion of the Poisson equation is based on a cascade of numerical operators, based on a set of progressively sparser grids. With errors in the solution of the order of 5% the computing time for the solver is of the same order as that to compute the numerical approximation $\underset{\sim}{L}F$ to the error the field equation $\underset{\sim}{L}\phi$.

Both the exact and the approximate elliptic solvers discussed above use operators which are only applicable when the operand is a periodic array. The physical problem is not periodic. Periodicity can be imposed on the near field plane by introducing reflections of the ϕ distribution in the working section walls.

APPENDIX II

THE NUMERICAL TREATMENT OF DISCONTINUITY FUNCTIONS

Let ϕ be Laplacian through each of a set of volumes separated by interfaces. By using Green's Theorem the value of ϕ can be expressed in terms of integrals with Green's function of the discontinuities in ϕ and $\partial\phi/\partial n$ across the interfaces

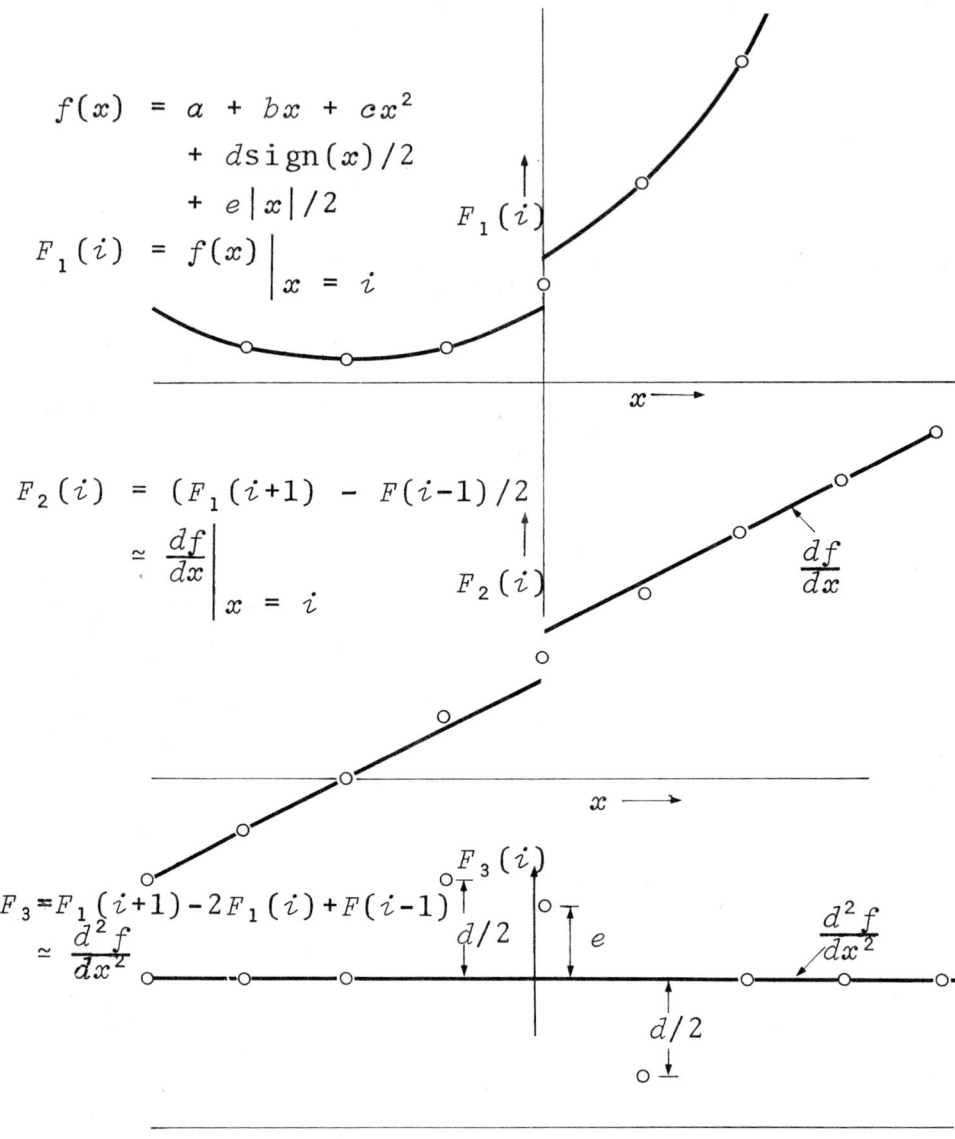

Sign(x)/2 is a doublet-like discontinuity
$|x|/2$ is a source-like discontinuity

Fig. 13. The errors in the numerical second derivatives are local to the discontinuity surface

between volumes.

In the discrete analogue the elliptic solver is used to invert $P \otimes F = G$ where G is zero except near interfaces. In special cases the result of this inversion can be computed analytically. The result can then be compared with the analytic integrals to obtain the error in computing the influence of discontinuity functions. (For first order numerical operators the inversion can be performed in physical space but for higher order operators the use of Fourier transform space is recommended.)

Let
$$\phi = (A + B \text{sign}(x_3))\cos(w_1 x_1)\cos(w_2 x_2) e^{-\mu|x_3|}$$
This represents a three-dimensional function with discontinuity distributions in the plane $x_3 = 0$:
$$\Delta\phi = 2B\cos(w_1 x_1)\cos(w_2 x_2)$$
$$\Delta\frac{\partial\phi}{\partial n} = -2\mu A\cos(w_1 x_1)\cos(w_2 x_2)$$
This ϕ distribution will satisfy $\nabla^2\phi = 0$ (except in $x_3 = 0$) when $\mu = \mu_1 = \sqrt{w_1^2 + w_2^2}$. This special value of μ is then the value for the analytic solution of the Laplace equation for the special discontinuity functions above.

Let F be the array of spot values of ϕ at integer values of x_1, x_2, x_3 for an arbitrary value of μ. Fig. 13 shows the effect of a discontinuity of ordinate and derivative on the numerical second derivative using a three point operator. The discontinuity in derivative produces an aberration at the discontinuity while the discontinuity of ordinate produces an aberration at two adjacent points.

Let $G = P \otimes F$ where $P \otimes$ is now the numerical equivalent in three-dimensions of the ∇^2 operator using 3 points in each direction. By substitution we obtain for $|x_3| > 1$ (*i.e.*, away from the region in which the operator straddles the discontinuity plane)

$$G = 4\{\sinh^2(\mu/2) - \sin^2(w_1/2) - \sin^2(w_2/2)\}F$$

We note that $G = 0$ for $|x_3| > 1$ when

$$\mu = \mu_2 = 2\sinh^{-1}(\sqrt{\sin^2(w_1/2) + \sin^2(w_2/2)})$$
$$= \sqrt{w_1^2 + w_2^2}\,(1 + \text{order } w_i w_j)$$

Thus for w_1 and w_2 both small (*i.e.*, with wavelengths of $\Delta\phi$ and $\Delta\partial\phi/\partial n$ large compared with the grid interval) the value of μ for which $P \otimes F = 0$ away from $x_3 = 0$ is nearly the value of μ for which $\nabla^2\phi = 0$ away from $x_3 = 0$. However, with $\mu = \mu_2$ we obtain:

$$G = -2A\sinh(\mu)\cos(w_1 x_1)\cos(w_2 x_2)$$
$$= \Delta\left(\frac{\partial\phi}{\partial n}\right)(1 + \text{order } w_i w_j) \qquad \text{at } x_3 = 0$$

and

$$G = -B\,\text{sign}(x_3)\cos(w_1 x_1)\cos(w_2 x_2)$$
$$= -\tfrac{1}{2}\text{sign}(x_3)\Delta\phi(1 + \text{order } w_i w_j)\,\text{at } x_3 = \pm 1$$

By scaling throughout to remove the factor near unity we then have

$$G = \Delta\left(\frac{\partial\phi}{\partial n}\right) \qquad \text{at } x_3 = 0$$
$$G = -\tfrac{1}{2}\text{sign}(x_3)\Delta\phi \qquad \text{at } x_3 = \pm 1$$
$$G = 0 \qquad \text{at } |x_3| > 1$$

The solution of $P \otimes F = G$ is then given in the form

$$F = \left\{\begin{array}{l} A(1 + \text{order } w_i w_j) \\ + B\,\text{sign}(x_3)(1 + \text{order } w_i w_j) \end{array}\right\}\cos(w_1 x_1) \times$$

$$\cos(w_2 x_2)e^{-\sqrt{w_1^2 + w_2^2}|x_3|}(1 + \text{order } w_i w_j)$$

Thus provided that wavelengths are large compared with the grid interval the solution $P \otimes F = G$ is a fair approximation to the analytic field induced by the same discontinuity functions $\Delta\phi$ and $\Delta\partial\phi/\partial n$.

We now consider the extension from the direct solution of the Poisson equation to the semi-direct solution of the nonlinear equation.

We start with the equation

$$P \otimes F_{t+1} = G_{t+1}$$

in which G_{t+1} is regarded as known. Subtracting $P \otimes F_t$ from both sides we obtain

$$P \otimes F_{t+1} - P \otimes F_t = - P \otimes F_t + G_{t+1}$$

which is equivalent to

$$P \otimes (F_{t+1} - F_t) = - \underset{\sim}{L} F_t$$

where for Laplacian fields

$$\underset{\sim}{L} F_t = P \otimes F_t - G_{t+1}$$

In the final steady state we have $F_{t+1} = F_t$ and the distinction between iteration steps disappears.

Then $\underset{\sim}{L} F_\infty = P \otimes F_\infty - G_\infty$

Now away from discontinuity surfaces $P \otimes F_\infty$ is the second order approximation to $\nabla^2 \phi$. The equation $LF_\infty = 0$ over the complete cyclic field is similarly the second order approximation to $\nabla^2 \phi = 0$ on both sides of the discontinuity surface. Thus LF_∞ may be regarded as the second order approximation to $\nabla^2 \phi$ in the presence of a discontinuity. The array $-G_\infty$ is then the correction for discontinuity effects to the discrete approximation to $\partial^2 \phi / \partial X_3^2$ used in computing $\nabla^2 \phi$.

The currently estimated discontinuity distributions of ϕ and $\partial \phi / \partial n$ are computed in the far field plane and are used to compute G_{t+1} representing the currently required discontinuity distributions in the near field plane. The action of the elliptic solver is then to add to F_t the Laplacian field $F_{t+1} - F_t$ induced by $G_{t+1} - G_t$. The result is that F_{t+1} is a Laplacian field with the required discontinuity distribution over $X_3 = 0$.

For the compressible flow equations we define $\underset{\sim}{L}F_t$ as the numerical approximation to $\underset{\sim}{L}\phi_t$ in which each first and second derivative has correction terms near discontinuity surfaces. These corrections are computed from the discontinuity functions required for F_{t+1}. The action of the elliptic solver is then to add a Laplacian field induced by $G_{t+1} - G_t$ in addition to the solution of the Poisson equation. Near the walls of the working section the equation $\underset{\sim}{L}\phi = 0$ is approximately the Prandtl-Glauert equation and near the walls the additional field does not disturb the field equation.

The action of the elliptic solver is the inverse of the action of the convolution operator $P \otimes$ applied over a complete plane. This is true whether or not $P \otimes$ is a good approximation to the operator ∇^2. The consequence of this property is that correcting the numerical derivatives used when representing $\underset{\sim}{L}\phi_t$ for the discontinuities required for ϕ_{t+1} is sufficient to cause the required discontinuities to appear in F_{t+1}.

A TRANSONIC HODOGRAPH THEORY FOR AEROFOIL DESIGN

J.W. Boerstoel

(National Aerospace Laboratory NLR, Amsterdam)

INTRODUCTION

To the aircraft industry aerofoil theory is useful for wing and helicopter rotor design. This is because the mathematical and computational means for analysis and design of two-dimensional flows are more developed than those for three-dimensional flows. Although it may be expected that for transonic wings three-dimensional effects are more important than for thinner wings with effectively subsonic flow (see the analysis of Cheng and Hafez [1]) aerofoil theory is still of great value for transonic wing design, in particular when the aspect ratio is large or moderate, the wing sweep not large and the size of the supersonic regions in the flow can be kept small.

The version of the hodograph method, developed during the last few years at NLR, is used as one of the means for transonic aerofoil design. Its advantage over so called direct methods (pressure distribution computed from given aerofoil geometry) is that the aerofoils computed always have low drag near a design condition (M_∞, C_L) that can be freely chosen. Its main drawback is that the aerofoil shapes and corresponding pressure distribution in the chosen design condition are not known in advance, but are found as a result of a computation.

A transonic aerofoil has to satisfy not only the requirement of low drag near a design point, but also requirements concerning the off-design behaviour (C_L-max, drag-rise and buffet boundaries). When a hodograph method is to be used one can try to translate the off-design requirements into desired properties of the design pressure distribution and the aerofoil shape, and to optimise the aerofoil and design pressure distribution.

The NLR hodograph method is sufficiently accurate and flexible that aerodynamic engineers can perform such optimisations without great effort.

Mathematically, the NLR hodograph method is interesting because it is based on the approximate solution of Tricomi boundary value problems for the mixed elliptic-hyperbolic hodograph equations. In this method approximation by a linear combination of solutions of the hodograph equations is applied. A special error minimisation principle was developed, which was the key to the desired flexibility needed for engineering optimisation of the aerofoils.

In the following sections the NLR hodograph method is outlined and recent computational results are presented.

INTRODUCTION TO HODOGRAPH THEORY

Transonic flows around shock-free aerofoils can be represented by solutions of a linear partial differential equation of elliptic-hyperbolic type for the stream function $\tilde{\Psi}$:

$$\tilde{L}\tilde{\Psi} \equiv PQ\tilde{\Psi}_{\tau\tau} + PQ_\tau \tilde{\Psi}_\tau - \tilde{\Psi}_{\theta\theta} = 0,$$

$$P = -2\tau(1-\tau)^{\gamma/(\gamma-1)}(1-\tfrac{\gamma+1}{\gamma-1}\tau)^{-1}, \quad Q = 2\tau(1-\tau)^{-1/(\gamma-1)} \tag{1}$$

where τ is a velocity parameter depending only on the local Mach number M:

$$\tau = \left\{1 + \frac{2}{(\gamma-1)M^2}\right\}^{-1} \tag{2}$$

and θ is the local flow angle. The differential operator $\tilde{L}$ is elliptic when $M<1$, and hyperbolic when $M>1$. Such solutions are defined on a two sheeted (τ,θ) surface with a point (τ^*,θ^*) as branch point. The (τ,θ) surface is called the hodograph surface.

First consider the structure of the solutions in terms of the hodograph variables and the relation between the physical plane of the flow around the aerofoil and the hodograph surface. (See Fig. 1).

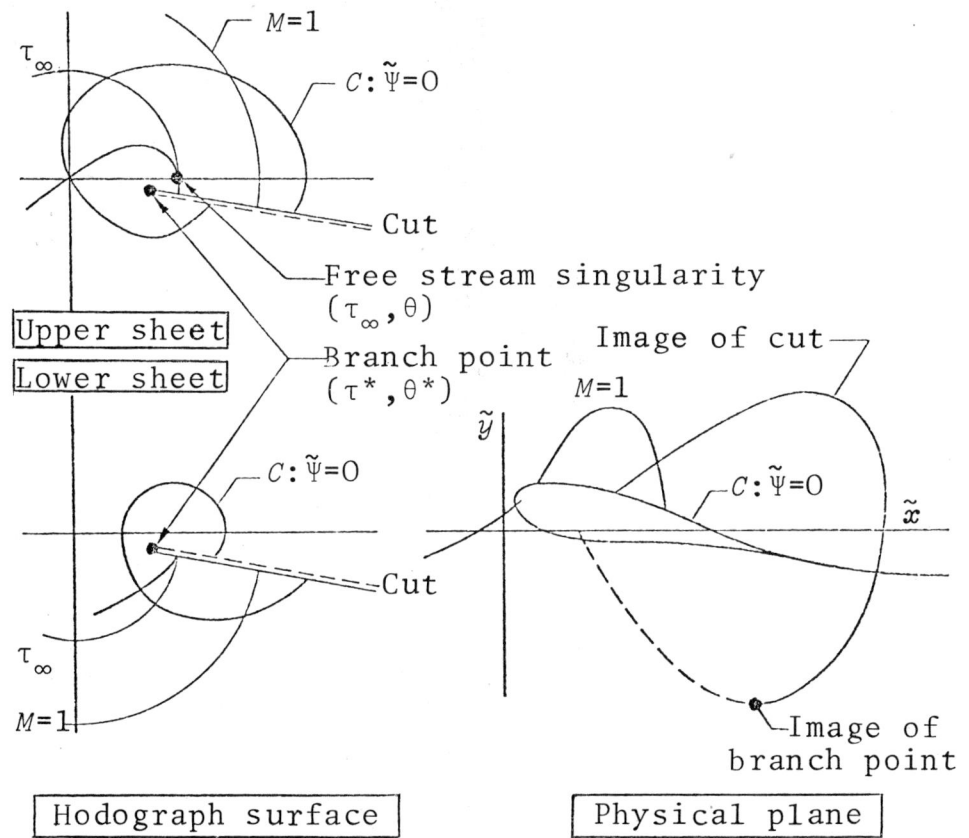

Fig. 1. Relation between hodograph surface and physical plane

(i) (τ,θ) are polar variables on the hodograph surface, with τ as radial variable.

(ii) The hodograph surface may be divided into two sheets by a cut that begins at the branch point (τ^*,θ^*) and extends outward along the radius $\theta = \theta^*$. The sheets are known as upper and lower sheets.

(iii) The image of the aerofoil on the hodograph surface is a closed curve C encircling the branch point, on which the stream function $\tilde{\Psi} = 0$. The exterior of the aerofoil in the physical plane maps onto the interior of the aerofoil image C on the hodograph surface.

(iv) The stream function of the flow has a free stream singularity of known type at infinity in the physical plane, and thus at a point $(\tau_\infty, 0)$ on one of the sheets of the hodograph surface. (τ_∞ is the value of the velocity parameter corresponding to the free stream Mach number M_∞.) The upper sheet is defined as the sheet containing the free stream point $(\tau_\infty, 0)$.

(v) The free streamline $\tilde{\Psi} = 0$ that extends in the physical plane from the front stagnation point to infinity upstream corresponds to a curve on the upper sheet from the origin to $(\tau_\infty, 0)$ where $\tilde{\Psi}$ has its free stream singularity. Similarly, the free streamline $\tilde{\Psi} = 0$ from the trailing edge downstream to infinity maps on to a curve connecting the trailing edge image on the lower sheet with $(\tau_\infty, 0)$ on the upper sheet.

(vi) When a stream function $\tilde{\Psi}$ has been defined, the mapping from the hodograph surface (τ, θ) to the physical plane $\tilde{z} = \tilde{x} + i\tilde{y}$ is given by a relation of the form

$$\tilde{z}(\tau, \theta) = \tilde{M}\tilde{\Psi}(\tau, \theta) \qquad (3)$$

where $\tilde{M}$ is a known linear operator.

When the stream function is given in operational form as a Mellin-Barnes contour integral in a complex ν-plane,

$$\tilde{\Psi}(\tau, \theta) = \text{Im} \int_{-i\infty}^{i\infty} F(\nu) f_\nu(\tau_\infty) \Psi_\nu(\tau) e^{-i\nu\theta - i\nu\pi} d\nu \qquad (4)$$

where:

$\Psi_\nu(\tau)e^{-i\nu\theta}$ is a particular solution of the hodograph equation (1), with $\Psi_\nu(\tau)$ the Chaplygin function [2];

$f_\nu(\tau_\infty)$ is a normalising function, depending on ν and on the value of τ_∞ corresponding through equation (2) with the free stream Mach number M_∞, and defined by Lighthill ([3] p.354);

$F(\nu)$ is the Mellin-Barnes transform of the stream function $\tilde{\Psi}$,

then the mapping from the hodograph surface to the physical plane $\tilde{z}(\tau,\theta) = \tilde{x}(\tau,\theta) + i\tilde{y}(\tau,\theta)$ is given by [4]:

$$\tilde{z} = \tfrac{1}{2}\left\{\frac{\tau_\infty}{\tau}\right\}^{\tfrac{1}{2}} Q \times$$

$$\left[\int_{-i\infty}^{i\infty} F(\nu)f_\nu(\tau_\infty)\left\{\Psi'_\nu(\tau)+\frac{\nu}{2\tau}\Psi_\nu(\tau)\right\}\frac{e^{-i(\nu-1)\theta}}{\nu-1}e^{-i\nu\pi}d\nu \right.$$

$$\left. + \overline{\int_{-i\infty}^{i\infty} F(\nu)f_\nu(\tau_\infty)\left\{\Psi'_\nu(\tau)-\frac{\nu}{2\tau}\Psi_\nu(\tau)\right\}\frac{e^{-i(\nu+1)\theta}}{\nu+1}e^{-i\nu\pi}d\nu}\right] \quad (5)$$

where the prime indicates differentiation with respect to τ and the bar that the complex conjugate of the expression under the bar has to be taken. The Mellin-Barnes integrals are found to exist usually only on a part of one of the sheets of the hodograph surface.

The Tricomi boundary value problem for transonic shock free flow can now be sketched. Assume that the aerofoil image C is given, together with the free stream Mach number M_∞ (and thus also τ_∞) and the location of the branch point (τ^*,θ^*). Assume also that the free stream singularity in

the stream function has been split off by putting
$$\tilde{\Psi} = \tilde{\Psi}_b + \tilde{\Psi}_a \tag{6}$$
where $\tilde{\Psi}_b$ is a given basic stream function with the desired free stream singularity of a subsonic free stream and satisfying the partial differential equation $\tilde{L}\tilde{\Psi}_b = 0$. The stream function Ψ_a must then also satisfy, everywhere inside the aerofoil image C, the partial differential equation
$$\tilde{L}\tilde{\Psi}_a = 0; \tag{7}$$
$\tilde{\Psi}_a$ must also satisfy on the image C the relation
$$\tilde{\Psi}_a + \tilde{\Psi}_b = 0 \tag{8}$$
This implies that a boundary value problem for the stream function $\tilde{\Psi}_a$ must be solved if $\tilde{\Psi}_a$ is not known. Ψ_a is known as the additional stream function.

The boundary value problem can be solved approximately by representing $\tilde{\Psi}_a$ by a finite sum of linearly independent solutions Ψ_{an}:

$$\tilde{\Psi}_a = \sum_{n=1}^{N} c_n \tilde{\Psi}_{an}(\tau, \theta), \tag{9}$$

$$\tilde{L}\tilde{\Psi}_{an} = 0, \quad n = 1(1)N. \tag{10}$$

The partial differential equations are automatically satisfied (by linearity of $\tilde{L}$) while the coefficients c_n are available to satisfy the boundary condition approximately.

When the coefficients c_n in the additional stream function have been found the physical plane variable $\tilde{z}$ can be found with the linear operator $\tilde{M}$:

$$\begin{aligned}\tilde{z} &= \tilde{M}\tilde{\Psi} \\ &= \tilde{M}\tilde{\Psi}_b + \sum_{n=1}^{N} c_n \tilde{M}\tilde{\Psi}_{an}\end{aligned} \tag{11}$$

To avoid overdetermination of the boundary value problem, the boundary condition $\tilde{\Psi}_a + \tilde{\Psi}_b = 0$ (see (8)) should not be prescribed on a segment of curve C in the supersonic part of the hodograph surface, that lies between two characteristics of the differential operator $\tilde{L}$ intersecting each other in an arbitrary point on the sonic line (see Fig. 2).

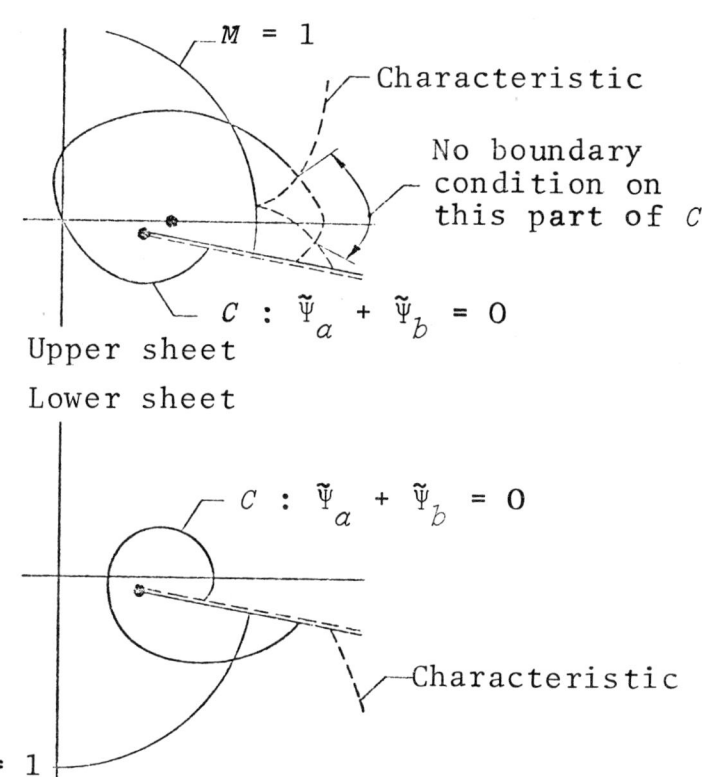

Given: M_∞, (M^*, θ^*), $\tilde{\Psi}_b$ and C

Required: $\tilde{L}\tilde{\Psi}_a = 0$

$\tilde{\Psi}_a + \tilde{\Psi}_b = 0$ on C

Fig. 2. Tricomi boundary value problem for additional stream function $\tilde{\Psi}_a$

On page 352 remarks are made as to how this condition is satisfied in practice.

CONSTRUCTION OF BASIC AND ADDITIONAL STREAM FUNCTIONS

The construction of the basic and additional stream functions is a lengthy mathematical process, which is not discussed in detail. Only the essential steps are outlined. A few details of the Lighthill-Nieuwland transformation technique for some of the functions $\tilde{\Psi}_{an}$ are given, however.

A basic solution $\tilde{\Psi}_b$ for transonic flows can be derived with the Lighthill-Nieuwland transformation technique [4] from a corresponding solution for incompressible flows. The basic solution Ψ_b for incompressible flows is defined as the sum of five terms:

$$\Psi_b \equiv \operatorname{Im} \phi_b \qquad (12)$$

$$\phi_b \equiv \frac{i\Gamma}{2\pi} \left\{ \tfrac{1}{2} \xi_\infty^{-2} \zeta^{*-1} \phi_{dip} + \phi_{ln} \right\} + \phi_e - \phi_{ec} - \phi_{bc} \qquad (13)$$

where Γ is the circulation of the flow, ζ^* the complex conjugate velocity in the branch point of the hodograph of the incompressible flow, and the other variables are defined below (ζ complex conjugate velocity of incompressible flow):

$$\xi = \left(1 - \frac{\zeta}{\zeta^*}\right)^{\frac{1}{2}} \qquad (14)$$

$$\xi_\infty = \left(1 - \frac{1}{\zeta^*}\right)^{\frac{1}{2}} \qquad (15)$$

$$\phi_{dip} = \left(1 - \frac{\xi}{\xi_\infty}\right)^{-1} \qquad (16)$$

$$\phi_{ln} = \int^{(-\zeta)} \phi_{dip}\, d(-\zeta) = -\ln\left(1 - \frac{\xi}{\xi_\infty}\right) - \frac{\xi}{\xi_\infty} \qquad (17)$$

$$\phi_e = \left(1 - \frac{\zeta}{\zeta^*}\right)^{\frac{1}{2}} \left(1 - \varepsilon\frac{\zeta}{\zeta^*}\right)^{-\frac{1}{2}} + \left(1 - \frac{\zeta}{\zeta^*}\right)^{-\frac{1}{2}} \left(1 - \varepsilon\frac{\zeta}{\zeta^*}\right)^{\frac{1}{2}} \quad (18)$$

$$\phi_{ec} = (1 - \varepsilon)^{\frac{1}{2}} \xi^{-1} \quad (19)$$

$$\phi_{bc} = \frac{i\Gamma}{2\pi\xi_\infty}(-\xi) \quad (20)$$

The terms with ϕ_{dip} and ϕ_{1n} are intended to guarantee that Ψ_b has the desired singular behaviour at the free stream singularity in incompressible flow. The other terms are chosen so that the basic solution becomes that of the flow over a (thin) ellipse-like shape. It can also be seen that the two sheeted nature of the hodograph surface in incompressible flow has been taken into account by removing the branch point by a conformal mapping from the hodograph surface to a ξ-plane.

The basic solution for transonic flows is derived from the incompressible singular solution using (with simplifying modifications) the Lighthill-Nieuwland transformation technique separately for each of the five terms in (13). The final result is a large number of complicated series representations for $\tilde{\Psi}_b$ and $\tilde{Z}_z$ which are one valued on the two sheeted hodograph surface for the transonic flows.

Similarly, the linearly independent regular solutions $\tilde{\Psi}_{an}$ for transonic flows that occur in the additional regular stream function $\tilde{\Psi}_a$ (see (9)) are defined with the Lighthill-Nieuwland transformation technique from corresponding regular solutions for incompressible flow. The incompressible regular solutions chosen are the real and imaginary parts of the expressions

$$\xi^m = \left(1 - \frac{\zeta}{\zeta^*}\right)^{\frac{1}{2}m}, \quad m = 0,1,2,\ldots \quad (21)$$

The most important details of the transformation technique are illustrated with an example.

When m is odd, (21) has the Mellin-Barnes integral representation [5]:

$$\phi_m = \left(1 - \frac{\zeta}{\zeta^*}\right)^{\frac{1}{2}m} = \frac{1}{2\pi i} \int_{-i\infty}^{i\infty} \frac{\Gamma(-\frac{1}{2}m+\nu)\Gamma(-\nu)}{\Gamma(-\frac{1}{2}m)} (\zeta^*)^{-\nu} \zeta^{\nu} e^{-i\nu\pi} d\nu \quad (22)$$

$$|\arg(\zeta/\zeta^*) - \pi| < \pi,$$

where Γ is the gamma function. This formula is first transformed into a solution of the hodograph equation (1) by replacing the factor ζ^{ν} by

$$e^{-\nu s_{\infty}} \Psi_{\nu}(\tau) e^{-i\nu\theta} \quad (23)$$

where s_{∞} depends on τ_{∞} by a formula given, for example, in [2] equation (6). The resulting integral

$$\tilde{\phi}_m = \frac{1}{2\pi i} \int_{-i\infty}^{i\infty} \frac{\Gamma(-\frac{1}{2}m+\nu)\Gamma(-\nu)}{\Gamma(-\frac{1}{2}m)} (\zeta^*)^{-\nu} e^{-\nu s_{\infty}} \Psi_{\nu}(\tau) e^{-i\nu(\theta+\pi)} d\nu \quad (24)$$

can be shown to converge if (use the asymptotic estimates for large $|\nu|$ of the Chaplygin functions, given in [2], and Stirling's asymptotic formula for the gamma functions):

$$|\theta - \theta^* + \pi| < \pi \iff \theta^* < \theta < 2\pi + \theta^*$$
$$\theta^* = -\arg \zeta^* \quad (25)$$

Now evaluate the Mellin-Barnes integral (24) by Mittag-Leffler's method (see, for example [5], section 7.4) into series expansions. The following three series are found:

$$S_m^{(1)} = \sum_{n=0}^{\infty} \frac{\Gamma(-\frac{1}{2}m+n)}{\Gamma(-\frac{1}{2}m)\, n!} \zeta^{*-n} e^{-n s_{\infty}} \Psi_n(\tau) e^{-in\theta}, \quad 0 \leq \tau \leq \tau^* \quad (26)$$

$$S_m^{(2)} = \sum_{n=0}^{\infty} \frac{\Gamma(-\frac{1}{2}m+n)}{\Gamma(-\frac{1}{2}m)\, n!} e^{-\frac{1}{2}m\pi i} (\zeta^* e^{s_{\infty}})^{-\frac{1}{2}m+n} \Psi_{-\frac{1}{2}m+n}(\tau) e^{-i(-\frac{1}{2}m+n)\theta}, \quad (27)$$
$$\tau^* \leq \tau < 1$$

$$S_m^{(3)} = \sum_{n=2}^{\infty} \frac{\Gamma(-\tfrac{1}{2}m-n)n!(-1)^n}{\Gamma(-\tfrac{1}{2}m)} \zeta^{*n} C_n \Psi_n(\tau) e^{in\theta}, \qquad 0 \leqslant \tau < 1 \tag{28}$$

where the C_n are real constants, depending only upon γ, and given in [2]. The convergence behaviour of the series can be analysed with the asymptotic estimates for the Chaplygin and gamma functions previously mentioned, and are found to be those indicated, with τ^* that value of τ which solves the equation

$$s(\tau) = s_\infty + \ln|\zeta^*|, \tag{29}$$

with the function $s = s(\tau)$ given in [2], equation (6). The Mellin-Barnes integral $\tilde{\phi}_m$ turns out to be equal to

$$\tilde{\phi}_m = + S_m^{(1)}, \qquad \theta^* < \theta < 2\pi + \theta^*, \qquad 0 \leqslant \tau \leqslant \tau^*$$
$$= + S_m^{(2)} - S_m^{(3)}, \qquad \theta^* < \theta < 2\pi + \theta^*, \qquad \tau^* \leqslant \tau < 1 \tag{30}$$

where the domain of θ has been restricted to that of the integral representation for $\tilde{\phi}_m$.

We now define the Mellin-Barnes integral to be valid in the sector $\theta^* < \theta \leqslant \pi$ on the lower sheet of the hodograph surface (first two lines of (31) below. Taking into account that the series $S_m^{(1)}$ and $S_m^{(3)}$ have a period 2π in θ and the series $S_m^{(2)}$ a period 4π, while their convergence behaviour does not depend on θ, we may continue $\tilde{\phi}_m$ analytically out of this domain into the subsequent domains as indicated.

From the construction of $\tilde{\phi}_m$ implied by these equations it is evident that $\tilde{\phi}_m$ is one valued near the branch point (τ^*, θ^*) on the hodograph surface.

In order to guarantee that a stagnation point exists on the upper sheet on the streamline $\tilde{\Psi} = 0$ (the aerofoil image C in Fig. 1), $\tilde{\Psi}$ should have a saddle point at $\tau = 0$. To create this saddle point two modifications were made to all the series representations for $\tilde{\Psi}_b$ and $\tilde{\Psi}_a$.

$$\left.\begin{array}{ll}\text{lower sheet, } 0\leqslant\tau\leqslant\tau^*, \\ \text{lower sheet, } \tau^*\leqslant\tau<1, \end{array}\right\} \left.\begin{array}{l}\theta^*\leqslant\theta\leqslant\pi: \\ \theta^*\leqslant\theta\leqslant\pi: \end{array}\right\} \begin{array}{l}\tilde{\phi}_m=+S_m^{(1)} \\ =+S_m^{(2)}-S_m^{(3)}\end{array}$$

$$\left.\begin{array}{ll}\text{upper sheet, } \tau^*\leqslant\tau<1, \\ \text{upper sheet, } 0\leqslant\tau\leqslant\tau^*, \end{array}\right\} \left.\begin{array}{l}-\pi<\theta\leqslant\theta^*: \\ -\pi<\theta\leqslant\theta^*: \end{array}\right\} \begin{array}{l}=+S_m^{(2)}-S_m^{(3)} \\ =-S_m^{(1)}-2S_m^{(3)}\end{array} \quad (31)$$

$$\left.\begin{array}{ll}\text{upper sheet, } 0\leqslant\tau\leqslant\tau^*, \\ \text{upper sheet, } \tau^*\leqslant\tau<1, \end{array}\right\} \left.\begin{array}{l}\theta^*\leqslant\theta\leqslant\pi: \\ \theta^*\leqslant\theta\leqslant\pi: \end{array}\right\} \begin{array}{l}=-S_m^{(1)}-2S_m^{(3)} \\ =-S_m^{(2)}-S_m^{(3)}\end{array}$$

$$\left.\begin{array}{ll}\text{lower sheet, } \tau^*\leqslant\tau<1, \\ \text{lower sheet, } 0\leqslant\tau\leqslant\tau^*, \end{array}\right\} \left.\begin{array}{l}-\pi<\theta\leqslant\theta^*: \\ -\pi<\theta\leqslant\theta^*: \end{array}\right. \begin{array}{l}=-S_m^{(2)}-S_m^{(3)} \\ =+S_m^{(1)}\end{array}$$

The first modification concerns the so called tail series (the terminology was introduced by Manwell[6] Ch.11, intro.). These series converge for all $\tau\in[0,1]$ and vanish in the limit of incompressible flow:

$$\tau_\infty \to 0, \quad \left(\frac{\tau}{\tau_\infty}\right)^{\frac{1}{2}} \to |\zeta|.$$

An example is the series $S_m^{(3)}$ occurring in $\tilde{\phi}_m$. The series themselves are solutions of the flow equation. Hence, it is permitted to modify $\tilde{\Psi}$ by adding multiples of these series to representations of $\tilde{\Psi}_b$ and $\tilde{\Psi}_a$. It was found necessary to remove all the tail series near the front stagnation point in the upper sheet in order to create the desired saddle point structure of $\tilde{\Psi}$ at higher τ_∞ values. This implied the function $\tilde{\phi}_m$ was redefined to:

$$\tilde{\phi}_m = -S_m^{(1)}, \qquad\qquad 0\leqslant\tau\leqslant\tau^*,\text{ upper sheet}$$

$$\left.\begin{array}{ll}-S_m^{(2)}+S_m^{(3)}, & \theta^*\leqslant\theta\leqslant\pi \\ +S_m^{(2)}+S_m^{(3)}, & -\pi<\theta\leqslant\theta^*\end{array}\right\} \tau^*\leqslant\tau<1,\text{ upper sheet} \quad (32)$$

$$\tilde{\phi}_m = +S_m^{(1)} + 2S_m^{(3)}, \qquad 0 \leq \tau \leq \tau^*, \text{ lower sheet}$$

$$\left.\begin{array}{l} +S_m^{(2)} + S_m^{(3)}, \quad \theta^* \leq \theta \leq \pi \\ -S_m^{(2)} + S_m^{(3)}, \quad -\pi < \theta \leq \theta^* \end{array}\right\} \tau^* \leq \tau < 1, \text{ lower sheet} \quad (32 \text{ cont.})$$

The second modification concerns the remaining series near the front stagnation point on the upper sheet. These series have the form

$$\tilde{\psi} = \mathrm{Im} \sum_{n=0}^{\infty} a_n \Psi_n(\tau) e^{-in\theta} \qquad (33)$$

with a_n complex coefficients. It can be shown that a_0 and a_1 should be zero if $\tilde{\psi}$ has a saddle point at $\tau = 0$. To create the saddle point the solution

$$a_0 + a_1 \Psi_1(\tau) e^{-i\theta} \qquad (34)$$

of the hodograph equation (1) was subtracted from all series with a_0 and a_1 not both zero. For ϕ_m this implied that $\tilde{\phi}_m$ was redefined to the final form:

$$\tilde{\phi}_m = -S_m^{(1)} \qquad +S_m^{(4)}, \qquad 0 \leq \tau \leq \tau^*, \text{ upper sheet}$$

$$\left.\begin{array}{l} -S_m^{(2)} + S_m^{(3)} + S_m^{(4)}, \quad \theta^* \leq \theta \leq \pi \\ +S_m^{(2)} + S_m^{(3)} + S_m^{(4)}, \quad -\pi < \theta \leq \theta^* \end{array}\right\} \tau^* \leq \tau < 1, \text{ upper sheet}$$

$$+S_m^{(1)} + 2S_m^{(3)} + S_m^{(4)}, \qquad 0 \leq \tau \leq \tau^*, \text{ lower sheet} \qquad (35)$$

$$\left.\begin{array}{l} +S_m^{(2)} + S_m^{(3)} + S_m^{(4)}, \quad \theta^* \leq \theta \leq \pi \\ -S_m^{(2)} + S_m^{(3)} + S_m^{(4)}, \quad -\pi < \theta \leq \theta^* \end{array}\right\} \tau^* \leq \tau < 1, \text{ lower sheet}$$

where (c.f., (26))

$$S_m^{(4)} = 1 + (-\tfrac{1}{2}m) \zeta^{*-1} e^{-s_\infty} \Psi_1(\tau) e^{-i\theta} \qquad (36)$$

From this final form we defined for odd $m \geq 3$:

$$\tilde{\psi}_{a(2m-3)} = \operatorname{Im} \tilde{\phi}_m$$
$$\tilde{\psi}_{a(2m-2)} = \operatorname{Re} \tilde{\phi}_m \tag{37}$$

while for $m = 1$:

$$\tilde{\psi}_{a1} = \operatorname{Im} \tilde{\phi}_1$$
$$\tilde{\psi}_{a2} = \operatorname{Re} \tilde{\phi}_1 \tag{38}$$

APPROXIMATE SOLUTION OF TRICOMI BOUNDARY VALUE PROBLEM

The Tricomi boundary value problem can be solved approximately by determining the coefficients in the representation (9) for the additional stream function Ψ_a.

While looking for a suitable method for the determination of the coefficients a problem arose, which is illustrated with results of model studies for incompressible flows. In Fig. 3 an acceptable pattern of the streamlines $\Psi = 0$ on a so called regularised hodograph surface is sketched.

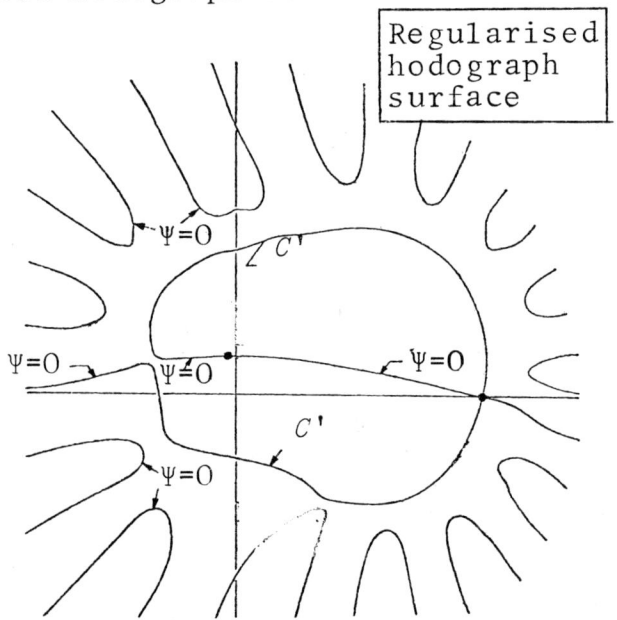

Fig. 3. Acceptable pattern of lines $\Psi = 0$ (incompressible flow)

(The regularised hodograph surface is obtained from the two sheeted hodograph surface by a mapping removing the branch point.) The aerofoil image C' is a closed curve $\Psi = 0$ (except for a small gap at the trailing edge which is unimportant from an engineering point of view.) Outside the aerofoil image C' are many other streamlines $\Psi = 0$, mapping into the interior of the aerofoil in the physical plane. Fig. 4, however, shows an unacceptable pattern of lines $\Psi = 0$: the aerofoil image is not a closed curve.

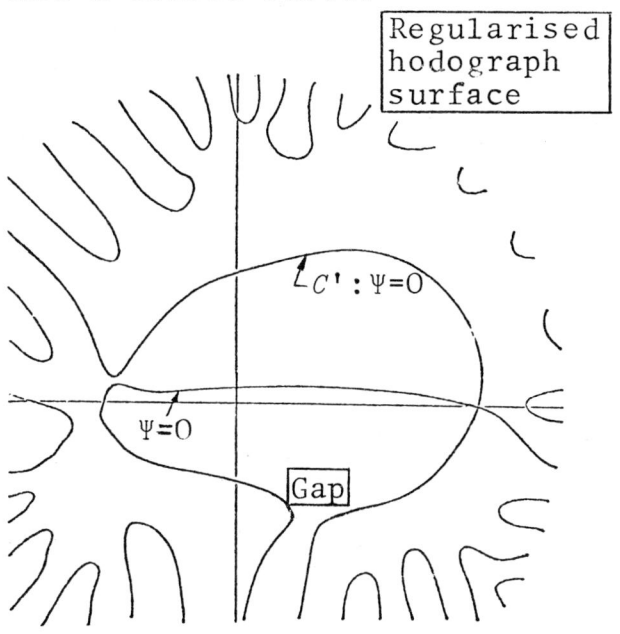

Fig. 4. Unacceptable pattern of lines $\Psi = 0$ (incompressible flow)

It was found that all available standard methods for the determination of the coefficients c_n failed, because usually the aerofoil images had at least one gap similar to that in Fig. 4.

To understand the solution it is important to observe that on the assumed aerofoil image C, $\tilde{\Psi}_a + \tilde{\Psi}_b$ are, in general, not exactly zero because of discretisation errors. It should also be noted that in a good approximate solution the curve C', where the condition $\tilde{\Psi} = \tilde{\Psi}_a + \tilde{\Psi}_b = 0$ is satisfied exactly, has to lie close to the assumed aerofoil image and to be a closed curve. It can be shown that the aerofoil, which corresponds to the curve C' with $\tilde{\psi} = 0$ (and not to the curve C' with $\tilde{\psi} \approx 0$) is then also a closed curve. If however C' is not a closed curve, the aerofoil is not a closed curve either.

A suitable method for the determination of the coefficients c_n giving (nearly always) closed aerofoil images C' near the assumed aerofoil image C consists of the minimisation of a special error norm that measures the difference between C and C'

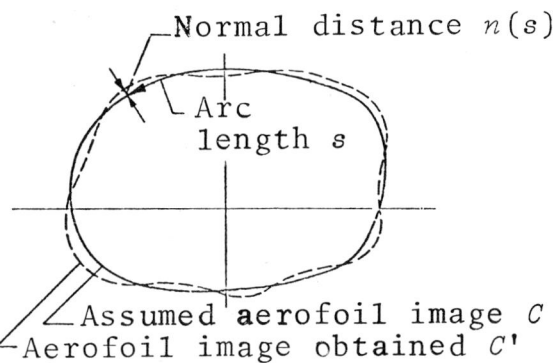

Mapping from hodograph surface to $\tilde{\xi}$-plane:

$$\tilde{\xi} = \left[1 - \left(\frac{\tau}{\tau^*}\right)^{\frac{1}{2}} e^{-i(\theta-\theta^*)}\right]^{\frac{1}{2}}$$

Fig. 5. Definition of normal distance $n(s)$ on regularised hodograph surface

(see Fig. 5; note the removal of the branch point of the hodograph surface by the mapping to a $\tilde{\xi}$-plane):

$$E = \oint_C w(s)\{n(s)\}^2 ds \qquad (39)$$

where $w(s)$ is a given positive weight function (usually $\equiv 1$), and $n(s)$ the distance between the curves measured along normals to the assumed aerofoil image C. It will be evident that gaps will be suppressed because in a gap $n(s)$ would be unbounded.

The error norm is discretised by the trapezoidal rule, and the distances $n(s)$ are computed from linear approximations to the stream function, as indicated in Fig. 6.

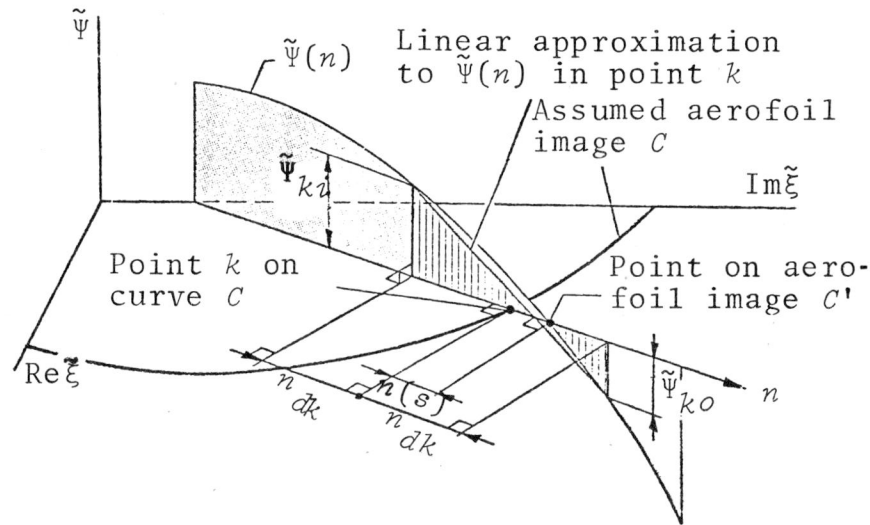

Fig. 6. Definition of variables in approximate error norm (40)

The result is

$$E \cong \sum_{k=1}^{K} \left(\frac{\tilde{\Psi}_{ko}+\tilde{\Psi}_{ki}}{\tilde{\Psi}_{ko}-\tilde{\Psi}_{ki}}\right)^2 n_{dk}^2 \tfrac{1}{2}(\Delta_k+\Delta_{k-1}) \qquad (40)$$

where Δ_k are the arc lengths between successive given points on C, n_{dk} are given distances on normals to C in the given points of C, and $\tilde{\Psi}_{ko}$ and $\tilde{\Psi}_{ki}$ are values of $\tilde{\Psi}$ indicated in Fig. 6.

The unknown coefficients c_n occur linearly in the $\tilde{\Psi}_{ko}$ and $\tilde{\Psi}_{ki}$. Hence, the approximation (40) to the error norm is a rational expression in the unknowns. The minimum is determined by a standard method [7].

One might ask in what sense a Tricomi boundary value problem is solved, because it is not the value of $\tilde{\Psi}$ on the assumed aerofoil image C that is minimised. At a first glance one would choose as error norm to be minimised:

$$E' = \oint_C w'(s)\{\tilde{\Psi}(s)\}^2 ds \qquad (41)$$

where $\tilde{\Psi}(s)$ is the value of $\tilde{\Psi}$ on the assumed aerofoil image and $w'(s)$ a non-negative weight function. (In fact this error norm was tried and found to be of no use.) However, for small $n(s)$ we may approximate $n(s)$ by

$$n(s) = -\frac{\tilde{\Psi}(s)}{\partial \tilde{\Psi}(s)/\partial n} + O(n(s))^2 \qquad (42)$$

where $\partial\tilde{\Psi}(s)/\partial n$ is the normal derivative of $\tilde{\Psi}$ on C. Hence, (41) can be approximated by

$$E' \cong \oint_C w'(s)\{\partial\tilde{\Psi}(s)/\partial n\}^2 \{n(s)\}^2 ds. \qquad (43)$$

The error norms are thus approximately equal if

$$w(s) \cong w'(s)\{\partial\tilde{\Psi}(s)/\partial n\}^2, \qquad (44)$$

so that error norm (39) solves the Tricomi boundary value problem, taking into account in a weight function the *a priori* unknown variations in normal derivative $\partial\tilde{\Psi}(s)/\partial n$ along C.

The existence and uniqueness of a solution of the Tricomi boundary value problem and the convergence of sequences of approximate solutions to an exact solution are not discussed here.

DETAILS ABOUT COMPUTATIONS

The series representing $\tilde{\Psi}_b$, $\tilde{\Psi}_{an}$, $\tilde{M}\tilde{\Psi}_b$ and $\tilde{M}\tilde{\Psi}_{an}$ have a poor convergence behaviour numerically in large parts of the hodograph surface. Their sum can be economically computed, however, by using the ε-algorithm as a convergence accelerator. (For details see, for example, annex D3 of Nieuwland's report [4].) We apply a complex version of the ε-algorithm, suitable for complex series.

The Chaplygin functions $\Psi_\nu(\tau)$ are products of hypergeometric functions with $\tau^{\frac{1}{2}\nu}$. The summation of hypergeometric series may present difficulties as a result of much cancellation of figures when summing the terms of the series. With the Chaplygin functions for $\nu = -99.5$ and $\tau = 0.32$ this cancellation is over 60 decimal places. Nieuwland found [4] that the only practical way of summing the terms of the series representation for $\tilde{\Psi}_\nu(\tau)$ was the application of multiple length arithmetic. We use an extensive table of the Chaplygin functions, stored on magnetic tape.

The aerofoil contour $\tilde{\Psi}(\tau,\theta) = 0$ can be computed (after having determined the coefficients c_n occurring in $\tilde{\Psi}_a$) by first solving (with the Regula Falsi for given τ the equation $\tilde{\Psi}(\tau,\theta) = 0$ for the value of θ satisfying this equation. Next these two values of τ and θ are substituted in series expansions for $\tilde{z}(\tau,\theta)$ and the aerofoil shape and pressure distribution are found.

Because the infinite series representations for $\tilde{\Psi}$ and $\tilde{z}$ are exact solutions of the hodograph equations the only errors in the final results arise from truncation of the Regula Falsi process and truncation of the series expansion when summing directly. When the series are summed by the ε-algorithm, errors arise which can be considered to be of stochastical nature; the magnitude of these errors can be easily estimated. Other data (Chaplygin functions, coefficients in series representations) are computed to such a high precision that they cannot affect the precision of the final data.

The accuracy of the aerofoil contour data is 10^{-4} to 10^{-5} of the chord (without smoothing interpolation operations). If desired, it is possible to compute coordinates, slopes and curvatures at about 400 points.

The computations are made with a system of nine Algol programs. A few programs mainly perform calculations, others are used for the management of data banks.

The computation of an aerofoil proceeds in three stages.

(i) First several parameters that determine the basic stream function $\tilde{\Psi}_b$ (τ_∞, τ^*, θ^* flow circulation Γ, ε, etc.) have to be determined so that the hodograph seems a reasonable starting point to obtain the desired aerofoil. This is done by computing lines $\tilde{\Psi}_b = 0$ in the regularised hodograph plane $\tilde{\xi}$ defined in Fig. 5; (The choice of $\tilde{\Psi}_b$ is sometimes checked by computing explicitly the aerofoil like shape defined by the lines $\tilde{\Psi}_b = 0$.)

(ii) An assumed image C of the aerofoil is then defined on the regularised hodograph surface $\tilde{\xi}$, the corresponding coefficients c_n are computed, and the aerofoil image C' where $\tilde{\Psi} = 0$ is estimated;

(iii) When the aerofoil image C' lies close enough to the assumed aerofoil image C, the aerofoil shape and other data corresponding to C' are computed accurately.

If, after the last step, the aerofoil shape and the pressure distribution in the design condition are not sufficiently good, an iterative cycle is set up by the aerodynamic engineer by specifying improved assumed aerofoil images C and restarting at point (ii). Sometimes it is necessary to choose new parameters for the basic stream function $\tilde{\Psi}_b$, when the desired aerofoil image deviates so much from that of $\tilde{\Psi}_b = 0$, that the capacity of the approximation method to correct for this is exceeded. The computation process for a specific aerofoil shape and design pressure distribution is represented schematically in Fig. 7.

The program system is operational on the NLR Cyber 72 computer (CDC 6200). A typical capacity claim in a computation, in which the program system is used once in all three modes mentioned

Hodograph Theory for Aerofoil Design

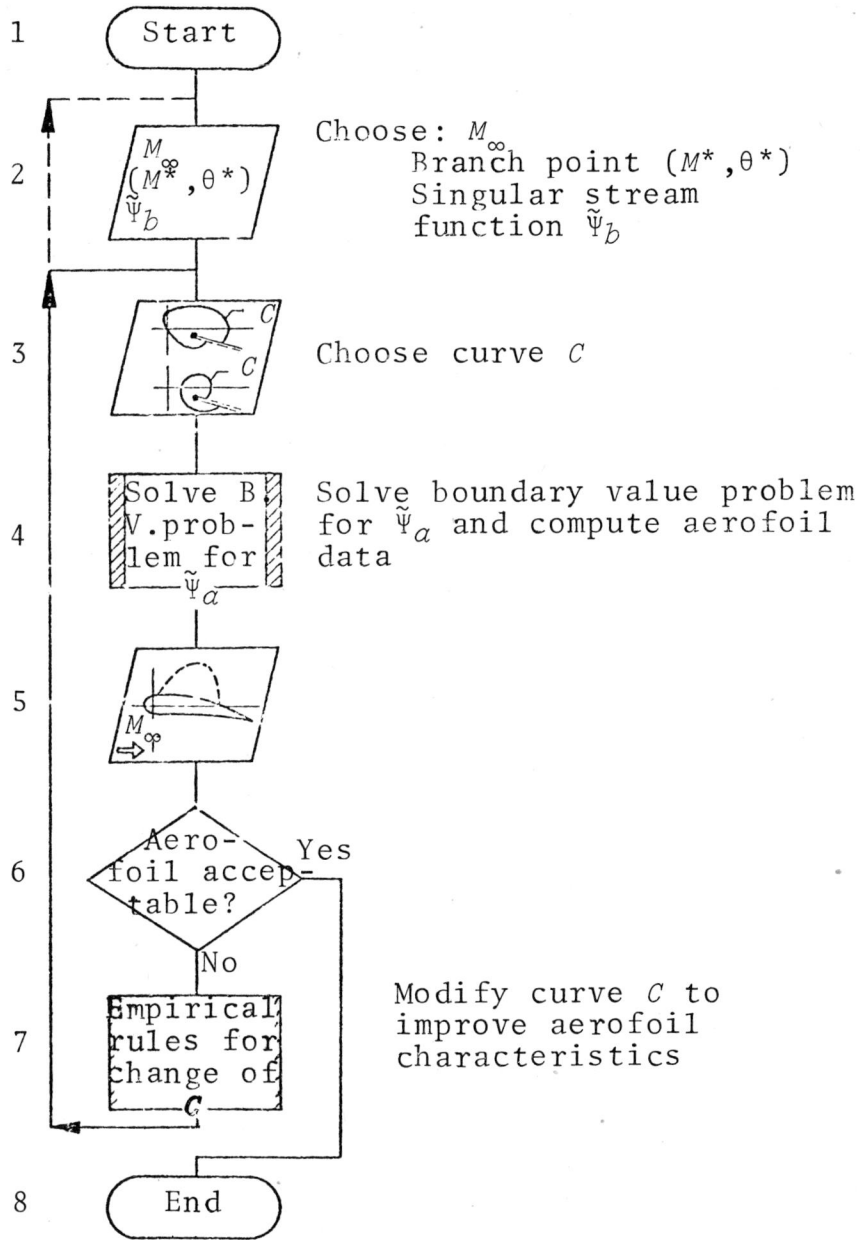

Fig. 7. Aerodynamic optimisation process of aerofoil shape and design pressure distribution

above, is about 4000 central processor seconds, 1600 input-output seconds and an average of 26 k core store. Two tapes and twenty disc files are used for data handling.

EXAMPLES OF COMPUTED AEROFOILS AND NUMERICAL RESULTS

Examples of recently computed aerofoils are given in Figs. 8, 9 and 10. Some of these are identical to those published in [8]. Older examples of aerofoils that are less interesting from an aerodynamic point of view are presented in [9].

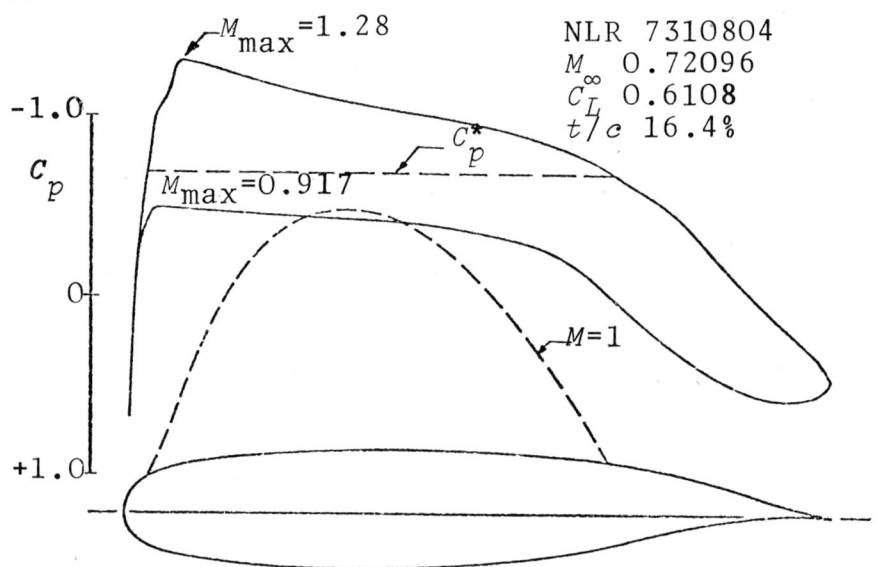

Fig. 8. Example of shock free aerofoil with design pressure distribution

The examples show that a large variation in aerofoil shapes and design pressure distributions can be realised.

Conclusions concerning the aerofoil shapes and design pressure distributions are the following. The peaks in the design pressure distribution at the nose correspond to curvature peaks in the aerofoil contour. The height of the super-

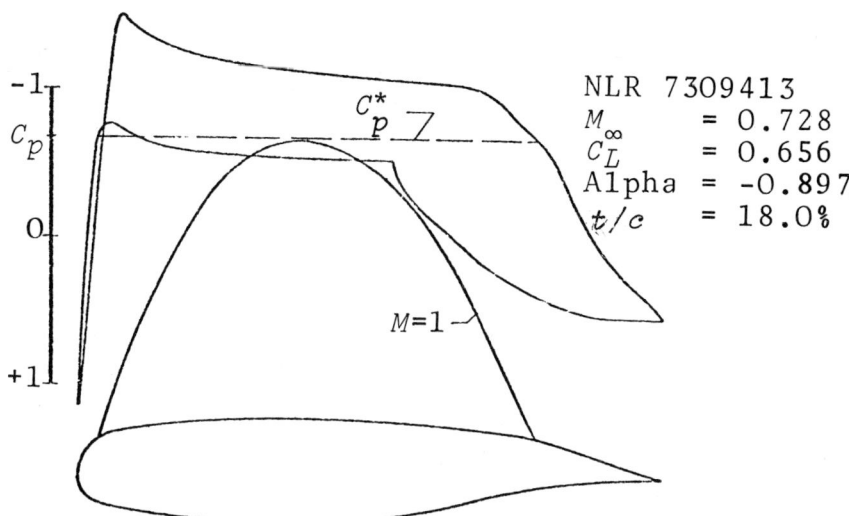

Fig. 9. Example of shock free aerofoil with design pressure distribution

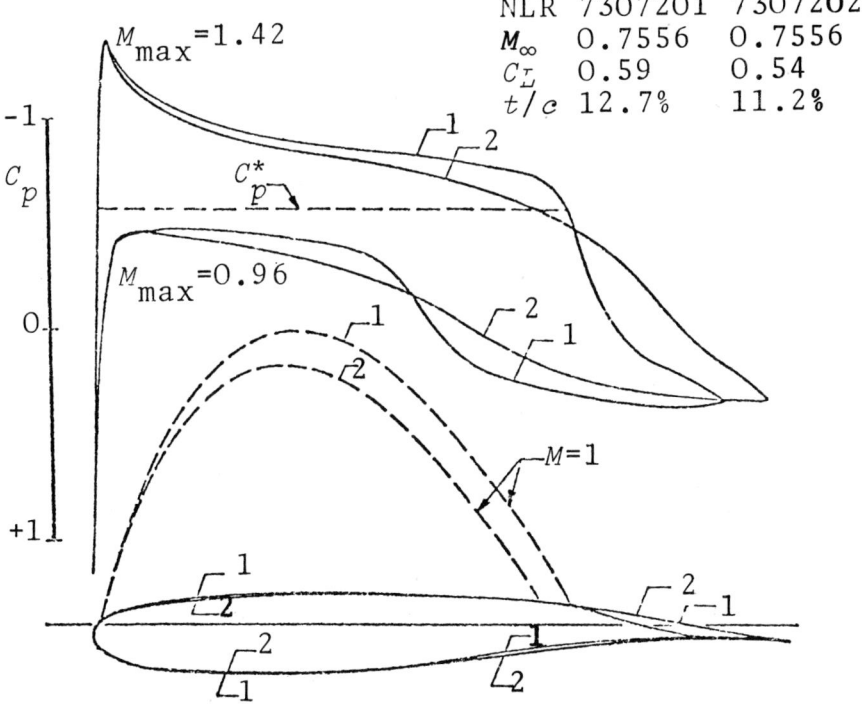

Fig. 10. Examples of shock free aerofoils with design pressure distributions

sonic zones is comparable to a few times the aerofoil thickness rather than the aerofoil chord. This can be important for favourable drag properties at off-design conditions (no drag creep). The aerofoil nose can be made blunt or shaped so that favourable low speed/high lift properties are realised (nose droop). At the rear of the aerofoil the pressure distributions can be modelled so that the boundary layer behaviour can be controlled. In particular, it is possible to design for Stratford type pressure recoveries, or to make pressure recoveries so steep that shock waves are simulated.

The computation of an aerofoil is illustrated in Figs. 10, 11 and 12. Results of a parameter study for the basic solution $\tilde{\Psi}_b$ are presented in Fig. 11. The drawn line represents the $\tilde{\Psi}_b = 0$ line of the basic solution ultimately chosen. Fig. 12 shows the optimisation in two steps to an assumed aerofoil image C giving the aerofoil shape 2 of Fig. 10.

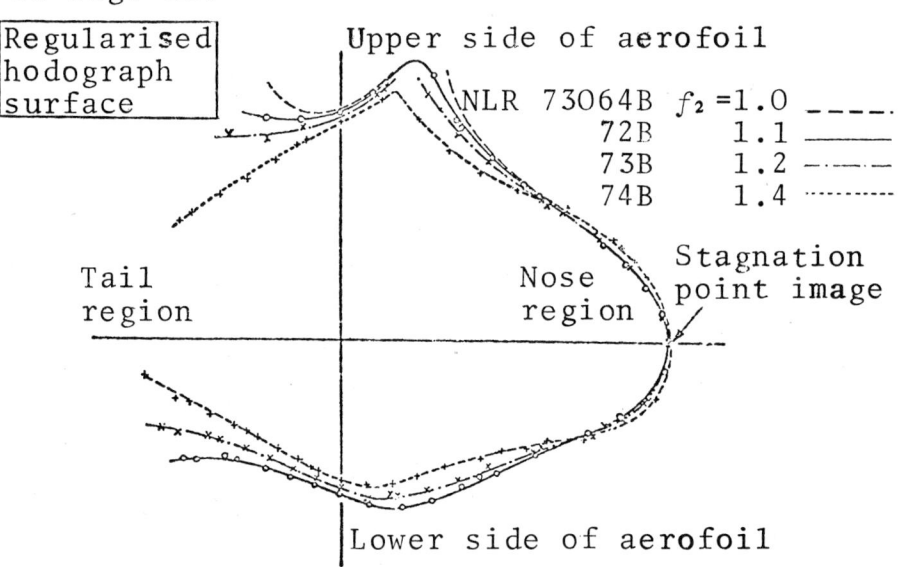

Fig. 11. Effect of variations of a parameter f_2 in $\tilde{\Psi}_b$ on the $\tilde{\Psi}_b = 0$ lines

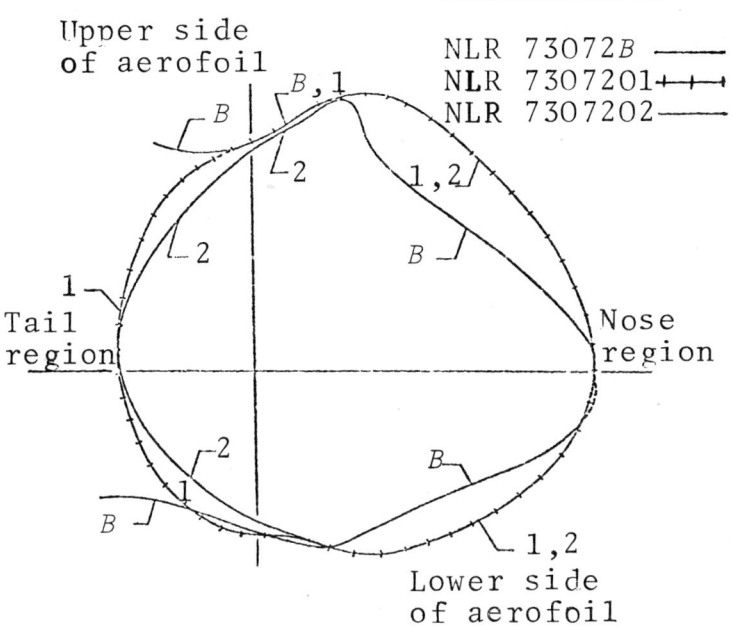

Fig. 12. $\tilde{\Psi}_b = 0$ lines and assumed aerofoil images of aerofoils of Fig. 10

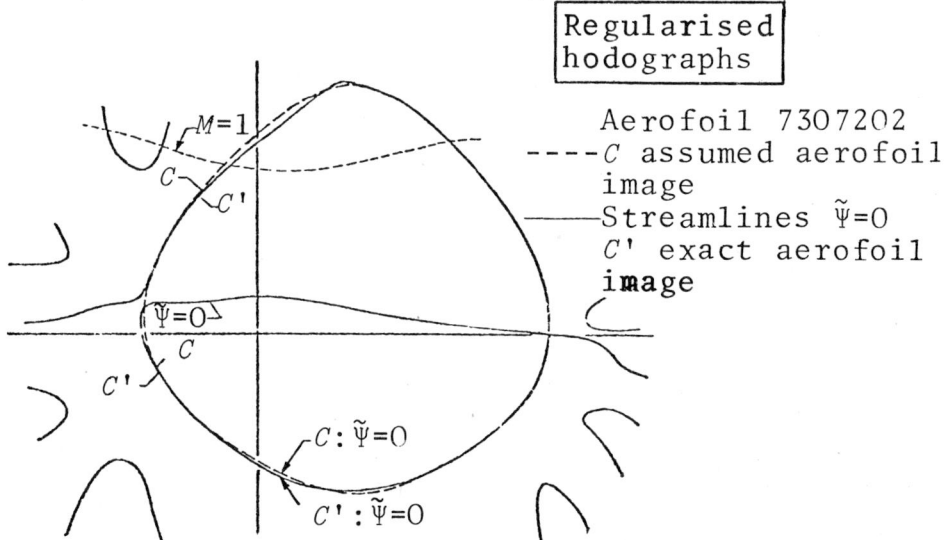

Fig. 13. Assumed aerofoil image, aerofoil image and streamlines $\tilde{\Psi}=0$ for aerofoil NLR 7307202 of Fig. 10

Some numerical details are shown in Fig. 13. First the small difference between the assumed aerofoil image C and the exact aerofoil image C' in the ξ-plane should be noted. Further it is seen that outside the aerofoil image C' where $\tilde{\Psi}=0$ there are other streamlines $\tilde{\Psi}=0$. These extra streamlines $\tilde{\Psi}=0$ map in the physical plane into the inside of the closed aerofoil.

At the end of "Introduction to Hodograph Theory" it was mentioned that mathematically the boundary condition $\tilde{\Psi}=0$ on the assumed aerofoil image C should not be prescribed on a certain segment of C in the supersonic region of the flow. This can be realised by choosing suitable weights w_k in expression (40) to be zero. In practice this is not necessary when the supersonic regions in the flow are not large, and the choice $w_k \equiv 1$ could then be made. This surprising procedure is possible for numerical reasons: in the supersonic regions $\tilde{\Psi}_b$ is "large" compared with $\tilde{\Psi}_a$. (In incompressible flow Ψ_b is large in corresponding regions because of the singularities of ϕ_e at $\zeta = \pm \zeta^*/\varepsilon$; in transonic flow these singularities disappear, but the general behaviour of $\tilde{\Psi}_b$ is similar to that for incompressible flow.) Hence, in the supersonic regions the general character of the profile is determined by $\tilde{\Psi}_b$. For this purpose too, ϕ_e (with its singularities at $\zeta = \pm \zeta^*/\varepsilon$ outside the aerofoil images) has been incorporated in Ψ_b ((12), (13)).

CONCLUSIONS

Starting from results in hodograph theory presented by Lighthill and Nieuwland an approximate method for the solution of Tricomi boundary value problems was developed. The solutions represent transonic flows around aerofoils. The approximation technique takes into account that the aerofoils and their images on the hodograph surface have to be closed curves. From an engineering point of view the range of aerofoils that can be computed seems to be sufficiently large.

The results published in this paper were obtained in investigations under contracts with the Netherlands Agency for Aerospace Programs (NIVR).

REFERENCES

1. Cheng, H.K. and Hafez, M.M., "Equivalence rule and transonic flow theory involving lift," *AIAA J.*, **11**, No. 8, 1210-1212 (1973).
2. Lighthill, M.J., "The hodograph equation in transonic flow. II. Auxiliary theorems on the hypergeometric functions," *Proc. Roy. Soc. A*, **191**, 341-351 (1947).
3. Lighthill, M.J., "The hodograph transformation in transonic flow. III. Flow round a body," *Proc. Roy. Soc. A*, **191**, 352-369 (1947).
4. Nieuwland, G.Y., "Transonic potential flow around a family of quasi-elliptical aerofoil sections," NLR TR-T 172 (1967).
5. Whittaker, E.T. and Watson, G.N., "A course of modern analysis," second edition, CUP (1915).
6. Manwell, A.R., "The hodograph equations," Oliver and Boyd, Edinburgh (1971).
7. Fletcher, R., "Generalised inverse methods for the best least squares solution of systems of non-linear equations," *Comp. J.*, **10**, 392-399 (1968).
8. Boerstoel, J.W. and Huizing, G.H., "Transonic shock-free aerofoil design by an analytic hodograph method," (presented at the Euromech 40 Colloquium, September 1973), NLR MP 73023 U (1973).
9. Boerstoel, J.W. and Huizing, G.H., "Transonic shock-free aerofoil design by an analytic hodograph method," AIAA paper 74-539 (1974); NLR MP 74025 U (1974).

SUPERSONIC FLOWS

F. Walkden

(University of Salford)

INTRODUCTION

This review paper is concerned with methods and problems associated with the computation of steady flowfields produced when a body is placed in a uniform supersonic stream. Attention is restricted to body shapes such that the effects of viscosity are confined to narrow layers lying close to body surfaces. In these cases good predictions of both surface pressures and flow outside the boundary layers (where viscous effects may not be neglected) can be obtained by solving equations of motion in which the effects of fluid viscosity are neglected.

The next section contains examples of some of the classes of body shapes and flows of interest in aeronautics. The mathematical problem which has to be solved is discussed in the following section. Distinctive features of numerical methods which have been used successfully to calculate a variety of steady inviscid supersonic flowfields are then examined. Some numerical results are presented in the next section and the final section contains some conclusions.

CLASSES OF PROBLEM

Cones with arbitrary cross-sections

The envelope of tangent planes at the pointed nose of an otherwise arbitrarily shaped body is a cone. This cone provides a close approximation to the body shape in the region close to the nose. By computing the supersonic flowfield produced by the cone, it is found that a good representation of disturbances produced by the actual body in the region close to its nose is obtained.

Conical body shapes are of interest because cone-fields (the name given to supersonic flow-fields produced by cones) yield the initial information needed to start calculations of flows associated with more complicated body shapes.

Apart from the special case of a circular cone at zero incidence, the calculation of cone-fields is not a trivial problem. Without going into great mathematical detail, this can be seen simply by noticing that:

(i) supersonic flow calculation methods that require information on an initial surface are often used to solve cone-field problems;

(ii) cone-fields have the property that quantities such as pressure, density and flow velocity magnitudes and directions are all constant along any ray through the vertex of the cone. Effectively, therefore, in a strict mathematical sense, this means that the unknown cone-field solution to be calculated is needed to provide starting data for its calculation.

Cone-field problems have been examined by many methods. In particular, work by South and Klunker [1] who used the method of lines and by Moretti [2] who used a conventional numerical marching method to obtain numerical solutions of certain cone-field problems asymptotically may be cited. Whilst both of these methods produce acceptable results for particular cases, they are not perfect for calculating general cone-fields, and it seems that more work is needed on this class of problem.

Fuselage shapes

Fig. 1 is a representation of a fuselage. Fuselage shapes for high speed aircraft are generally three-dimensional because designers indent fuselages to obtain smooth variation of the total aircraft cross-sectional area with distance along the aircraft. Indentation of the fuselage is needed to compensate for area taken up by wings and cockpit canopies when these components are attached to the fuselage.

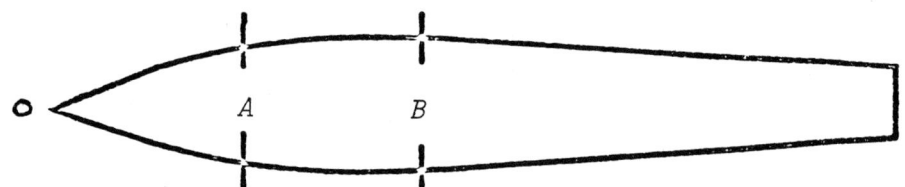

(a) Longitudinal section of a fuselage which is axisymmetric except in the region $A - B$:

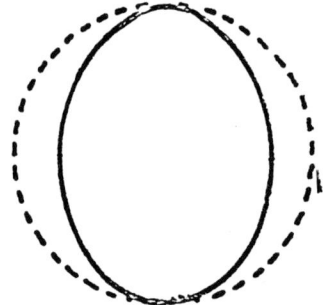

(b) Typical cross-section illustrating indentation and departure from axial-symmetry in the region $A - B$

Fig. 1

The technique for obtaining smooth variations in total cross-sectional areas is known as "area-ruling" and an "area-ruled" configuration may be expected to have a lower wave-drag than a "non area-ruled" configuration.

Numerical solutions of nonlinear equations of motion for flowfields associated with fuselage shapes are required in order to obtain accurate predictions of wave-drag for given shapes. Flow velocity magnitudes and directions on fuselage surfaces are also required to initiate viscous boundary layer calculations.

Wings and combinations of fuselages, cockpit canopies and wings

In addition to fuselages alone, of course, accurate wave-drag estimates are required for wings alone and for all the different possible combinations of components like fuselages, cockpit canopies and wings.

Intakes

Fig. 2 represents a simple intake attached to the surface of a wing. Aircraft designers are interested in obtaining details of the flow around the outside of intakes. Aero-engine designers, on the other hand, are interested in details of the flow inside the intake. Fig. 3 shows how complicated these internal flow structures may be.

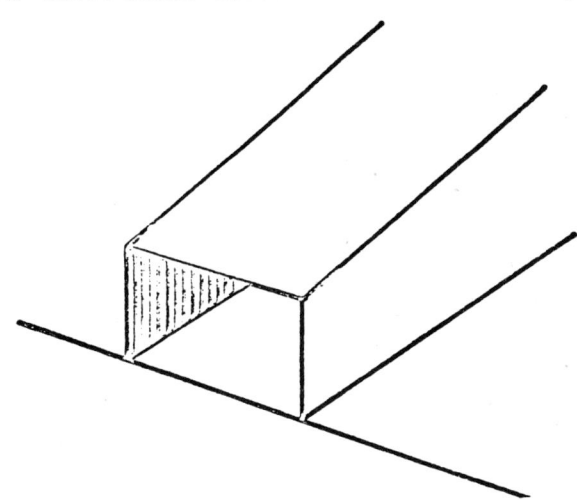

Fig. 2. Representation of an intake attached to a wing

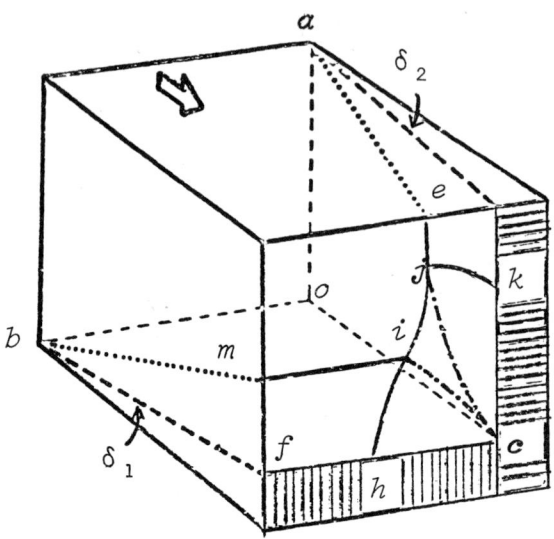

Fig. 3. A representation of flow inside an engine air intake shock surfaces: $aejo$, oji, $boim$, oih, ojk; vortex sheets: ojc, oic. The point c is a vortical singularity

The flow patterns shown in Fig. 3 are representative of flow inside the class of intakes which are constructed so that, at the design point, the flow in the intake is two-dimensional. Complicated shock interactions involving three-dimensional flows may occur, for example, when an aircraft is manoeuvering. The problem of calculating flows of the type shown in Fig. 3 has been considered by Kutler using a method similar to that described in [8]. Work is being carried out by BAC at Filton using Kutler's method to compute intake flows similar to that illustrated in Fig. 3.

Mixed flow problems

For the body shapes described in pp. 354 - 358, the flow speed may be supersonic throughout the flowfield. However, for some shapes, under certain conditions, locally subsonic regions can and do occur in practice. Although the problem of computing mixed flows is not considered here, some examples of ways in which mixed flows may occur are given to complete the discussion of classes of supersonic flows which are of interest.

Fig. 4 is a schematic diagram showing a subsonic region of flow associated with spillage from an intake mounted on a fuselage.

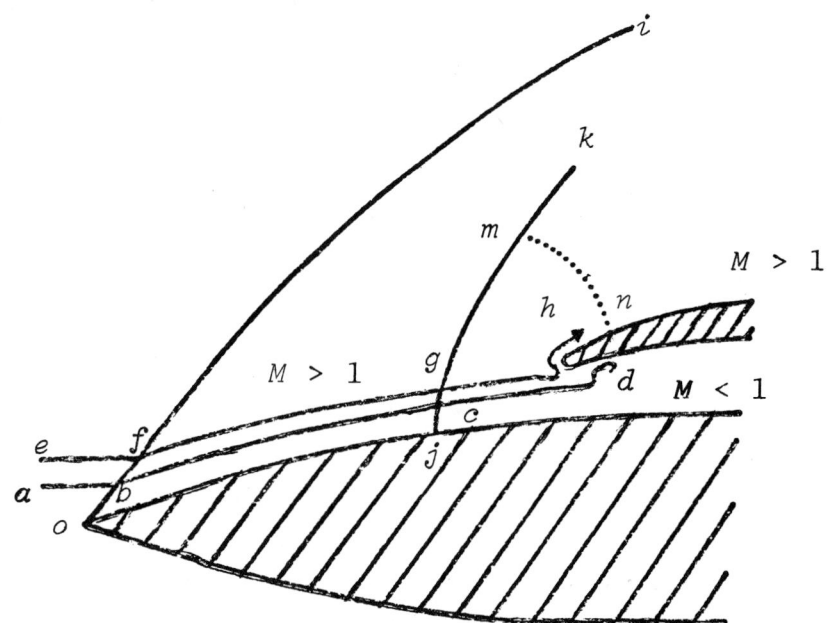

Fig. 4. A representation of the flowfield produced by air spillage from an engine intake attached to a fuselage. $obfi$ and $jgmk$ are shock waves, mn is a sonic line separating regions in which the flow is subsonic ($M < 1$) from regions in which the flow is supersonic ($M > 1$). $ofgh$ and $abcd$ are streamlines, oj is the fuselage surface

Embedded subsonic regions may also occur at junctions such as those formed where cockpit canopies and wings join a fuselage. In these situations, inviscid theory would be expected to predict a flow pattern qualitatively similar to that in Fig. 5(a). In practice, however, the pressure jump across the shock wave would probably cause a boundary layer separation. In this case, resulting inviscid/viscous flow interactions might lead to a flow pattern similar to that in Fig. 5(b)

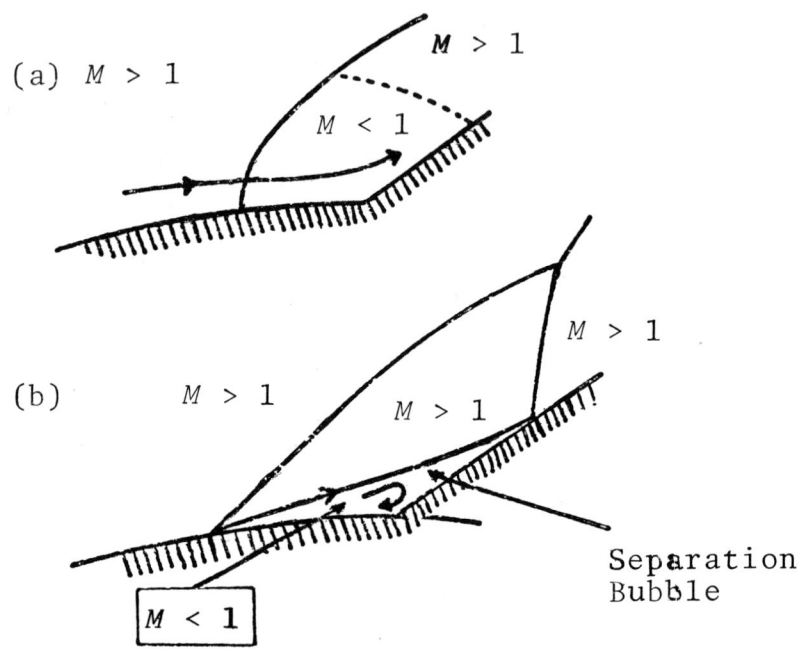

Fig. 5. Representations of mixed subsonic ($M < 1$) and supersonic ($M > 1$) flows

Mixed flows involving embedded subsonic regions or viscous/inviscid flow interactions present difficult theoretical problems. Much research remains to be done in developing efficient techniques for obtaining reliable numerical solutions for problems of this type.

THE MATHEMATICAL PROBLEM

For calculations of flowfields associated with conventional aircraft, air may be assumed to be an ideal gas with constant specific heats. Then, if viscous effects are neglected, equations of steady motion relating pressure p, density ρ and the velocity vector $\underset{\sim}{v}$ may be expressed in the form

$$\mathrm{div}(\rho \underset{\sim}{v}) = 0 \tag{1}$$

$$\gamma p/(\gamma-1)\rho + 0.5(\underset{\sim}{v}\cdot\underset{\sim}{v}) = \text{constant} \qquad (2)$$

$$\rho \text{grad}(0.5\underset{\sim}{v}\cdot\underset{\sim}{v}) - \rho\underset{\sim}{v}\times\text{curl}\underset{\sim}{v} = -\nabla p \qquad (3)$$

where γ is a constant equal to the ratio of specific heats. For air $\gamma = 1.4$. Equations (1) to (3), respectively, express conservation of mass, conservation of energy and conservation of momentum.

This paper is concerned only with steady flowfields in which local fluid speeds are always supersonic. In such cases, equations (1) to (3) form a hyperbolic system, and for each point in the flowfield a real Mach cone generated by rays inclined at the local Mach angle $\mu = \sin^{-1}[(\gamma p/\rho|\underset{\sim}{v}|^2)^{\frac{1}{2}}]$ to the local velocity vector $\underset{\sim}{v}$ can be constructed. Now definitions of time-like and space-like surfaces are needed. A surface is said to be time-like if the interiors of Mach cones associated with all the points on the surface lie on one and the same side of the surface. A surface is space-like if no part of the surface, however small, forms a time-like subsurface. Thus in a uniform supersonic flow, a plane perpendicular to the uniform flow direction will be a time-like surface. Any surface composed of streamlines forms a space-like surface in both uniform and non-uniform flows. Surfaces of bodies through which there is no fluid flow are streamsurfaces because, if $\underset{\sim}{n}$ represents a vector normal to the body surface, then

$$\underset{\sim}{v}\cdot\underset{\sim}{n} = 0 \qquad (4)$$

at every point on the surface. It follows, therefore, that body surfaces in flowfields which are supersonic everywhere are also space-like surfaces.

For hyperbolic partial differential equations, mixed initial and boundary value problems are ones in which the differential equations have to be solved either in a part or in the whole of some region bounded by a given time-like surface and a space-like surface which intersects the time-like surface. In general, for such problems, the dependent variables have to be specified over the time-like surface and certain boundary conditions have to be specified on the space-like surface to determine a unique solution of given partial differential equations.

The classes of problems described in pp. 354 - 358 are those in which some aerodynamic shape is placed in a uniform supersonic flow. Since, here, attention is restricted to shapes and uniform flow conditions such that flows are supersonic everywhere, it follows that, in regions where the flow has to be calculated, equations (1) to (3) will be hyperbolic. In general, the mathematical problem which has to be solved to obtain theoretical solutions for the classes of problems listed in pp. 354- 358 will be a mixed initial and boundary value problem governed by equations (1) to (3).

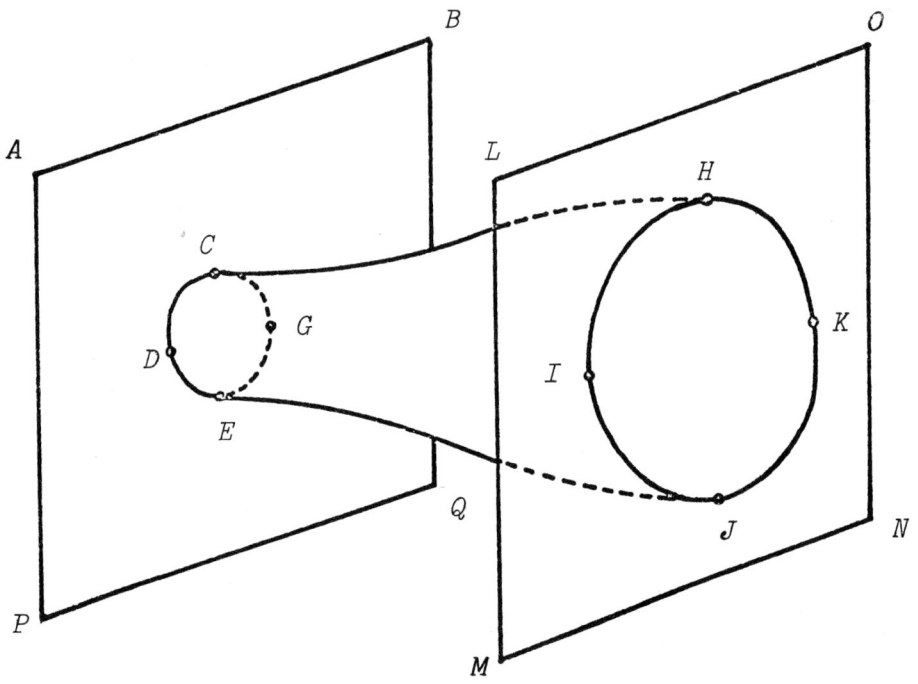

Fig. 6. An exterior flow problem

Fig. 6 is an illustration of an exterior flow problem. *ABPQ* is a time-like surface, probably extending to infinity in all directions. The surface *CDEHIJ* represents a given body surface which is space-like. The problem generally is to obtain values of p, ρ and $\underset{\sim}{v}$ in the region downstream of *ABPQ* and outside the body surface *CDEHIJ*. In practice, values of p, ρ and $\underset{\sim}{v}$ are often required only in relatively small subregions of this space, regions consisting of points at which the flow is influenced by some part of the given body surface. Depending on whether the solution of the equations of motion has to be calculated in the whole or in part of the region downstream of *ABCD*, specified values of p, ρ and $\underset{\sim}{v}$ have to be supplied on either the whole or part of *ABCD* and a boundary condition, which is usually an equation equivalent to (4), has to be specified on the body surface. Then that solution of equations (1) to (3) which satisfies the prescribed initial and boundary conditions has to be obtained. For most practical problems *ABCD* may be taken to be a plane perpendicular to the uniform flow direction. This is impossible only if the flow normal to this plane is less than the speed of sound. As far as specifying the values of p, ρ and $\underset{\sim}{v}$ is concerned, the usual situation is that values at points of *ABCD* that are remote from the body are simply known constant uniform stream values. Closer to the curve *CDEG* in the surface *ABPQ* the values of p, ρ and $\underset{\sim}{v}$ may be determined by the uniform stream condition and the shape of some forebody upstream of *ABPQ*. It has to be assumed that these values are either known or can be obtained, for example, by other calculations. Typically, if the forebody is some kind of cone, or can be approximated closely by a cone, then solving a cone-field problem (see p. 354) would yield the necessary information.

Since equations (1) to (3) are nonlinear, solutions of mixed initial and boundary value problems generally have to be constructed numerically. This aspect of the problem is considered in the next section.

NUMERICAL METHODS FOR MIXED INITIAL AND BOUNDARY VALUE PROBLEMS

The equations of motion given in the last section are hyperbolic when the flow is supersonic. Consequently, numerical solutions of mixed initial and boundary value problems associated with supersonic flow can generally be obtained simply by solving finite difference representations of the equations of motion, in a step-by-step fashion, at finite difference mesh points throughout the region of interest. Although the occurrence of shock waves is a complication, it is one which, for many problems at least, can be dealt with effectively.

In order to solve the equations of motion numerically, they must, of course, be written as differential equations instead of in the vector form given in the previous section. It is clear that there are several ways in which this can be done. Moreover, when the problem is considered seriously, it soon becomes apparent that depending on the choice of coordinate surfaces ($i.e.$, the independent variables in the equations of motion), the method of treating shock waves, the method of constructing discrete representations of equations of motion, the accuracy of the discrete representations and whether operator splitting is used or not, then, theoretically at least, many different numerical methods can be constructed for any given supersonic flow problem.

In practice, a fairly wide variety of methods has been developed and applied successfully to particular problems. These methods are not discussed here in detail, but, references to original papers are given and, for cited methods, particular distinguishing features from the following table are mentioned.

Table I
Possible Features for Numerical Methods

F1 Coordinate surfaces
- (*a*) Fixed curvilinear surfaces in space
- (*b*) Stream surfaces

F2 Equations of Motion
- (*a*) In divergence form
- (*b*) Some equations expressed as characteristic relations

F3 Shock waves
- (*a*) Detected and fitted as discontinuities
- (*b*) "Captured" by finite difference equations in which case shock waves appear as rapid but smooth changes in the dependent variables
- (*c*) Bow shocks detected and fitted but other shock waves captured

F4 Discrete representations of equations of motion (at points in field, *i.e.*, at points which are not boundary points)
- (*a*) Finite difference representations of conservation laws
- (*b*) Finite difference representations of characteristic relations
- (*c*) Explicit use made of characteristic relations

F5 Order of accuracy of numerical scheme
- (*a*) Order 1
- (*b*) Order 2

F6 Operator splitting used
- (*a*) Yes
- (*b*) No

The features mentioned in Table I do not represent an exhaustive list. There are other distinctive features. For example, in many cases, different methods may be further distinguished by the way in which boundary conditions are applied. The restricted list of possible features in Table I is sufficient for the discussion here. Although the list is restricted, it is clear that many different methods can be constructed.

It is generally agreed that it is important to carry out computations using coordinate systems so that boundary surfaces are coordinate surfaces. The methods cited here satisfy this requirement.

Methods described by Moretti [3] and Marconi and Salas [4] have features F1(a), F2(a), F3(a), F4(a), F5(b), F6(b). Reference [3] describes a possible method but it does not contain any significant numerical results.

Methods described by Walkden and Caine [5] and Walkden, Laws and Caine [6] have features F1(b), F2(b), F3(b), F4(b), F5(a), F6(b). Reference [7] describes the derivation of semi-characteristic forms of equations of motion used in [6].

Kutler, Lomax and Warming [8] describe a method which has features F1(a), F2(a), F3(b), F4(a), F5(b), F6(b).

Chu [10] has described a method having features F1(b), F2(b), F3(c), F4(c), F5(b), F6(b).

The methods cited represent only a small sample of the total number in the scientific literature. All the methods listed here have been used successfully to treat certain practical supersonic flow problems, and are methods I recommend as worthy of consideration by anyone who has to select a method for computing supersonic flows and who has limited experience in the field of computational fluid dynamics.

In the next section, graphs illustrating some results are presented. Certain of the numerical results shown have been obtained using one of two numerical methods developed at Salford. One of the methods, known as the new bow-shock fitting operator-splitting method, has features F1(a), F2(b), F3(a), F4(b), F5(a), F6(a). The other method, the new shock-capturing operator-splitting method, has features F1(a), F2(b), F3(c), F4(b), F5(a), F6(a).

Detailed descriptions of both of these new methods should be available soon in separate reports [12], [13].

NUMERICAL RESULTS

This section contains some graphs of numerical results which have been obtained at Salford for flow past a delta wing, a fuselage cockpit canopy combination and a simple intake.

Delta wing

Some computed results for flow past the symmetric delta wing with rounded leading edges described in [11] are presented here. The results are for the case when the wing is placed at zero incidence in a uniform stream of Mach number 3.5. At this Mach number the wing leading edges are sonic. Over the first 20% of the wing, cross-sections perpendicular to the main-stream direction are circles associated with an axisymmetric conical nose extending over the front 20% of the length of the wing. Over the rear 80% of the wing, the cross-sections are ellipses such that the ratio of minor to major axes varies from 1 at the 20% station to zero at the wing trailing edge.

Detailed knowledge of the wing shape is not required here. The results are provided only to illustrate differences that might be expected in numerical solutions near a shock wave when the wave is "captured" instead of being fitted as a discontinuity.

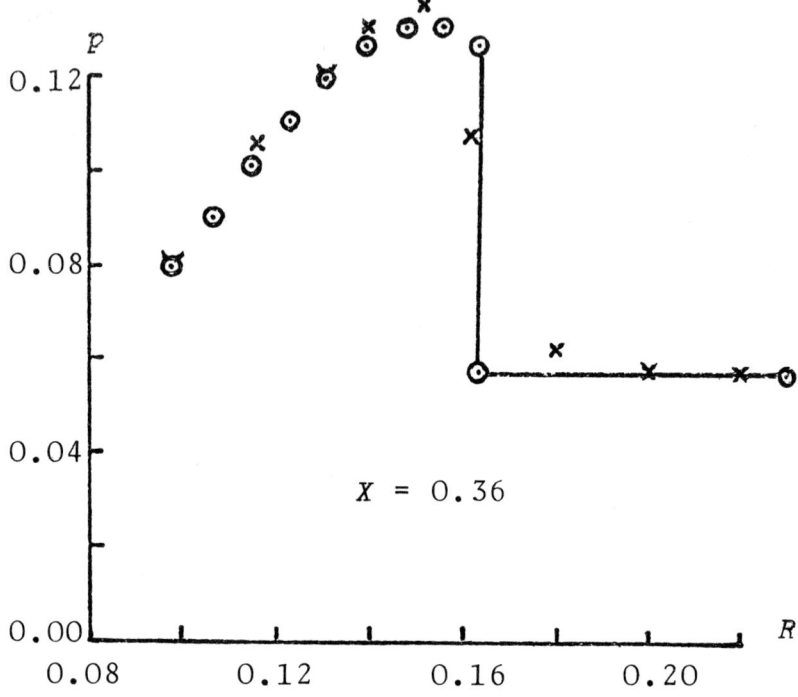

Fig. 7. Variation of pressure with distance from a delta wing in the symmetry plane through the top centre line at a point 36% of the centre chord length from the nose. x results obtained using shock-capturing method [6]. ⊙ results obtained using bow-shock fitting method [13]

Fig. 7 shows the variation of pressure with distance from the wing surface in the wing symmetry plane through the top centre line. The plot is at a point which is 36% of the wing length from the nose.

The circles in Fig. 7 represent results obtained with the bow-shock fitting operator-splitting method mentioned earlier. The crosses represent results obtained using a shock-capturing method similar to that described in [6].

Although the shock discontinuity has been smoothed in the shock-capturing method results, it should be noted that the pressures predicted by the two methods are in good agreement in regions away from the shock wave. In particular, the predicted surface pressures are in good agreement.

Figs. 8 and 9 show similar results at two other stations along the length of the wing. The graphs in Figs. 8 and 9 show that "captured" shock waves are propagated through the flowfield correctly.

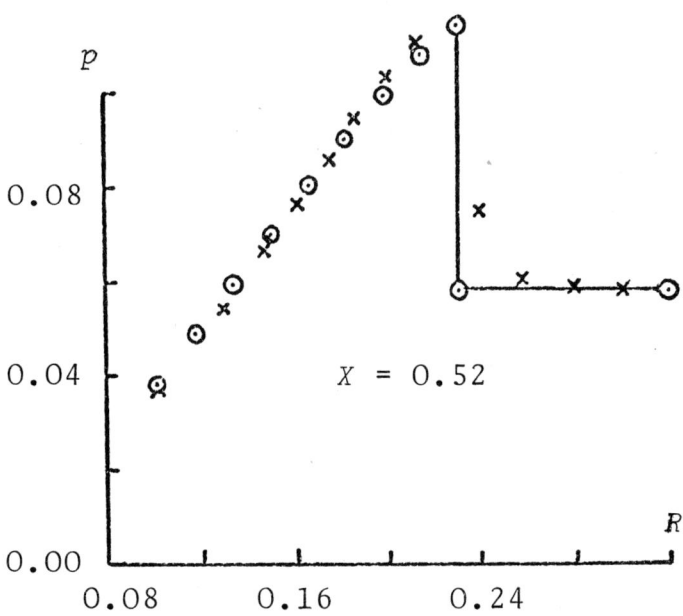

Fig. 8. Variation of pressure with distance from a delta wing in the symmetry plane through the top centre line at a point 52% of the centre chord length from the nose. x results obtained using shock-capturing method [6]. o results obtained using bow-shock fitting method [13]

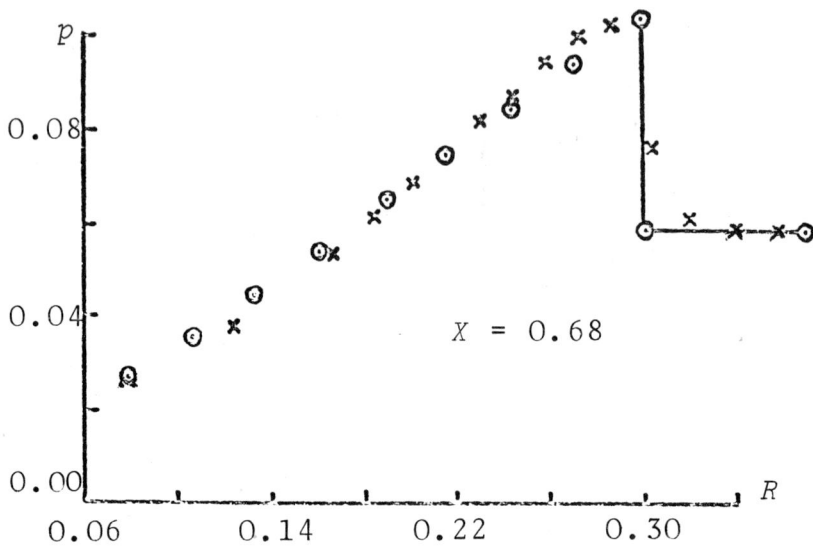

Fig. 9. Variation of pressure with distance from a delta wing in the symmetry plane through the top centre line at a point 68% of the centre chord length from the nose. x results obtained using shock-capturing method [6]. ⊙ results obtained using bow-shock fitting method [13]

Computing times needed to obtain the delta wing results were as follows

Shock-capturing method : 200 seconds on a CDC 7600 computer

Bow-shock fitting method: 62 seconds on a CDC 7600 computer

These results do not indicate that, in the delta wing study, bow-shock fitting is a significantly faster procedure than shock-capturing. The primary reason for the differences in computing times quoted here is the fact that the basic finite difference procedure used in the fitting method is much faster than the older one used in the shock-capturing method.

Supersonic Flows

Fuselage-cockpit canopy combination

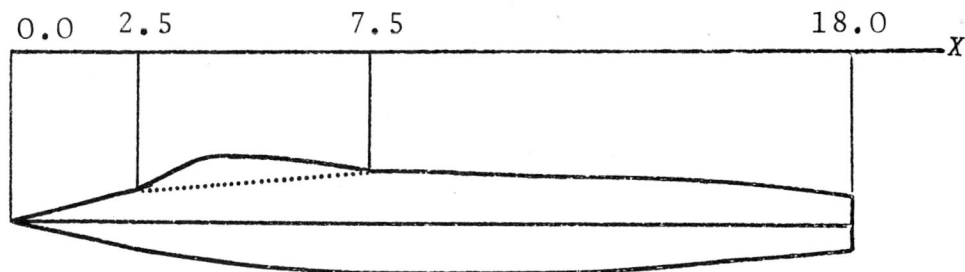

Fig. 10. Longitudinal cross-section of a cockpit canopy/fuselage combination

Fig. 10 shows the cross-sectional shape of an axisymmetric fuselage combined with a cockpit canopy. The section is through the combination's plane of symmetry. Two shock-capturing methods have been used to calculate the flowfield produced when the shape shown in Fig. 10 is placed in a uniform stream which has a Mach number of 2.0 and which is directed along the axis of symmetry of the fuselage. One method is similar to the method described in [6]. This method uses stream surfaces as coordinate surfaces. The other is the operator-splitting shock-capturing method [11] mentioned earlier. This method uses given fixed geometric surfaces in space as coordinates. The form of coordinate surfaces used to obtain results shown here is given in Appendix I.

In Fig. 11, for the top centre line surface pressures predicted by the stream-surface method are compared with corresponding values predicted by the geometric-surface method.

The two sets of theoretical results shown in Fig. 11 are in good agreement. This suggests that the quoted numerical results are free from truncation error. When, as in the present case, experimental results are not available, assessments of the accuracy of numerical results have to be made using a variety of techniques. These techniques include comparisons between theoretical results obtained by different methods.

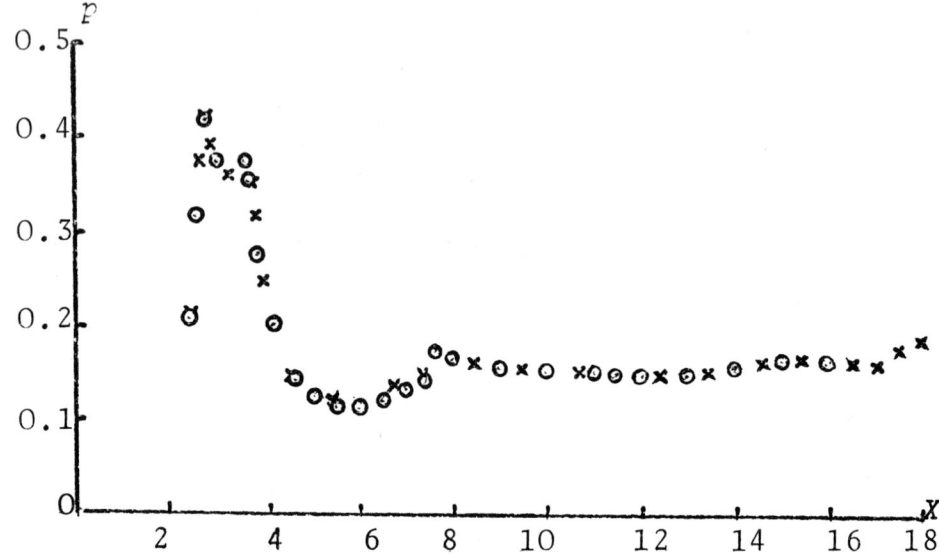

Fig. 11. Variation of surface pressure with distance along the top centre line of the fuselage/cockpit canopy combination shown in Fig. 10. x results obtained by a method which uses stream-surfaces as coordinates [6]. ⊙ results obtained by a method which uses fixed geometric surfaces in space as coordinates [11]

Figs. 12 and 13 show comparisons of variations of surface pressure with polar angle θ for two values of x along the length of the combination. For each value of x, θ = 0 at the top centre line point. The angle θ is measured at the appropriate point on the axis of symmetry of the fuselage.

Computing times:

stream-surface method : 438 seconds on CDC 7600
fixed geometric surface method : 1078 seconds on CDC 7600

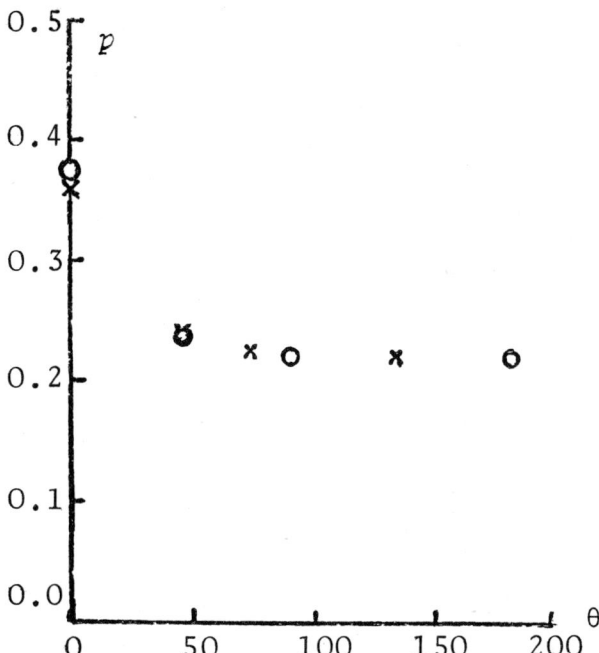

Fig. 12. Variation of pressure on the surface of a cockpit canopy combination (Fig. 10) with polar angle θ in a plane normal to the axis of symmetry of the fuselage at $X = 3.0$

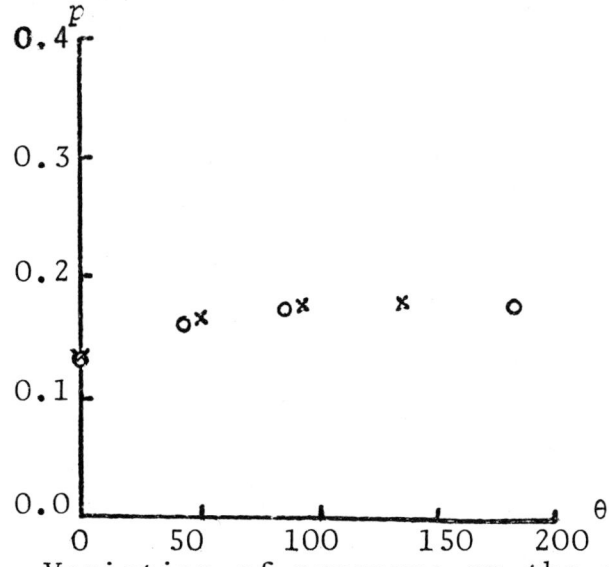

Fig. 13. Variation of pressure on the surface of a cockpit canopy combination (Fig. 10) with polar angle θ in a plane normal to the axis of symmetry of the fuselage at $X = 5.0$

Intakes

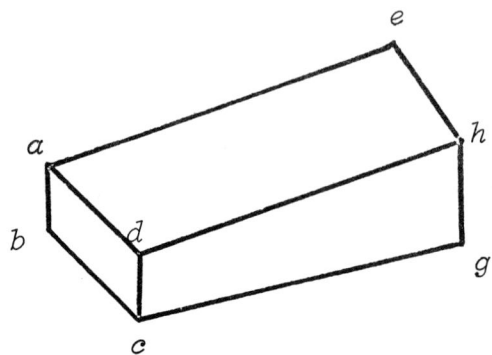

Fig. 14. Sketch of an engine air intake with parallel plane side-walls. $ad = 2b$, $ab = 2h$ (see Appendix). The lines dh and cg are inclined at 7.6° to each other. The planes $dhgc$, aeb are parallel and form the side-walls of the intake

Fig. 14 is a sketch of an intake which has two parallel plane side-walls. The top and bottom surfaces of the intake are planes inclined at an angle of $\theta = 7.6°$ to each other. In this section, results are shown for a particular intake ($b/h = 102$) placed in a uniform supersonic stream (Mach number = 4.0) which is parallel to the intake sidewalls and is directed so that it is inclined at an angle of 3.8° to both the top and the bottom surfaces of the intake.

For computational purposes, some modelling of the given intake shape was carried out. Near its leading edge, the top of the model surface points in the stream direction. The model top surface is then faired into the given intake top surface over four finite difference mesh lengths. In addition, the sharp corner at the junction of the given intake top and right side surfaces is rounded in the model.

The region in which the flowfield is calculated may be restricted by making use of two symmetry properties associated with the intake problem described here.

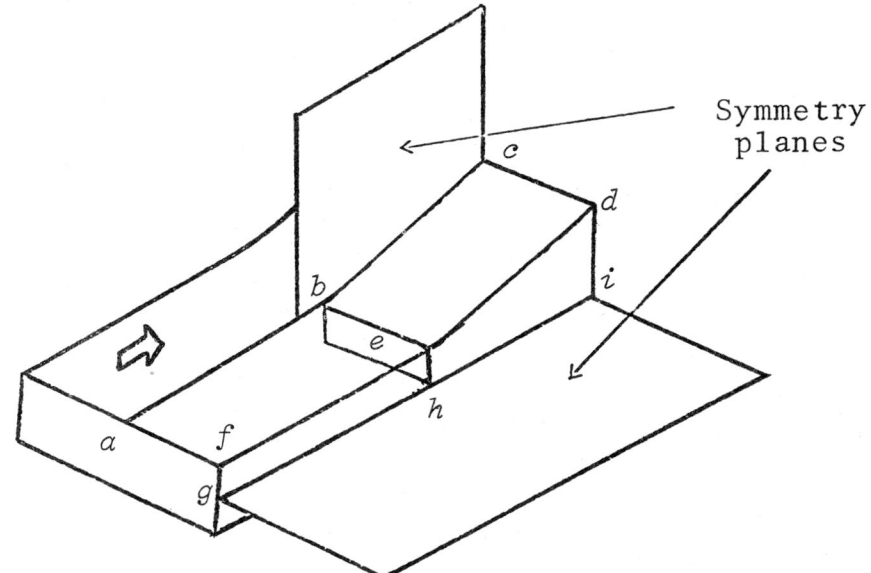

Fig. 15. Sketch illustrating the model of the intake used for computation

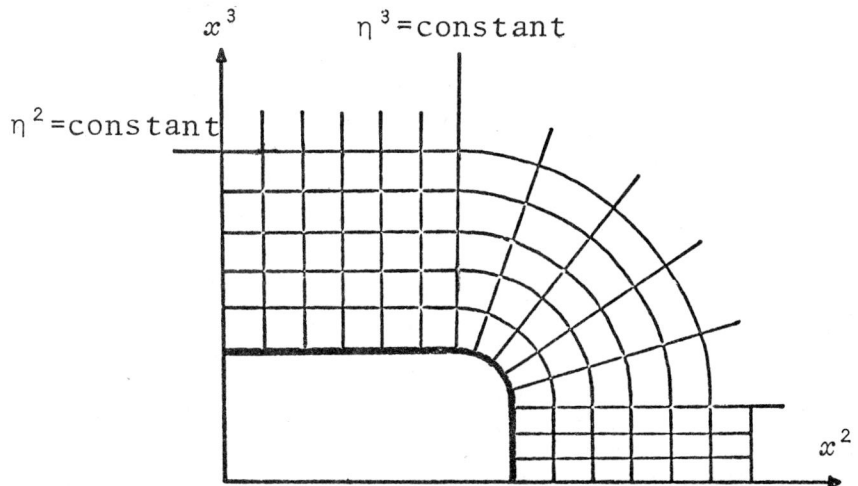

Fig. 16. Illustration of curves of intersections of surfaces η^2 = constant and η^3 = constant with a plane η^1 = constant in Cartesian space (see equations in Appendix I)

Fig. 15 is a sketch illustrating the modelling of the intake surface and the symmetry planes for the flowfield produced by the intake. The flow is undisturbed at points in that plane which is perpendicular to the uniform stream and which also passes through the leading edge of the model of the intake. The form of equations defining co-ordinate surfaces η^1 = constant, η^2 = constant and η^3 = constant for the intake model are given in Appendix I. Fig. 16 illustrates the curves of intersection of surfaces η^2 = constant and η^3 = constant with a plane η^1 = constant. The intake model surface is η^2 = 0.

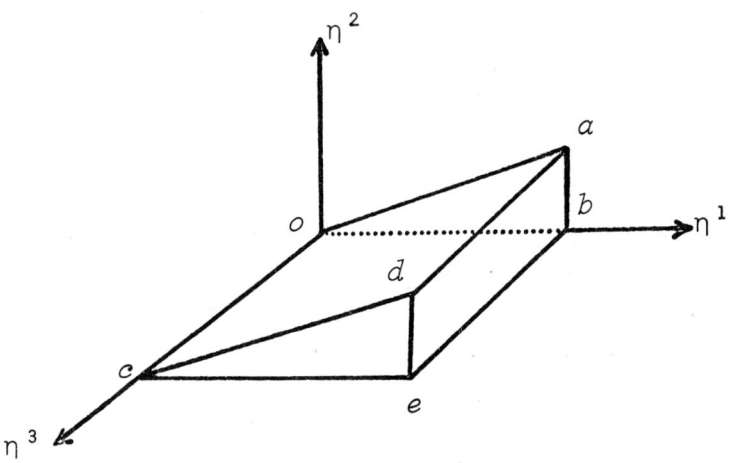

Fig. 17. Working space in which computation of the flowfield produced by an intake model like that shown in Fig. 15 was carried out. Symmetry conditions applied on planes *oab* and *cde*. Boundary condition applied on plane *obec* which is the image in η-space of part of the intake surface

Fig. 17 illustrates the computational region in (η^1, η^2, η^3) space within which disturbances produced by the intake have to be calculated.

Figs. 18, 19 and 20 show comparisons between calculated and measured pressures on the top surface of the intake. Strictly, the measured pressures are for an equivalent intake wedge, but the

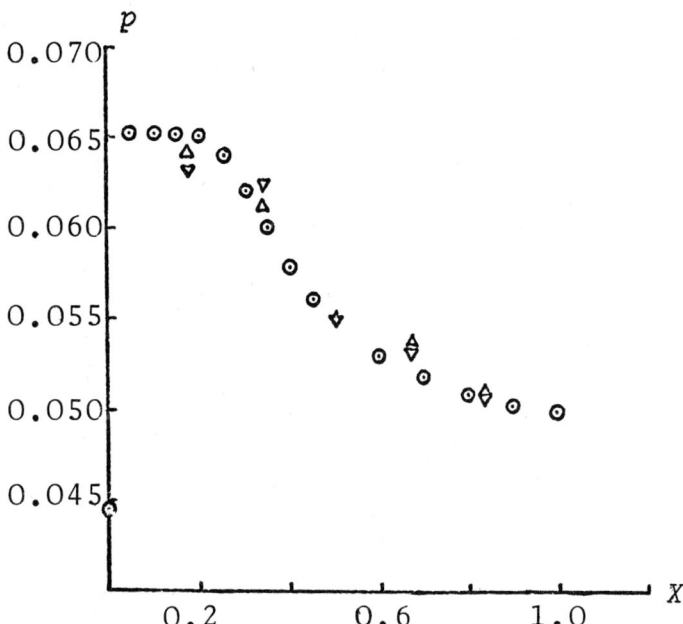

Fig. 18. Variation of pressure on the intake surface top centre line with distance in the uniform stream direction. ⊙ computed pressures, △ measured pressures on the upper surface, ▽ measured pressures on the lower surface

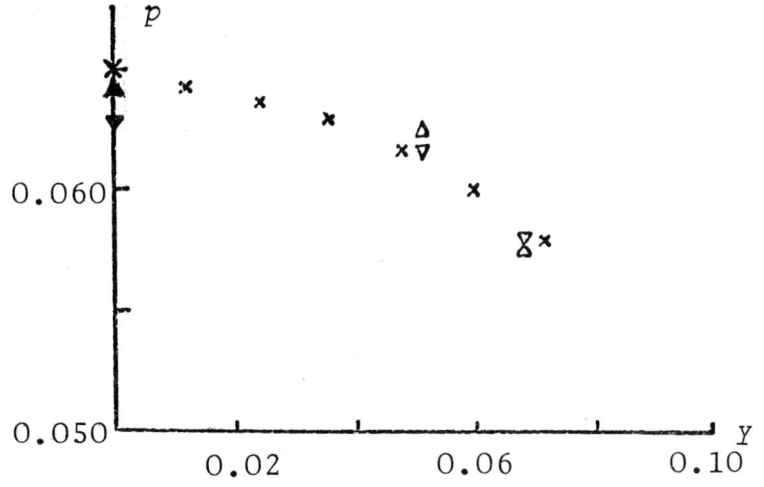

Fig. 19. Variation of pressure on the intake surface with distance from the top centre line in the plane $X = 0.166$

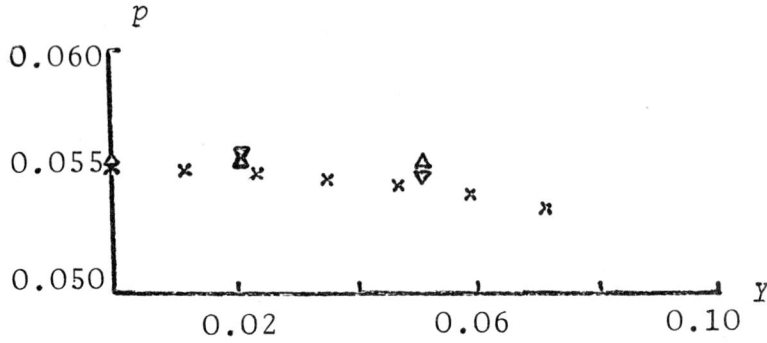

Fig. 20. Variation of pressure on the intake surface with distance from the top centre line in the plane $X = 0.50$

model intake may be regarded as a model of the intake wedge. The close agreement between the numerical and experimental results suggests that the theoretical results are satisfactory and that the intake model used is a satisfactory model of the actual shape used in the experiments.

Computation time for the intake problem was 288 seconds on a CDC 7600.

CONCLUSIONS

The results given in the previous section indicate that it is possible to treat complete flows or parts of flows associated with general body shapes. It has been shown that reasonably accurate results can be obtained in cases when the flow remains supersonic everywhere. For shapes which are much more complicated than those considered in the last section, the amount of computation needed to obtain accurate results may increase significantly beyond the times quoted. However, provided necessary computer time is available, there is no reason why such problems should present special difficulties.

The results for the delta wing (p. 367) indicate that it is not necessary to treat all shock waves as discontinuities in order to obtain

accurate estimates of surface pressures. Although
I have not yet encountered any, there may be some
problems which will be solved satisfactorily only
if all shock waves in the solution are detected
and fitted as discontinuities. If this has to be
done, then I am sure that it will be possible to
produce necessary algorithms and computer programs.
Moretti [3] is already making good progress in
this direction.

Important, outstanding, difficult, problems
are mixed flow problems such as the intake-spill-
age problem and the viscous/inviscid flow interaction problem mentioned earlier. It is impossible to guess how long it will take before mixed
flow calculations can be done routinely. The
practical importance of these problems cannot be
overstated and there can be little doubt that in
the immediate future a great deal of research
effort will be devoted to various problems of this
type.

ACKNOWLEDGEMENTS

All the program development and small scale
testing of the new methods mentioned in this paper
were carried out on the SRC's IBM 370/195 computer
at Chilton. Without these facilities, it would
have taken years rather than months to obtain
results such as those presented here. I grateful-
ly acknowledge the SRC's support.

I am also grateful to Mr. E.L. Goldsmith of
RAE, Bedford, who provided the intake problem and
supplied the experimental results used to validate
the numerical method. Finally, I acknowledge also
the work of Mr. P. Caine and Mr. G.T. Laws who
have been responsible for much of the detailed
work, including the program handling, necessary to
obtain the numerical results presented here.

REFERENCES

1. South, J.C. and Klunker, E.B., "Methods for calculating non-linear conical flows," *in* "Analytic Methods in Aircraft Aerodynamics," NASA SP-228, 131-158 (October 1969).

2. Moretti, G., "Inviscid flowfield about a pointed cone at an angle of attack," *AIAA J.*, **5**, No. 4, 789-791 (April 1967).

3. Moretti, G., "Three-dimensional, supersonic, steady flows with any number of imbedded shocks," AIAA Paper No. 74-10 (May 1974).

4. Marconi, F. and Salas, M., "Computation of three-dimensional flows about aircraft configurations," *J. Computers Fluids*, **1**, 185 (1973).

5. Walkden, F. and Caine, P., "Application of a pseudo-viscous method to the calculation of the steady supersonic flow past a waisted body," *Int. J. Num. Meth. Eng.*, **5**, 151-162 (1972).

6. Walkden, F., Laws, G.T. and Caine, P., "Shock capturing numerical method for calculating supersonic flows," *AIAA J.*, **12**, No. 5, 642-647 (May 1974).

7. Walkden, F., "A form of the supersonic flow equations for an ideal gas," ARC Report 34160 (1972).

8. Kutler, P., Lomax, H. and Warming, R.F., "Computation of space shuttle flow fields using noncentered finite difference scheme," AIAA Paper No. 72-193 (December 1972).

9. Rakich, J.V. and Kutler, P., "Comparison of characteristics and shock capturing methods with application to the space shuttle vehicle," AIAA 10th Aerospace Sciences Meeting, San Diego, California (January 1972).

10. Chu, C.W., "Compatibility relations and a generalized finite difference approximation for three-dimensional steady supersonic flow," *AIAA J.*, **5**, No. 3 (March 1967).

11. Butler, D.S., "The numerical solution of hyperbolic systems of partial differential equations," *Proc. Roy. Soc. A.*, **255**, 232-252 (1960).

12. Walkden, F., Laws, G.T. and Caine, P., "Operator-splitting shock-capturing method for supersonic flows," *in the press*.
13. Walkden, F. and Caine, P., "Operator-splitting bow shock-fitting method for supersonic flows," *in the press*.

APPENDIX I

Coordinate Systems for Computation

1. *Fuselage/Cockpit Canopy Combination*

$$x^1 = \eta^1$$
$$x^2 = r_{ob}(\eta^1,\eta^3)\eta^2 + r_b(\eta^1,\eta^3)(1-\eta^2)$$
$$x^3 = \eta^3$$

x^1, x^2 and x^3 are cylindrical polar coordinates. $r = r_b(x^1,x^3)$ is the equation of the surface of the combination in cylindrical polar coordinates. $r = r_{ob}(x^1,x^3)$ is a given function such that $r_{ob} > r_b$.

2. *Intake*

$$\left. \begin{array}{l} x^1 = \eta^1 \\ x^2 = \eta^3(h(\eta^1)-\delta)/(\eta^3)_1 \\ x^3 = b + \eta^2 \end{array} \right\} 0 \leq \eta^3 \leq (\eta^3)_1$$

$$\left. \begin{array}{l} x^1 = \eta^1 \\ x^2 = h(\eta^1) - \delta + (\eta^2+\delta)\sin\left[\dfrac{((\eta^3)-(\eta^3)_1)\frac{\pi}{2}}{(\eta^3)_2-(\eta^3)_1}\right] \\ x^3 = b - \delta + (\eta^2+\delta)\cos\left[\dfrac{(\eta^3)-(\eta^3)_1\frac{\pi}{2}}{(\eta^3)_2-(\eta^3)_1}\right] \end{array} \right\} (\eta^3)_1 \leq \eta^3 \leq (\eta^3)_2$$

$$\left. \begin{array}{l} x^1 = \eta^1 \\ x^2 = \eta^2 + h(\eta^1) \\ x^3 = \left[\dfrac{(\eta^3)-(\eta^3)_3}{(\eta^3)_2-(\eta^3)_3}\right](b-\delta) \end{array} \right\} (\eta^3)_2 \leq \eta^3 \leq (\eta^3)_3$$

where

$$h = h_0 \qquad \eta^1 < (\eta^1)_1$$

or

$$h = h_0 + \frac{0.5\tan\theta_b(\eta^1-(\eta^1)_1)^2}{(\eta^1)_2-(\eta^1)_1} \qquad (\eta^1)_1 \leq \eta^1 \leq (\eta^1)_2$$

or

$$h = \eta^1 \tan\theta_b \qquad \eta^1 \geq (\eta^1)_2$$

θ_b, $(\eta^1)_1$, $(\eta^1)_2$, δ, b, $(\eta^3)_1$, $(\eta^3)_2$ and $(\eta^3)_3$ are constants. x^1, x^2 and x^3 are Cartesian coordinates.

APPLICATIONS OF LINEARISED SUPERSONIC WING THEORY TO THE CALCULATION OF SOME AIRCRAFT INTERFERENCE FLOWS

M. Purshouse and R.K. Nangia

(British Aircraft Corporation Limited)

PREAMBLE

This paper illustrates the way in which a simple numerical formulation of linearised supersonic wing theory can be used to calculate some typical aircraft interference flows.

The Mach box method is used, in which disturbance regions in a wing plane are overlaid by a grid of constant source strength Mach boxes. In the real plane these boxes are rectangular, having Mach lines as diagonals, but become square when a coordinate transformation to the equivalent problem at Mach 1.414 is introduced. The problem is solved in this transformed plane.

Flow perturbations produced at a point in the downstream zone of action from a source are calculated by using the appropriate Aerodynamics Influence Coefficient (AIC) relationships. The AIC's are evaluated from analytical expressions. The source strengths of boxes on the wing plane are determined at their centres by the local wing slope on the wing itself. Off the wing, the appropriate condition on the velocity potential is applied. An empirical area factor correction is used to assign a representative source strength to boxes cut by a subsonic leading edge. Some applications of the method are given.

The first is in the calculation of an underwing flowfield at a typical intake location. The requirements for good intake performance are reviewed, and the extent to which they can be met by underwing installation is considered. In particular, the effect of leading edge droop on the intake flowfield is discussed in its implications for intake performance.

The second is the calculation of the flowfield at a fin

location arising from asymmetric elevon deflection. From the calculated interference field, the side force on fin and rudder can be estimated.

Finally, consideration is given to a close coupled wings configuration. This problem is important in examining the possible use of additional lifting surfaces on slender winged aircraft. It also arises in studies of controlled configured vehicles and low sonic-boom supersonic transports. Here we have calculated the interference field produced by one wing on another at Mach 2 for an arbitrary configuration: two geometrically similar deltas of 65° sweep. The variables considered are vertical height spacing between the wings and relative wing sizes.

Comments are made on the consequences of the interference field for the planform design of the interfered wing; however no attempt has been made to optimise the wing shapes.

INTRODUCTION

A knowledge of the interference flows induced by a wing or wing-body combination has always been of interest in the design of supersonic aircraft. Consideration of such flows is important in deciding, for example, the size and/or positioning of engine nacelles, controls and lifting and non-lifting surfaces.

The equations governing these flows are, of course, nonlinear but, because of the difficulty of solving them for the more general configurations within a reasonable computer time, much reliance is still placed on the simpler linear theory backed up by experiment.

Many of the flows about an aircraft, however, are described quite well by the linear theory, and here we present a theoretical approach for calculation of the wing-induced interference flows encountered in some typical problems of current aeronautical interest.

METHOD AND COMPARISONS WITH CONICAL THEORY

There are many ways in which steady and unsteady linearised supersonic wing theory can be formulated numerically. One way is to consider the disturbance regions in the wing plane (including the diaphragm ahead of subsonic leading edge and the wake) as a source sheet with correct boundary conditions [1], [2], [3], [4]. An approximate solution is obtained by overlaying the source sheet with a suitable grid of elementary areas such as characteristic boxes (Mach lines forming the edges), Mach boxes (Mach lines forming the diagonals) [5] and triangles [6] - Fig. 1.

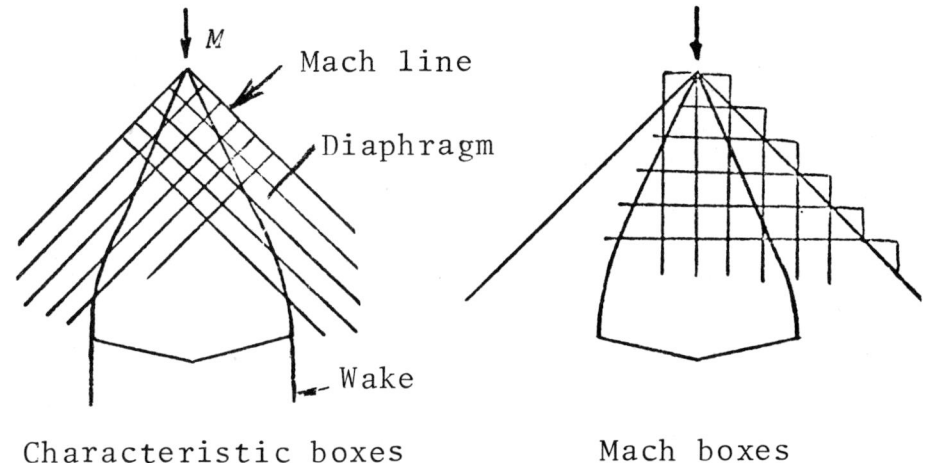

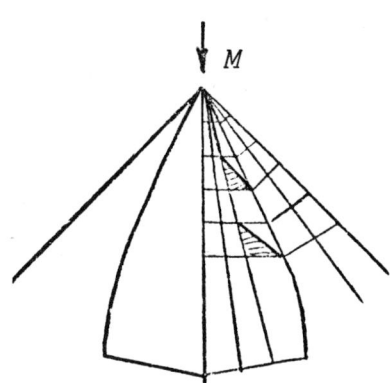

Fig. 1. Grids used in linearised theory

Of these grids, because of their relation to the Mach lines, the first two lend themselves to simple Aerodynamic Influence Coefficient (AIC) approaches, and the effect of one element on another becomes purely a function of relative position. In the case of triangular elements this is not possible and the shape of the element has to be taken into account.

It has been shown that the accuracy with which flow properties may be described in any of these approaches depends mainly on the number of elements in the grid chosen and the way in which any singularities are dealt with.

However, whichever method is chosen, advantage can be taken of the assumed linearity of the problem so that for a wing of given planform the effects of incidence, camber and thickness can be calculated independently and then superposed.

In this paper, the Mach box technique [7] to [11] has been used because it is well suited to the calculation of out-of-plane flows.

A brief discussion is presented in Appendix A.

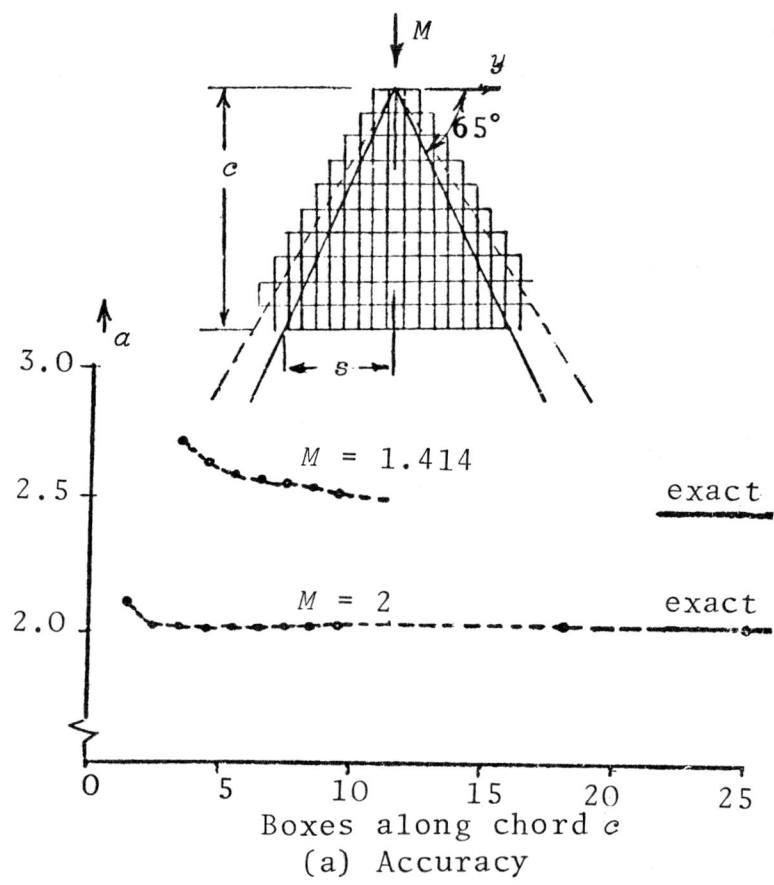

(a) Accuracy

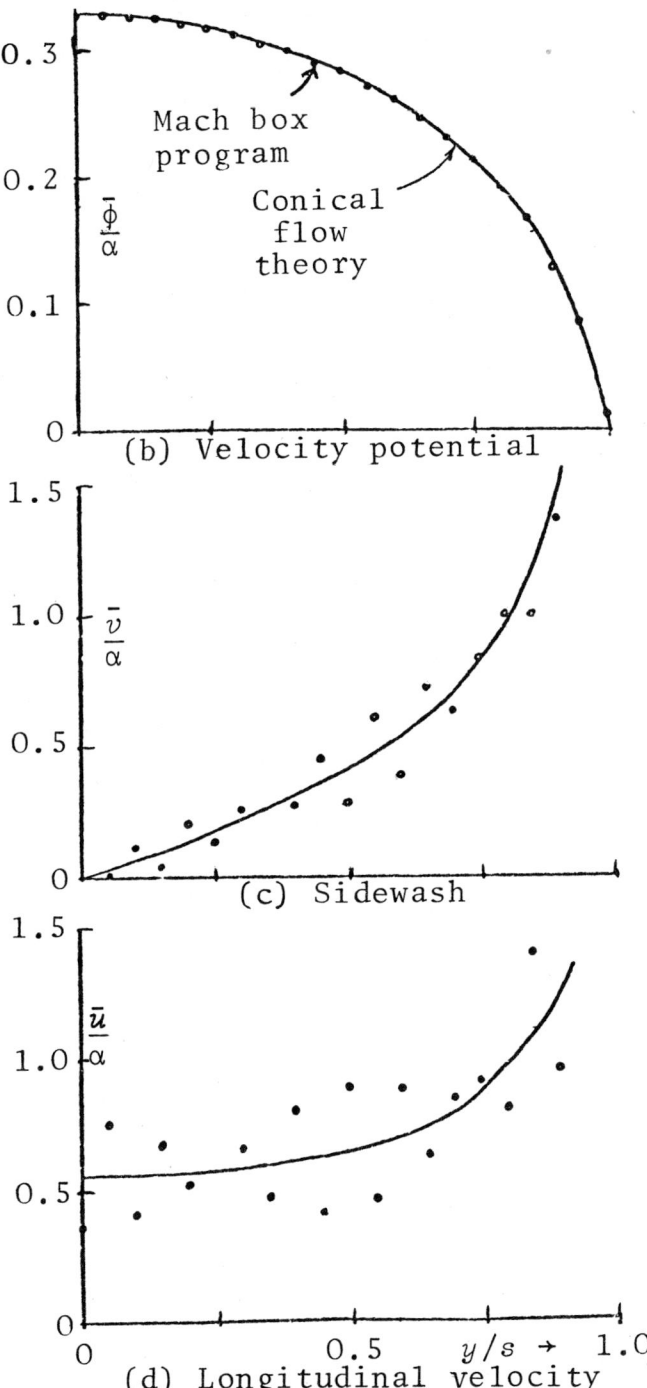

Fig. 2. Comparisons with conical wing theory

Figs. 2(a) to (d) show how the results from the method, obtained for a planar delta of 65° sweepback at incidence, compare with analytical results from conical flow theory.

Fig. 2(a) shows the computed lift curve slope a at two Mach numbers as a function of the number of boxes along the chord. Naturally, accuracy improves with an increase in the number of boxes but even with only 5 boxes along the chord, a is only in error by 2%.

Figs. 2(b) to (d) show the computed non-dimensional distributions of velocity potential, sidewash and longitudinal velocity in the wing plane. These are compared with the exact conical flow theory predictions. The computations involved the use of a Mach grid with 25 boxes along the centre-line chord.

The distribution of velocity potential is given accurately by the present method; the sidewash and longitudinal velocity distributions exhibit oscillations which are a result of the simple representation of the downwash singularity at the subsonic leading edge. Their amplitude cannot be reduced merely by using a finer grid [10]: more refined techniques on the lines of [8] are applicable.

However, as can be seen, the oscillations are stable and the mean lines agree well with the conical flow theory. The severity of this type of oscillation is reduced when more practical wing cambers, which incorporate drooped leading edges, are considered, or when the wing has a supersonic leading edge.

APPLICATIONS

Flowfield at a typical intake location

The design of a supersonic intake, in attempting to satisfy the two important requirements of (*i*) low external drag and (*ii*) good internal performance, must take account of the aircraft flowfield in which it is to operate. For example, in underwing installations, the external drag is minimised [12] by aligning the nacelles at half the mean intake face sidewash angle (Fig. 3).

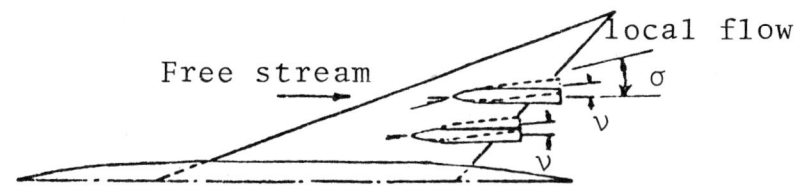

Fig. 3. Nacelles aligned at half local sidewash angle to reduce external drag [12]

The internal performance of the intake depends on the mean incident Mach number and the over-all level of non-uniformity of the flow. In general, reduced entry Mach number tends to improve the over-all pressure recovery achieved, whilst increased flow distortions tend to worsen it. The intake internal performance is also affected by improper alignment in nominally uniform flow: for example, in two-dimensional intakes, certain conditions of sideslip can lead to the intake shocks impinging downstream of the cowl-lip - the resulting reflected shocks system producing losses of pressure recovery and unwanted additional flow distortions [13]. Fig. 4 shows experimental results illustrating the sensitivity of typical two-dimensional and axisymmetric intakes [14] to sideslip.

An understanding of the interference flows generated by the aircraft is therefore essential if optimum intake performance is to be realised. Flowfield surveys are normally carried out to aid this understanding but such surveys are expensive and do not necessarily provide a physical feel for the way in which the over-all intake flowfield picture is influenced by small changes in, for example, wing leading edge camber. It is here that the linear theory can be useful.

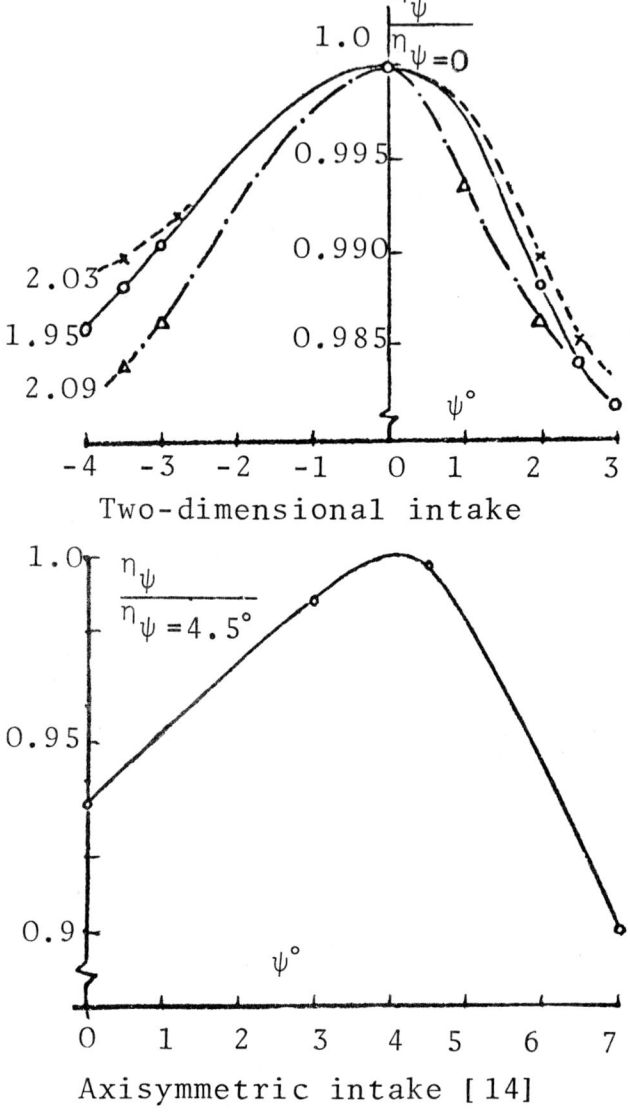

Fig. 4. Effect of yaw on intakes

We illustrate this by referring to an intake location under a wing with subsonic leading edges. This type of intake location has several advantages [14]: one is that the initial compression in the wing flowfield is near isentropic and thus the reduced local Mach numbers under the wing are achieved for a very small stagnation loss. This leads to an over-all increase in intake pressure recovery. A further consequence of using the compression under the wing is that the intake frontal area can be reduced for a given capture mass flow thus giving a useful weight saving.

We take the example of a wing with leading edges swept back 65° in Mach 2 flow (Fig. 5).

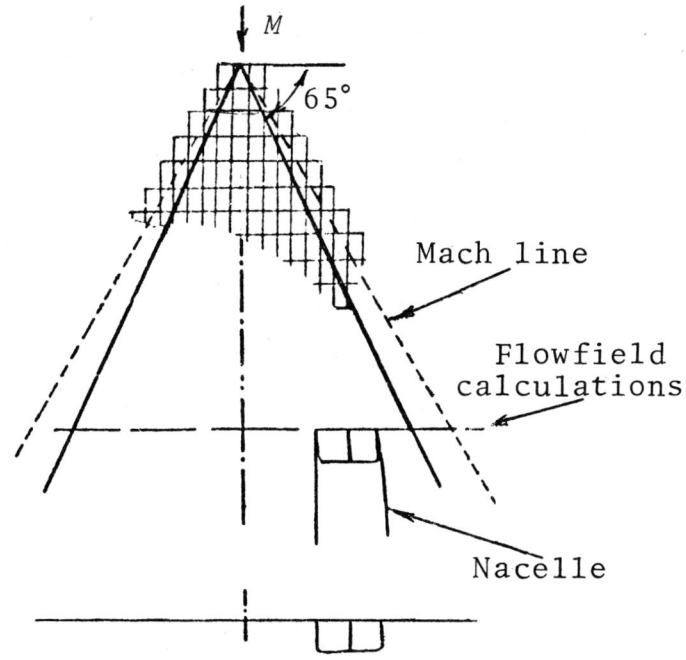

Fig. 5. Wing nacelle geometry

The calculations have used 25 boxes to the nacelle plane. Fig. 6 shows the upwash, sidewash and longitudinal "wash" contour distributions under a flat wing; the analytical results for upwash by Nielson and Perkins [15] are shown for comparison. In the nacelle region, the analytical and calculated results agree well.

Besides illustrating simply the flowfield for a flat wing, Fig. 6 shows how the flowfield under a wing of arbitrary thickness or camber, with stated planform, would respond to a small incidence change; it shows that the flowfields in the

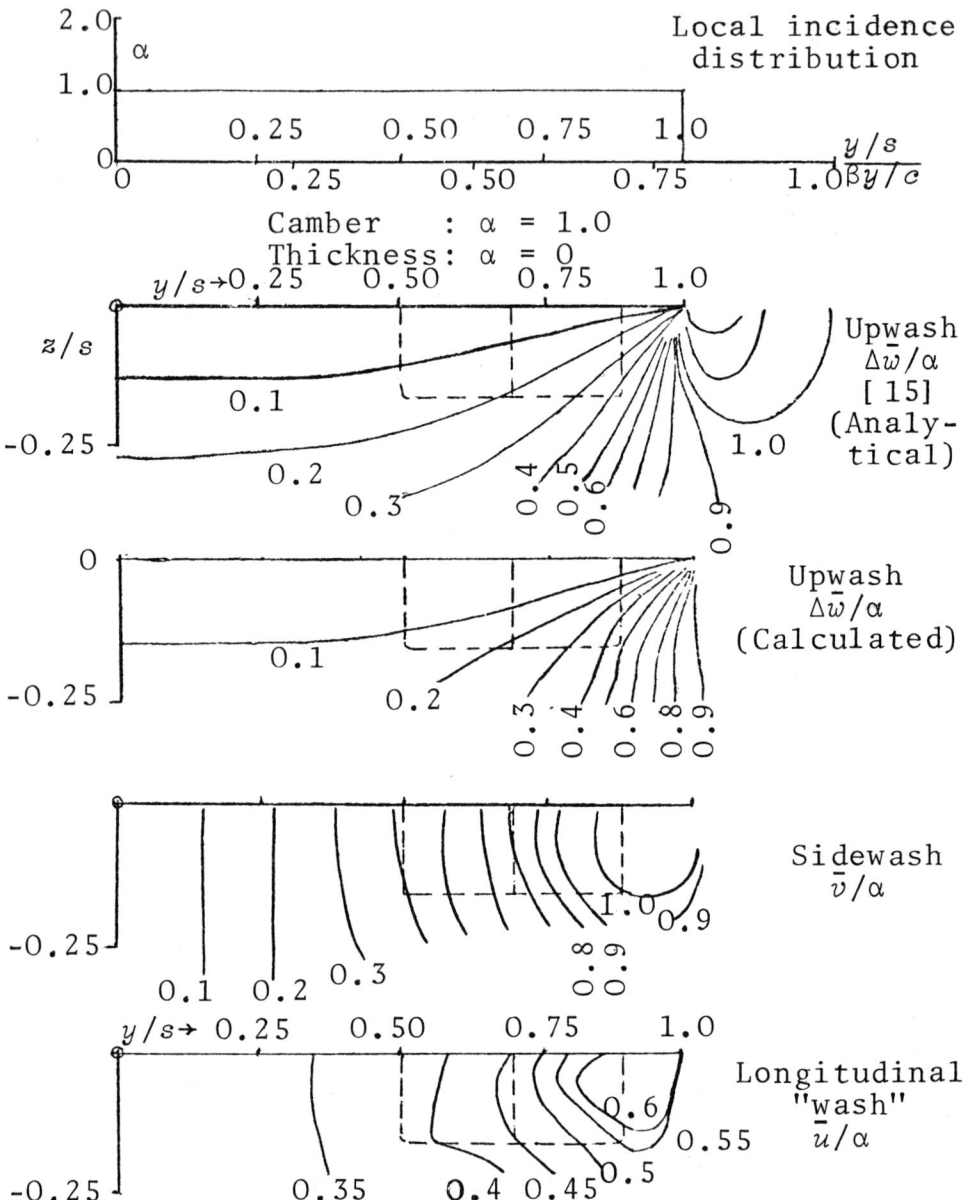

Fig. 6. Underwing flowfield - planar wing

region well inboard of the leading edges are relatively insensitive to incidence changes and are therefore attractive from the point of view of nacelle location. Fig. 7 illustrates a supersonic transport (SST) design incorporating this idea [16].

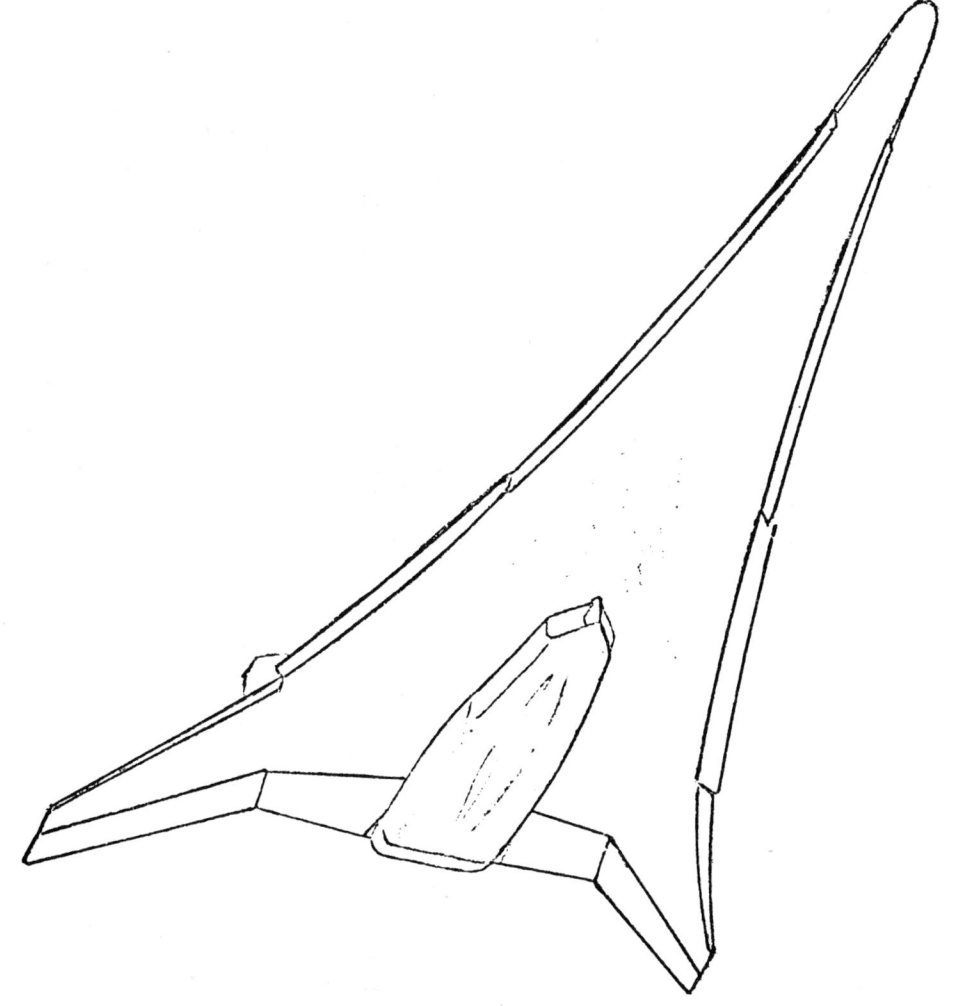

Fig. 7. SST configuration with underwing nacelles along centre [16]

Fig. 8 shows the effect of an 8th power conical camber giving zero local incidence at the leading edge.

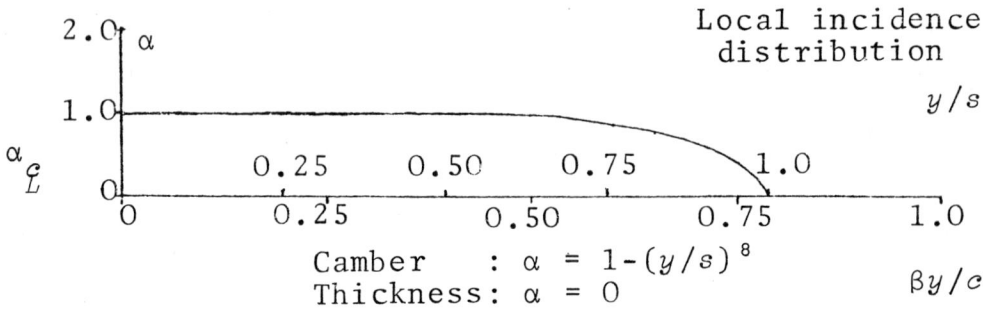
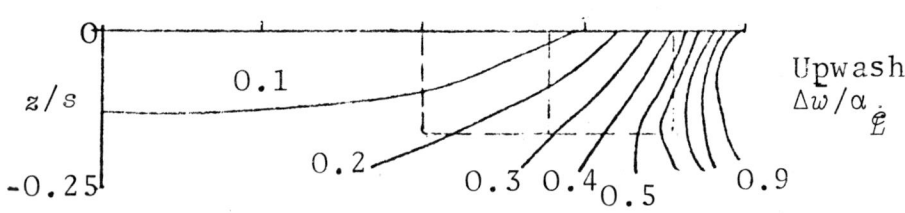
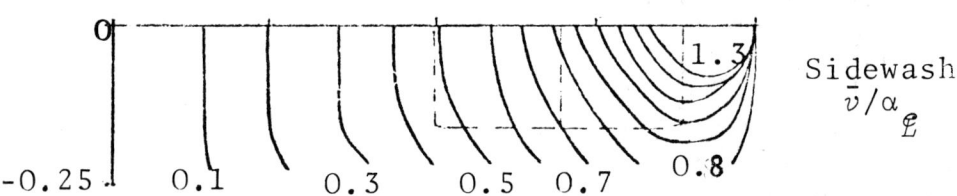
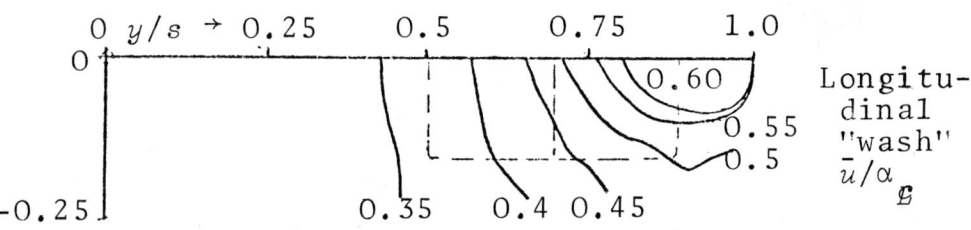

Fig. 8. Underwing flowfield - cambered wing

Comparison of Figs. 6 and 8 illustrates the effect of leading edge droop at the same centre-line incidence. Not surprisingly the effect is to spread out over a large area those rapid changes in flow properties normally occurring in a very small region near the leading edge. It is also seen that the spatial separation of the contour lines has changed not only in a spanwise direction but in a direction normal to the wing. The upwash under the cambered wing is greater at all points of interest. The values of sidewash and longitudinal "wash," although greater at points nearer the wing, tend to fall away more rapidly with distance normal to the wing. This point will be important in examining later cases which include larger and more representative leading edge droops and a thickness distribution.

In the cruise condition, however, the effects on the field arising from thickness, camber and incidence are of comparable magnitude. We therefore illustrate the effects of simple thickness distribution to put these effects in their proper perspective.

Fig. 9 shows maps of the three non-dimensional components of the perturbation velocity. These relate to a conical thickness distribution (quadratic in the spanwise direction) per unit wing centre-line thickness to chord ratio calculated at the intake face.

These thickness effects must be taken into account in determining the absolute values and changes of Mach number.

Finally, Figs. 10 and 11 compare the flow-fields under a wing of the given planform for two different camber distributions. Both include the effects of the thickness distribution described, corresponding to a 4% ratio of centre-line thickness to chord at the intake face.

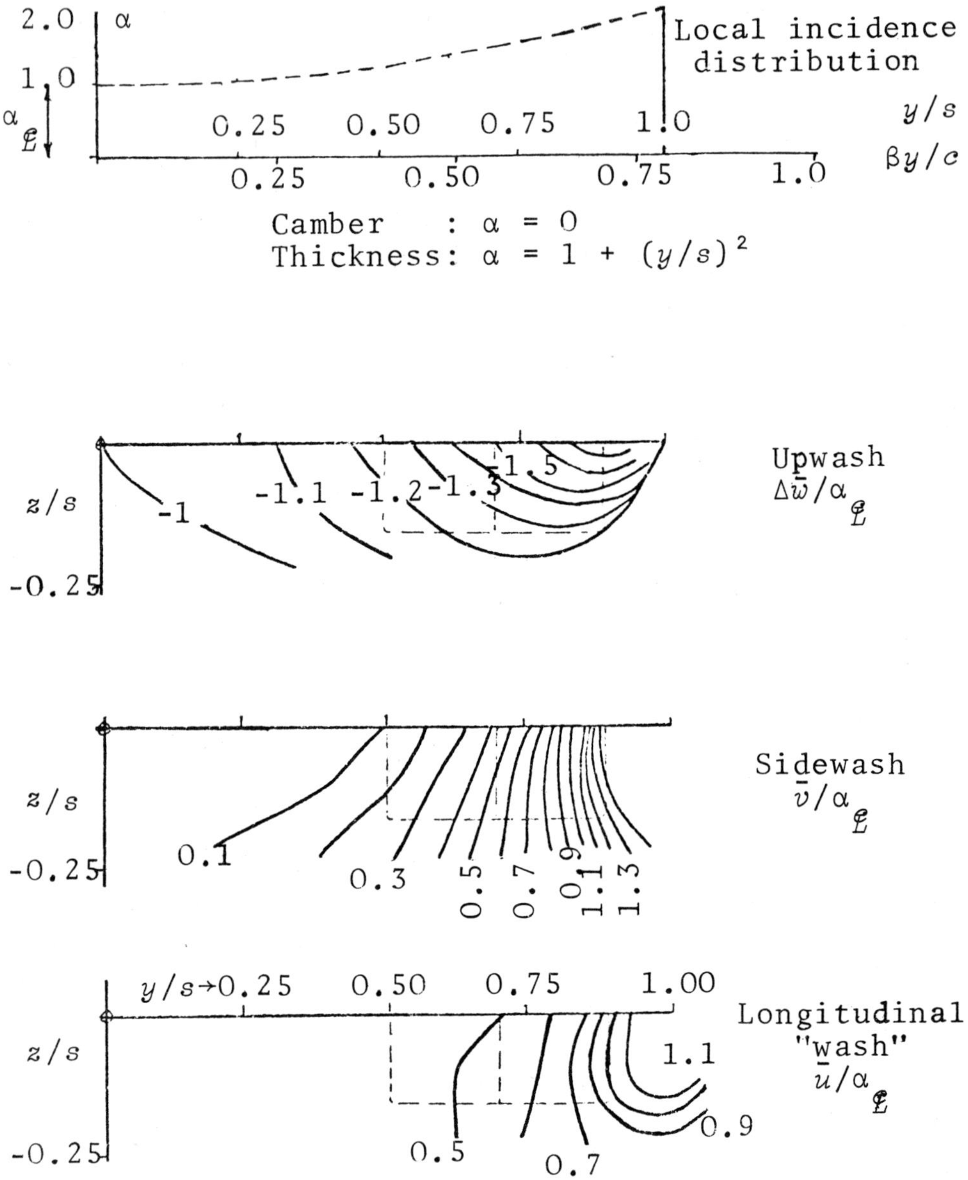

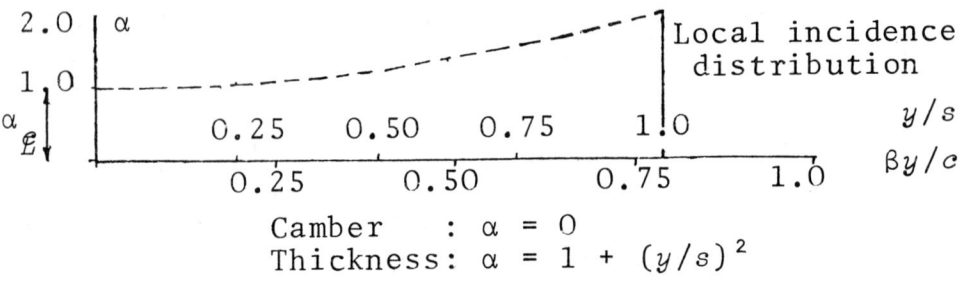

Fig. 9. Underwing flowfield - thickness effects

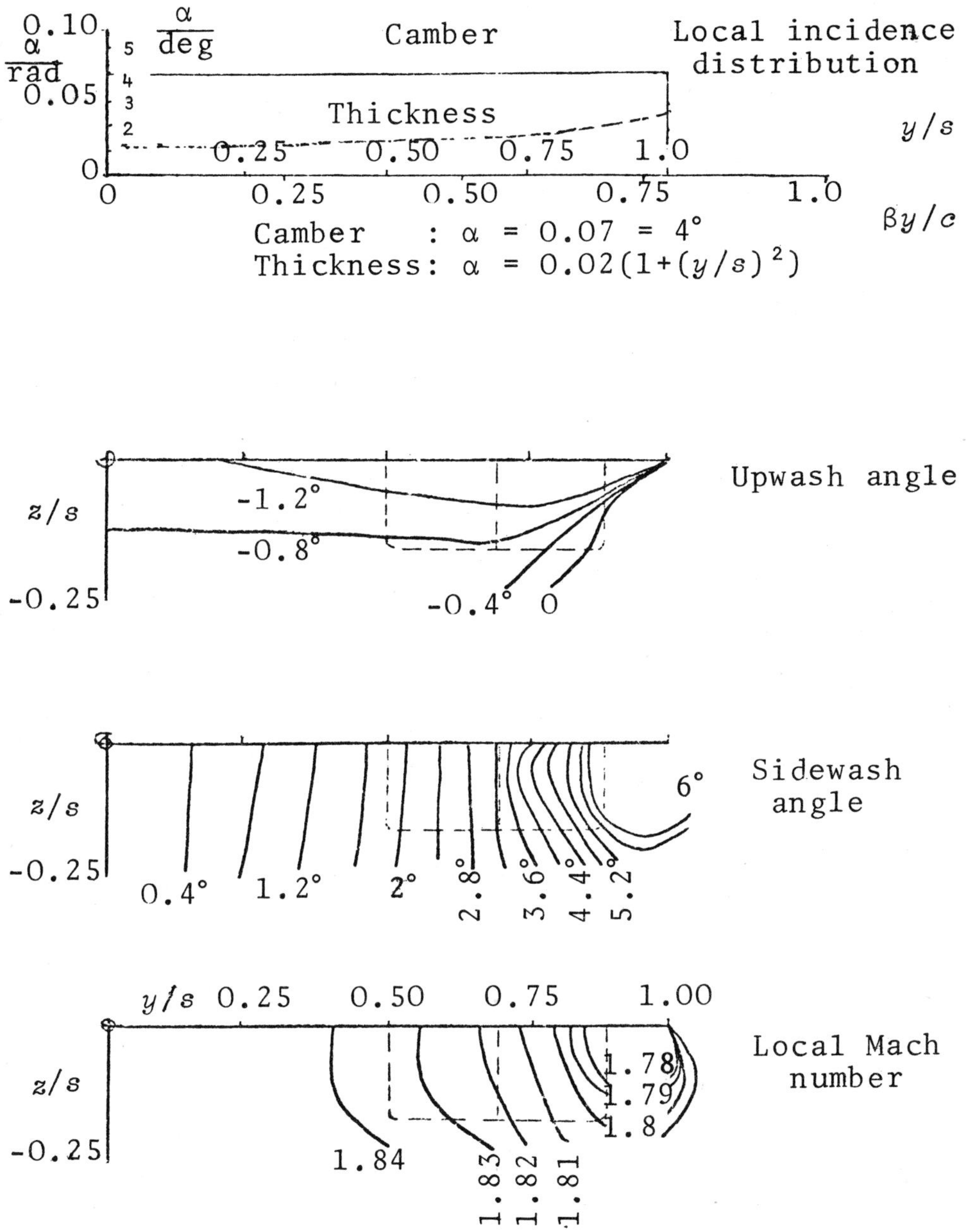

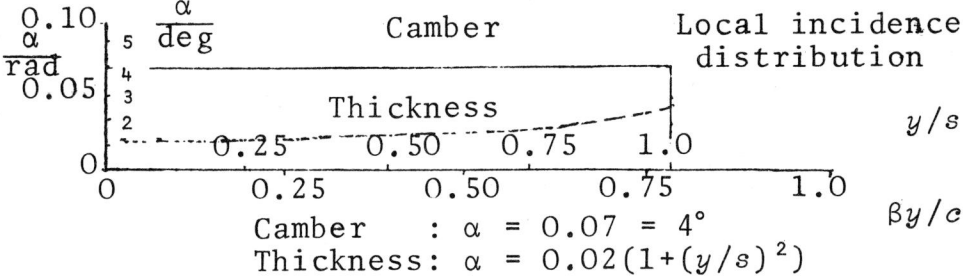

Fig. 10. Underwing flowfield. Representative wing 4° incidence, thickness zero droop leading edge

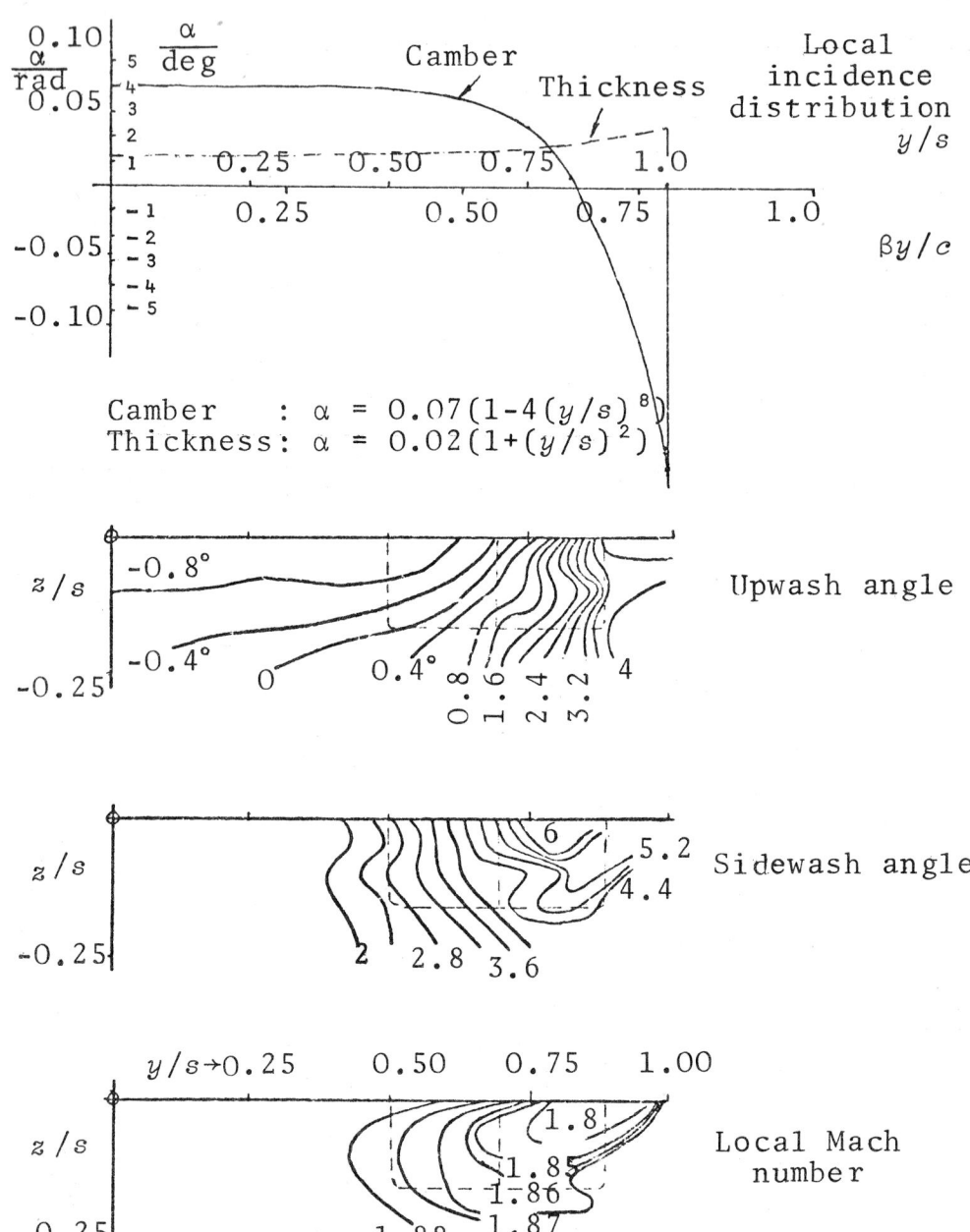

Fig. 11. Underwing flowfield. Representative wing 4° incidence at centre-line, thickness and leading edge droop

Fig. 10 shows results for a wing with zero mean camber at an incidence of 4°. Fig. 11 shows results for an 8th power mode of conical droop camber with 4° centre-line incidence and -12° incidence at the leading edge. It can be seen that the changes to the flowfield are considerable. The increased inboard penetration of the leading edge effect is evident in the cambered case. The variation in upwash is also considerably more severe near an outboard intake; the sidewash and Mach number distributions are also significantly modified.

A more satisfactory approach would, of course, be to compare the flowfields at the same c_L rather than the same incidence, but as this would require more information about the aircraft concerned (e.g., centre of gravity position and control powers)- it has not been attempted here.

Accurate information of this type would be of great value in the design of intakes and subsequent estimation of their performance.

No direct experimental evidence is available for comparison, but the flowfield maps presented in Fig. 12 (from [17]) for a wing of more complicated geometry provide encouragement, and it is hoped to proceed with more calculations on a wing of this type.

Consideration will probably need to be given to improving the representation of the subsonic leading edge.

Elevon/fin-rudder interaction

The deflection of a control surface causes additional perturbations in the flowfield about an aircraft. In particular, a sidewash field so produced can lead to substantial side forces on fin or rudder surfaces in the interference region.

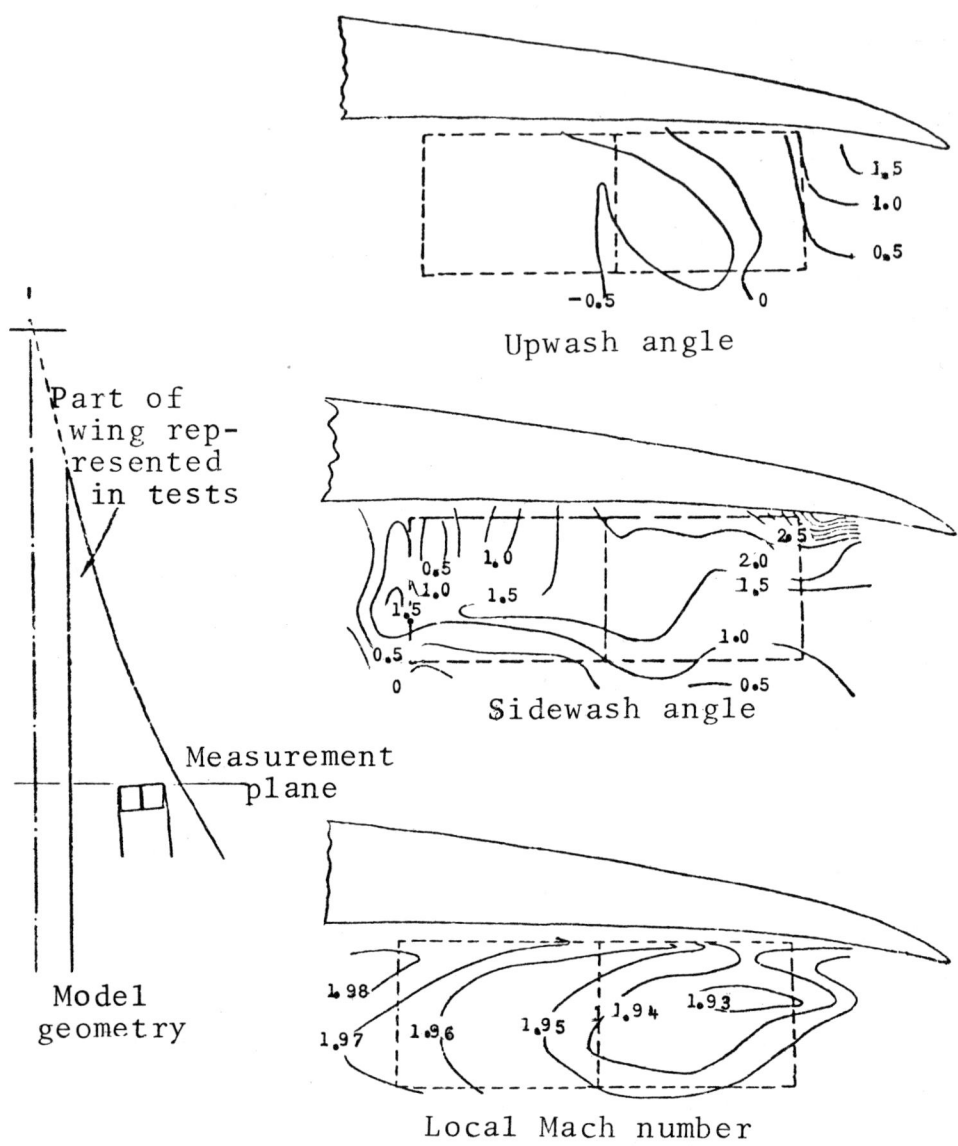

Fig. 12. Underwing flowfield studies [17]. Free stream Mach 2, incidence 3.5° (body datum)

Linearised Supersonic Wing Theory

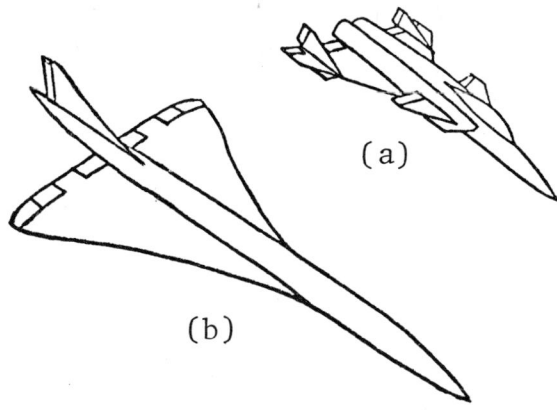

Fig. 13

This sort of problem is likely to arise in modern configurations of the type shown in Fig. 13(a) but also occurs in conventionally finned slender winged supersonic aircraft with trailing edge controls (Fig. 13(b)). Here the condition arises as a result of asymmetric control surface deflections during roll manoeuvres, or during the corrective action following an engine failure.

In aircraft of this type, the half span can form a single control surface, or may be split into two or more such surfaces, which can then either move independently, or be geared to move in fixed ratios to one another.

Clearly the area of the fin within the interference region of each control surface depends on the flight Mach number.

Two idealised cases involving asymmetric elevon deflection are illustrated here: first, a representative half span control surface at Mach 1.414. At this Mach number a large part of the fin area is in the interaction region. Fig. 14 shows the computed sidewash field per unit control deflection ($\bar{v}/\delta a$); values of the order 1 to 1.5 are calculated.

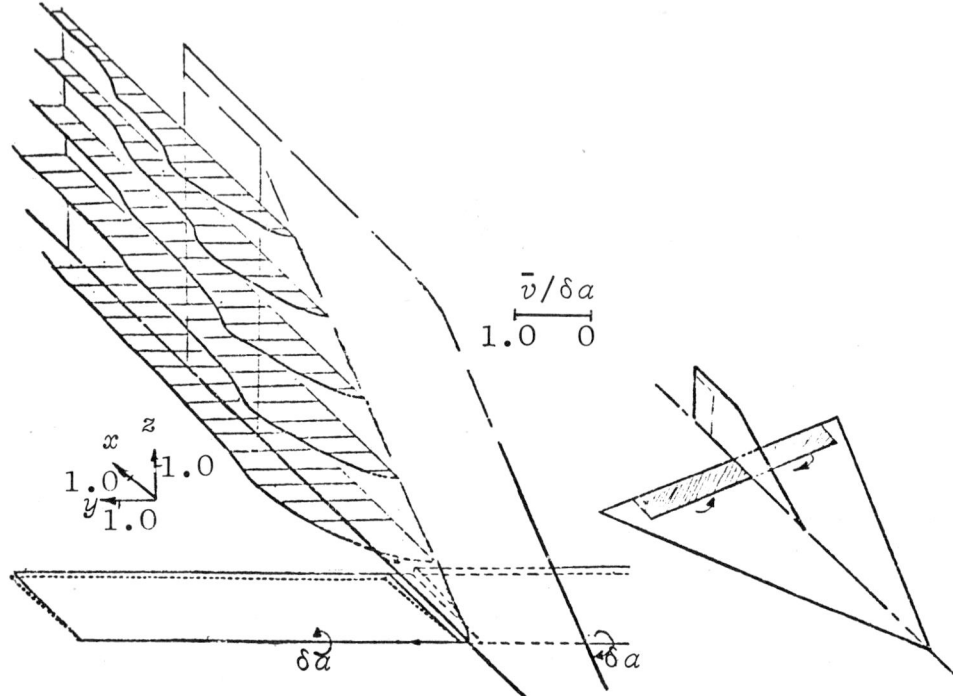

Fig. 14. Sidewash at fin due to half span control deflection Mach 1.414

Second, a part span elevon case at Mach 2 is considered. Here, only the inboard control surface is felt at the fin. Fig. 15 shows the computed sidewash field per unit control deflection $(\bar{v}/\delta a)$; values of the order of 0.6 arise.

Further work is required to corroborate the results with experiment.

Parallel interfering wings

Ideas for the use of additional (non-retractable) lifting surfaces on slender-winged aircraft with a view to improvement of low speed characteristics, as well as the more general need for studies of the control configuration vehicle concept [18], have highlighted the interfering wings problem (Fig. 16). Recent research on second generation and low sonic boom supersonic transports has shown that there are benefits to be obtained by distributing the aerodynamic lift over a longer length [19] to [25] and aircraft with up to three lifting surfaces (Fig. 17) have been considered.

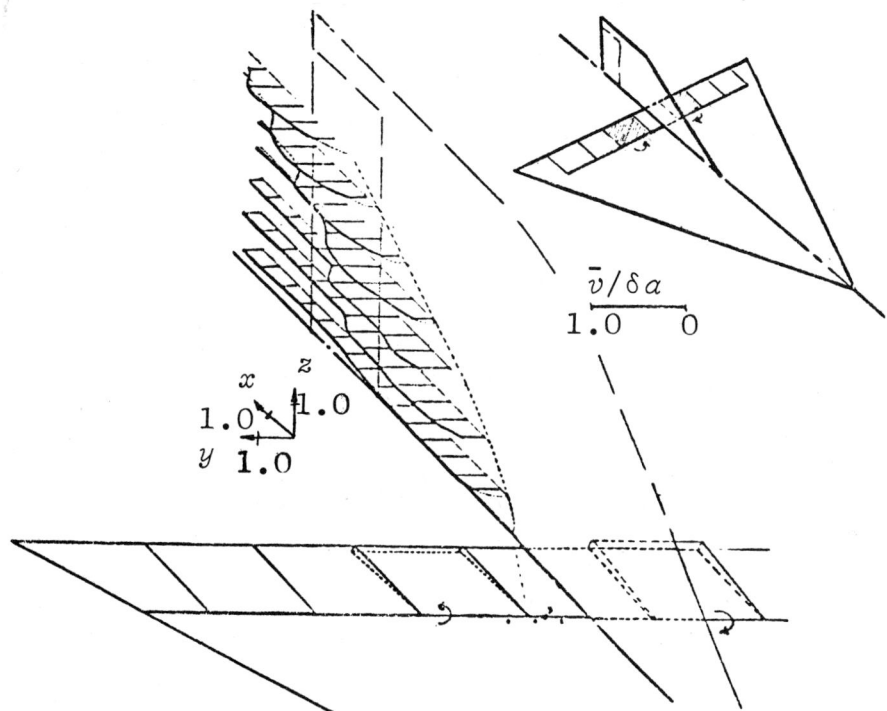

Fig. 15. Sidewash at fin due to part span control deflection Mach 2

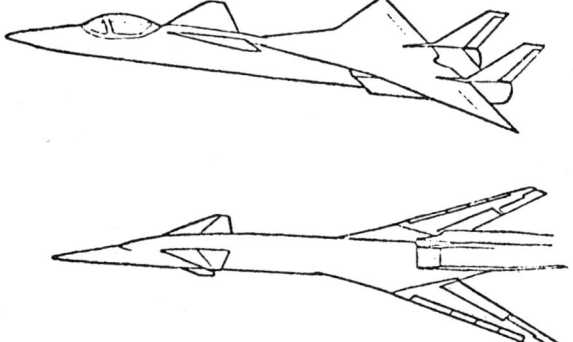

Fig. 16. Control configuration vehicle concepts [18]

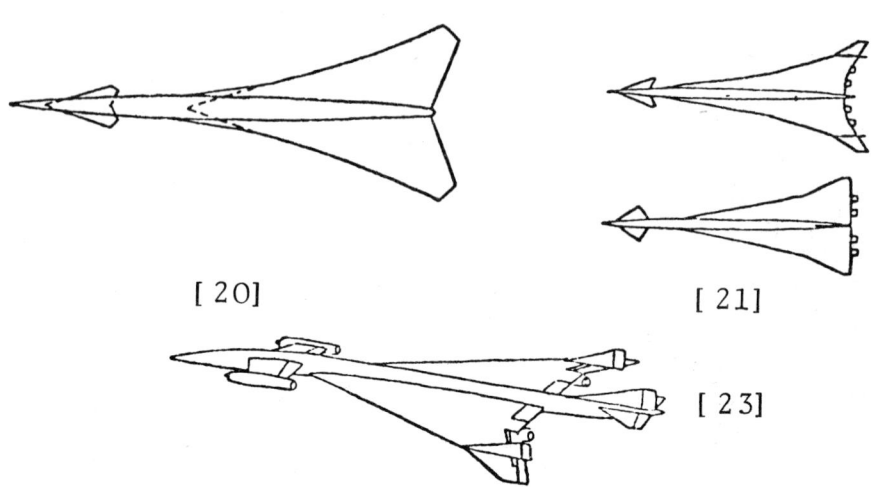

Fig. 17. Low sonic boom supersonic transports

So, to provide some understanding of the interaction between wings, we have looked at an arbitrary configuration with two geometrically similar 65° deltas at Mach 2 (see notation in Fig. 18); in particular at the effect of relative areas and the vertical spacing h.

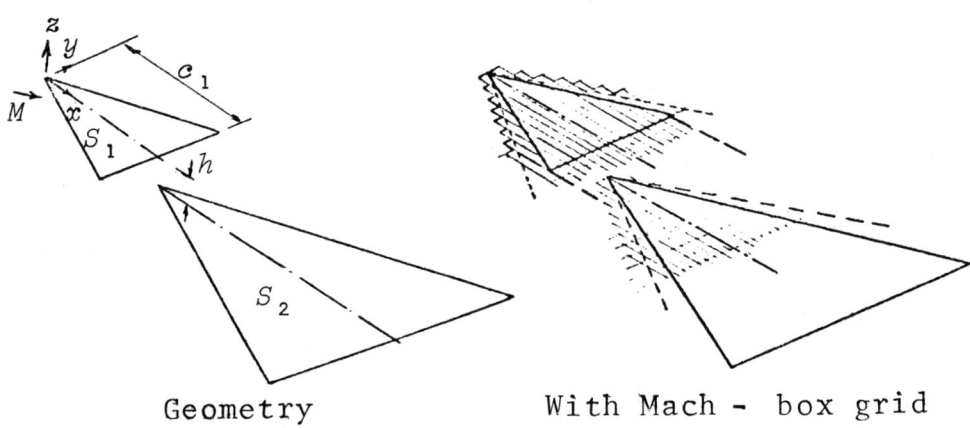

      Geometry   With Mach - box grid

Fig. 18. Parallel interfering wings

The first lifting (canard) wing produces a perturbed flow in its downstream zone of influence; if the second (main) wing is within this region a loading will be induced on it. The method has been formulated to estimate this interference loading.

The upwash field behind the canard is calculated at the Mach box centres on the wing. These upwash values then become the "cancelling" source distribution required to ensure satisfaction of the tangential flow boundary condition there [6] to [11].

Fig. 19 gives an idea of the vertical velocity field induced by the first wing in the plane containing the main wing, for three values of spacing h. As expected, with h increasing, the wake edge discontinuities in velocity are smoothed out. The central region in all cases shows downwash and the outer region upwash.

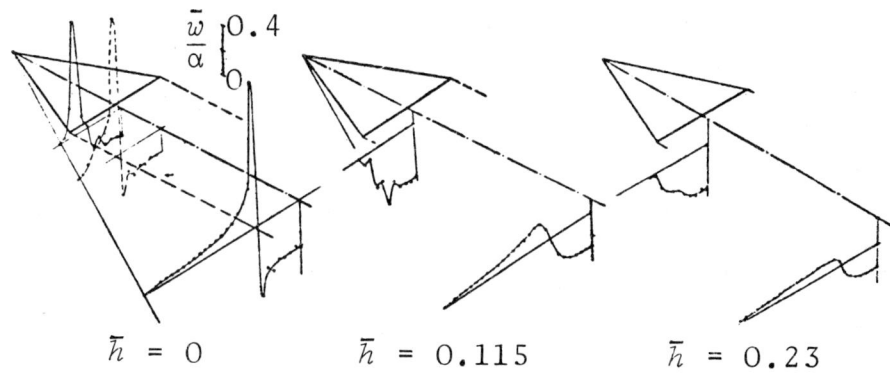

$\bar{h} = 0$ $\bar{h} = 0.115$ $\bar{h} = 0.23$

Fig. 19. Upwash behind canard at three different heights $\bar{h}$

Fig. 20(a) shows the centre plane ($y = 0$) upwash at three heights compared with the exact values from [3]. It is noted that the numerical results oscillate about the exact values. By simple surface fitting of the data locally, improvements as shown in Fig. 20(b) can be achieved.

Fig. 21 gives an idea of the mutual interference between the two wings. The effects of varying the relative size of the two surfaces (S_2/S_1) and the vertical separation h are considered. Fig. 21(a) shows the interference lift slope a_{12}. It is noted that a_{12} reduces with increasing S_2/S_1 and increasing h. This follows

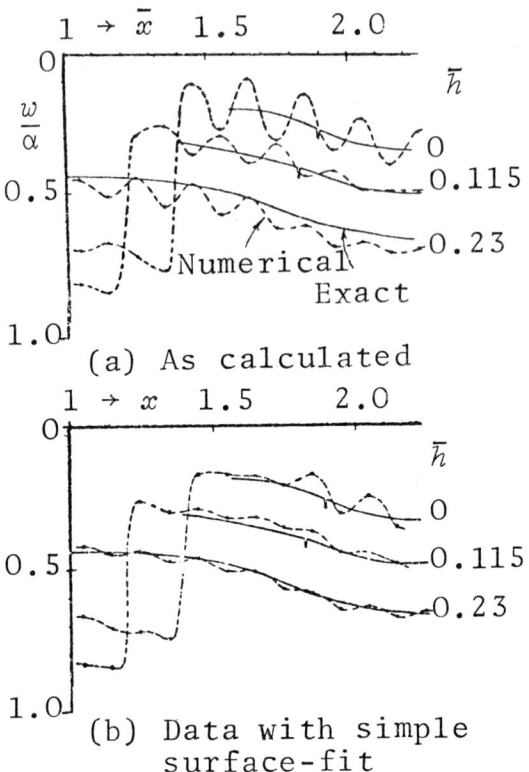

Fig. 20. Centre-plane upwash at three different heights $\bar{h}$

from the upwash pictures of Fig. 19. In Fig. 21(b) the variation of the centre of interference lift is shown. The main wing centre of lift is also shown for comparison. The oscillatory nature of these curves for small S_2/S_1 can be explained on the basis that wings located at varying heights beneath the canard experience the trailing edge and tip effects from the canard at different points on their surfaces.

Some idea of the trimmed longitudinal characteristics of the two wing configuration can be gained by examining Fig. 22, which shows to a base of centre of gravity position the induced drag and trimmed lift coefficient relationship for a particular case (S_2/S_1 = 3.06, $\bar{h}$ = 0.0144). Neutral stability exists at $\bar{x}_{cg}$ = $\bar{x}_{ac}$ = 1.773, and from

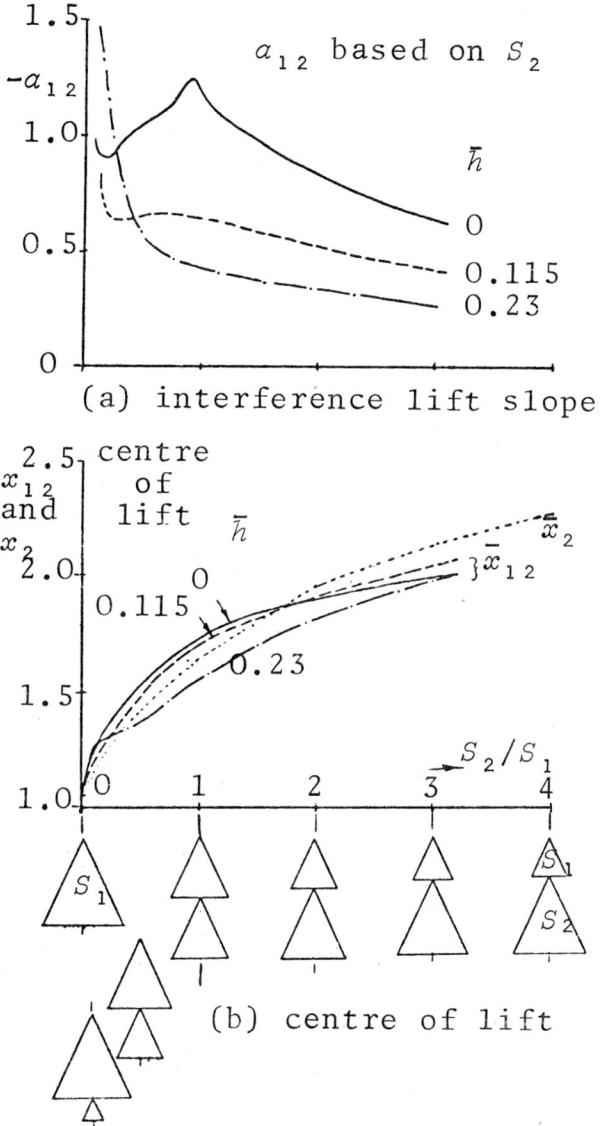

Fig. 21. Mutual interference between two wings 65° deltas at Mach 2

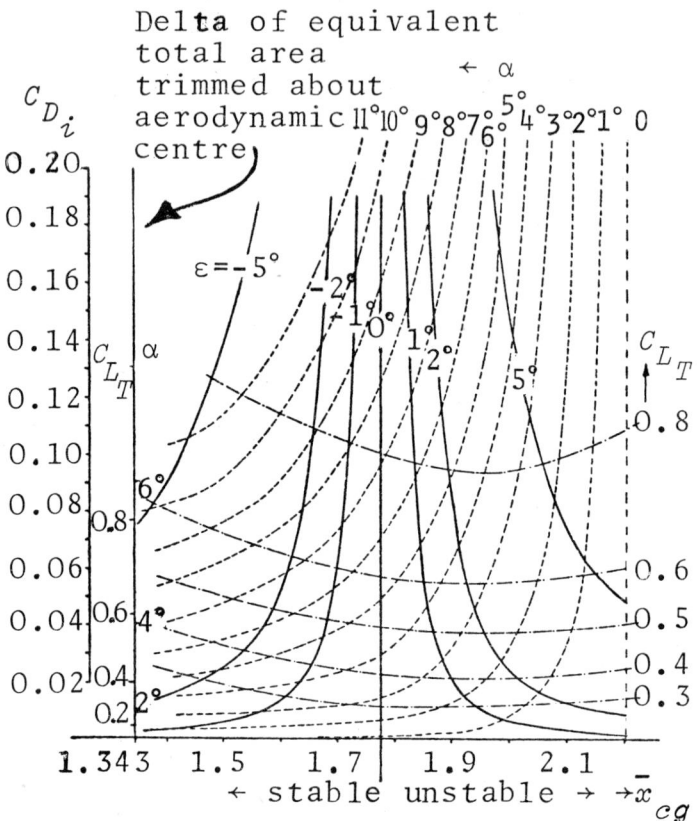

Fig. 22. Trimmed characteristics - effect of varying ε

the plot we see that for a given C_{L_T} value we may compare the induced drag of the two-delta configuration with that of a single delta of the same total area, and sweepback, trimmed about its aerodynamic centre. At low values of C_{L_T} (0.3, 0.4) the two-wing configuration has only marginally higher induced drag, but the difference becomes larger at values of C_{L_T} of 0.7, 0.8.

It will be interesting to see how other factors, e.g., leading edge camber and planform shape, will modify the lift/drag characteristics. Further systematic studies are therefore required that consider the effects of camber, planform and longitudinal and vertical spacing. From limited experience with these studies, however, there seems to be a good case for removing parts of the wing from the central downwash area to the more efficient upwash regions near the tips as illustrated in Fig. 23.

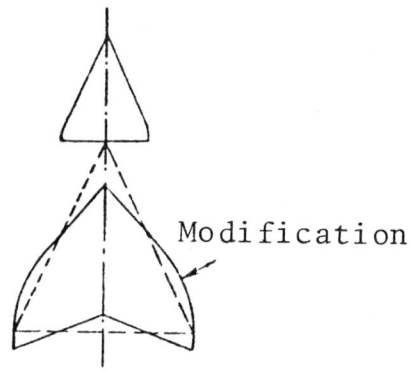

Fig. 23. Modification of main wing planform

CONCLUDING REMARKS

This paper illustrates the use of Mach box technique of linearised supersonic wing theory for calculation of interference flows. Where comparison has been made with other independent analytical solutions generally good agreement is shown. It is felt that the method can contribute significantly to knowledge and handling of interference flows especially where full nonlinear solutions are not easily available. Some consideration however needs to be given to improving the representation of subsonic leading edges.

We acknowledge the support given by the British Aircraft Corporation (Filton) to the work presented in this paper. Any opinions expressed are, however, our own.

NOTATION

a lift curve slope

a_∞ velocity of sound

c wing chord or chord length to nacelle plane $(= c_1)$

C_L	lift coefficient
C_{D_i}	induced drag coefficient
C_m	pitching moment coefficient about first wing apex
f	α_1/α_2
h	separation distance between two wings
$\bar{h}$	h/c_1
H	source strength
ℓ	box length
L	lift
M	Mach number
M_L	local Mach number
m	pitching moment
q	dynamic pressure
s	semi-span
S	area
t	thickness
x,y,z	rectangular Cartesian coordinate system
$\bar{x},\bar{y},\bar{z}$	x/c, y/c, z/c
X,Y,Z	x, βy, βz
u,v,w	perturbation velocities in x,y,z directions
$\bar{u},\bar{v},\bar{w}$	u/U_∞, v/U_∞, w/U_∞
U,V,W	velocity influence coefficients (VIC) in x,y,z directions
U_∞	free stream velocity
Δw	$w - U_\infty \sin\alpha$
α	angle of attack
β	$(M^2 - 1)^{\frac{1}{2}}$
δ_a	antisymmetric control deflection angle
ε	$\alpha_1 - \alpha_2$

η	intake pressure recovery
ξ, η, ζ	coordinates along x, y, z axes defining source position
$\bar{\xi}, \bar{\eta}, \bar{\zeta}$	$x-\xi_*$, $y-\eta_*$, $z-\zeta_*$
ξ_*, η_*, ζ_*	$\xi, \beta\eta, \beta\rho$
ν	nacelle orientation angle ⎫ see Fig. 3
σ	local flow sidewash angle ⎭
ϕ	perturbation potential
$\bar{\phi}$	$\phi/(U_\infty \cdot c)$
Φ	Velocity Potential Influence Coefficient (VPIC)
ψ	angle of yaw (or negative sideslip)

Subscripts

1	first wing
2	second wing
12	effect of first wing on the second
ac	aerodynamic centre
cg	centre of gravity
$\mathcal{L}$	centre line
P	at point P
u	upper surface
ℓ	lower surface
ψ	at angle of yaw

REFERENCES

1. Evvard, J.C., "Use of source distributions for evaluating theoretical aerodynamics of thin finite wings at supersonic speeds," NACA Rep. 951 (1949).

2. Stewart, H.J., "A review of source superposition and conical flow methods in supersonic wing theory," *J. Aero. Sciences*, **23**, No. 5, 507-516 (May 1956).

3. Donovan, A.F. and Lawrence, H.R., *Editors*, "Aerodynamic Components of Aircraft at High Speeds," **VII**, Oxford University Press (1957).

4. Donovan, A.F., Lawrence, H.R., Goddard, F.E. and Gilruth, R.R., *Editors*, "High Speed Problems of Aircraft and Experimental Methods," **VIII**, Oxford University Press (1961).

5. Zartarian, G. and Hsu, P.T., "Theoretical studies on the prediction of unsteady supersonic airloads on elastic wings," Parts I and II WADC Tech Rep. 56.97 (1955).

6. Appa, K. and Smith, G.C.C., "Development and applications of supersonic unsteady consistent aerodynamics for interfering parallel wings," NASA CR-2168 (1973).

7. Moore, M.T. and Andrew, L.V., "Unsteady aerodynamics for advanced configurations," Part IV., FDL-TDR 64-152 (1965).

8. Andrew, L.V. and Moore, M.T., "Further developments in supersonic aerodynamic influence coefficient methods," Proceedings of the AIAA Symposium on Structural Dynamics and Aeroelasticity, AIAA, New York (1966).

9. Ashley, H., "Some considerations relative to the preduction of unsteady air loads on lifting configurations," *J. Aircraft*, **8**, No. 10, 747-56 (October 1971).

10. Li, J.M., "A refined prediction method for the unsteady aerodynamics of supersonic elastic aircraft," *J. Aircraft*, **9**, No. 1, 61-68 (January 1972).

11. Li, J.M. and Rowe, W.S., "Unsteady aerodynamics of non-planar wings and wing-tail configurations of elastic flight vehicles in supersonic flight," *J. Aircraft*, **10**, No. 1, 19-27 (January 1973). [See also AIAA Paper 72-378.]

12. Landrum, E.J., "Effect of nacelle orientation on the aerodynamic characteristics of an arrow wing-body configuration at Mach number 2.03," NASA TN D-3284 (1966).

13. Nangia, R.K., "Three-dimensional wave interactions in supersonic intakes," 2nd International Symposium in Air Breathing Engines, Sheffield (March 1974).
14. Reinhart, W.A. and Tjonneland, E., "Inlet flow field studies for the supersonic transport," Soc. of Automotive Engineers, National Aeronautical Meeting, Washington, D.C. (April 1965).
15. Nielson, J.N. and Perkins, E.W., "Charts for the conical part of the downwash field of swept wings at supersonic speeds," NASA TN 1780 (December 1948).
16. "Langley tests supersonic transport design," *Aviation Week and Space Technology*, (January 1974).
17. Dobson, M.D., "Test at a Mach number 2.0 on a rectangular, twin duct air intake with variable geometry, situated in the flow field of a slender wing," RAE Tech Report 68285 (December 1968).
18. Yaffee, M.L., "New controls to shape future aircraft," *Aviation Week and Space Technology*, (October 1974).
19. Ferri, A. and Ismail, A., "Effects of Lengthwise lift distribution on sonic boom of SST configurations," *AIAA J.*, 7, 1538-41 (August 1969).
20. Ferri, A., Wang, Huai-Chu and Sorenson, H., "Experimental verification of low sonic boom configuration," NASA CR-2070 (June 1972).
21. Carlson, H.W., Barger, R.L. and Mack, R.J., "Application of sonic boom minimisation concepts in supersonic transport design," NASA TN D-7218 (June 1973).
22. Kane, E.J., "A study to determine the feasibility of a low sonic boom supersonic transport," NASA CR 2332 (December 1973).

23. Brown, D.A., "Advanced SST concepts studied," *Aviation Week and Space Technology* (January 1972).

24. Loftin, L.K., "Towards a second generation supersonic transport," *J. Aircraft*, **11**, No. 1, 3-9 (January 1974).

25. Goodmanson, L.T. and Gratzer, L.B., "Recent advances in aerodynamics for transport aircraft," Part 1, *Astro and Aero.* (December 1973); Part 2, *Astro and Aero.* (January 1974).

APPENDIX A

LINEARISED THEORY USING MACH BOX TECHNIQUES

As the Mach box method has been described by several authors [7] to [11] a brief summary only is presented here.

The method is an approximate means of solving the linearised governing equation

$$(M^2-1)\phi_{xx} - \phi_{yy} - \phi_{zz} = 0$$

for supersonic flow. This equation describes the small perturbation flow in the neighbourhood of a thin wing at small incidence, subject to the usual restrictions on the value of M (*i.e.*, $M > 1.2$ approximately).

In simple isolated lifting wing problems the $z = 0$ plane, in which the wing lies approximately, is imagined to divide the flowfield into two half spaces. Sources are then distributed in the disturbance region of the upper and lower surfaces of this plane to satisfy the appropriate boundary condition there. As the upper and lower half spaces are thus separated, any calculations to determine the flow perturbations at a point have to include contributions from all the sources in the Mach forecone from the point.

The expression for the velocity potential ϕ at a point P in Fig. A1 can be written

Linearised Supersonic Wing Theory

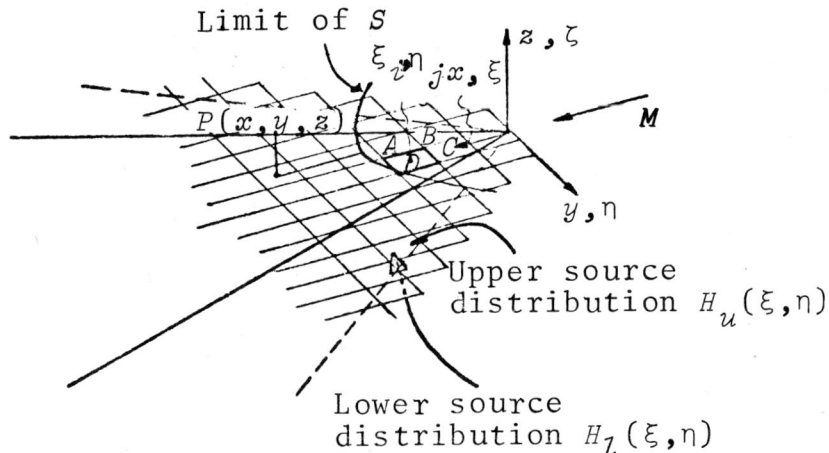

Physical plane

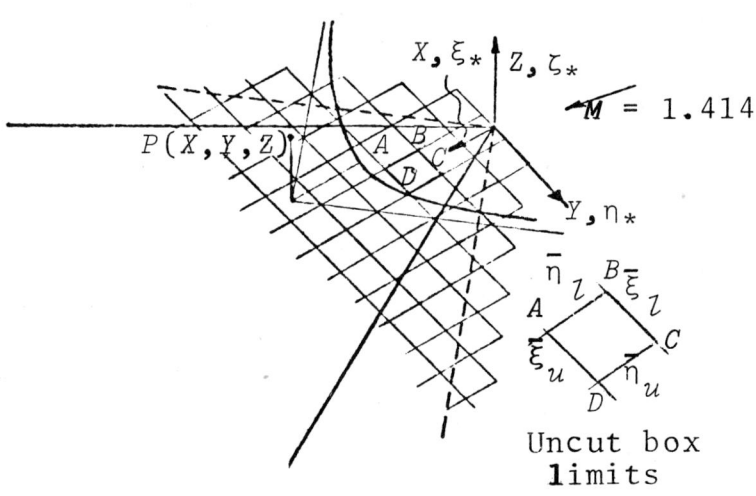

Equivalent problem - transformed plane Mach 1.414

Fig. A1. Mach box method

$$\phi_P(x,y,z) \propto \iint_S \frac{H_u(\xi,\eta)\,d\xi\,d\eta}{\left\{(x-\xi)^2 - \beta^2(y-\eta)^2 - \beta^2 z^2\right\}^{\frac{1}{2}}} \quad (A1)$$

where $H_u(\xi,\eta)$ is the upper source strength per unit area, *i.e.*, the source *intensity*.

Stewart [2] has shown that if the constant of proportionality in (Eq. A1) is $1/\pi$, the local source intensity equals the local downwash and thus has a simple physical interpretation.

This constant can be used by rewriting (Eq. A1), *i.e.*,

$$\phi_P(x,y,z) = \frac{1}{\pi}\iint_S \frac{H_u(\xi,\eta)\,d\xi\,d\eta}{\left\{(x-\xi)^2 - \beta^2(y-\eta)^2 - \beta^2 z^2\right\}^{\frac{1}{2}}} \quad (A2)$$

DETERMINATION OF SOURCE INTENSITIES

(*i*) For a wing of given planform, the effects of incidence, camber and thickness can be calculated independently and then superimposed.

(*ii*) On the wing, the source intensity equals the local wing surface slope [2].

(*iii*) Off the wing, in wakes or regions between a subsonic leading edge and the leading Mach lines, *i.e.*, in the so called diaphragms, the source intensity at all points must ensure that the no load condition is satisfied. Thus for incidence and camber problems, where flow properties on either side of the wing are asymmetric, ϕ must remain constant in the x direction [7]. In the symmetrical thickness problem, the source strength must be zero in these regions by symmetry.

The source sheet is approximated by a grid of rectangular boxes. The required source intensity is determined at the centre of each box and is assumed uniform over the box area. In the present calculation, boxes lying partly in the wing and partly in the diaphragm are assigned a source strength determined by the area fraction of the box which lies in each region [8].

The contribution to ϕ at P arising from the source distribution over the Mach box $ABCD$, centre ξ_i, η_j, normally written

$$\phi_P(x,y,z) = \frac{1}{\pi} \iint_{ABCD} \frac{H_u(\xi_i,\eta_j)\,d\xi\,d\eta}{\left\{(x-\xi)^2 - \beta^2(y-\eta)^2 - \beta^2 z^2\right\}^{\frac{1}{2}}} \quad (A3)$$

can now be rewritten.

First, however:

(i) all lengths are non-dimensionalised with respect to the box length ℓ

(ii) a coordinate transformation to the equivalent problem at $M = 1.414$ is introduced: (Fig. A1)

$$X = x; \quad Y = \beta y; \quad Z = \beta z$$
$$\xi_* = \xi; \quad \eta_* = \beta\eta; \quad \zeta_* = \beta\zeta$$

(iii) two new variables are introduced

$$\bar{\xi} = X - \xi_*; \quad \bar{\eta} = Y - \eta_*$$

Thus (Eq. A3) becomes

$$\phi_P(x,y,z) = \frac{1}{\pi\beta\ell} \iint_{ABCD} \frac{H_u(\xi_i,\eta_j)\,d\xi_*\,d\eta_*}{(\bar{\xi}^2 - \bar{\eta}^2 - Z^2)^{\frac{1}{2}}}$$

$$= \frac{1}{\pi\beta\ell} H(\xi_i,\eta_j) \iint_{ABCD} \frac{d\bar{\xi}\,d\bar{\eta}}{(\bar{\xi}^2 - \bar{\eta}^2 - Z^2)^{\frac{1}{2}}}$$

$$\phi_P(x,y,z) = \frac{1}{\beta\ell} H(\xi_i,\eta_j) \, \Phi(\bar{\xi}_i,\bar{\eta}_j,Z) \quad (A4)$$

where $\Phi(\bar{\xi}_i,\bar{\eta}_j,Z)$ is the velocity potential influence coefficient.

The full expression for $\phi(x,y,z)$ becomes:

$$\phi(x,y,z) = \frac{1}{\beta\ell} \sum_i \sum_j H(\xi_i,\eta_j) \; \Phi(\bar{\xi}_i,\bar{\eta}_j,Z)$$

where i and j are chosen to include all the boxes wholly or partially within the Mach forecone from P.

Velocity potential influence coefficient is then obtained by integrating the function over the area of each box

$$\Phi(\bar{\xi},\bar{\eta},Z) = \frac{1}{\pi} \iint\limits_{ABCD} \frac{d\bar{\xi}\,d\bar{\eta}}{(\bar{\xi}^2-\bar{\eta}^2-Z^2)^{\frac{1}{2}}} \qquad (A5)$$

Similar influence coefficients for u,v,w velocities can also be calculated by differentiating the function (A3) in the appropriate direction.

ANALYTICAL EXPRESSIONS FOR THE INFLUENCE COEFFICIENTS

In steady flow, analytical expressions may be derived for the influence coefficients for velocity potential and for the three components of velocity. These results are summarised here for a box uncut by the Mach hyperbola. Application of the formulae to boxes with one or more corners outside the hyperbola results in improper arguments of the functions concerned. This problem can be overcome by assigning the correct boundary value to such functions.

(i) *Velocity potential*

$$\Phi(\bar{\xi},\bar{\eta},Z) = \frac{1}{\pi} \iint \frac{d\bar{\xi}\,d\bar{\eta}}{((\bar{\xi}^2-\bar{\eta}^2-Z^2)^{\frac{1}{2}}}$$

Integrating first with respect to $\bar{\eta}$ gives

$$\Phi(\bar{\xi},\bar{\eta},Z) = \frac{1}{\pi} \int_{\bar{\xi}_l}^{\bar{\xi}_u} [\sin^{-1} \bar{\eta}/(\bar{\xi}^2-Z^2)^{\frac{1}{2}}]_{\bar{\eta}_l}^{\bar{\eta}_u} d\xi \quad (A6)$$

Integration by parts, after some manipulation then gives

$$\Phi(\bar{\xi},\bar{\eta},Z) = \frac{1}{\pi} \left[\left[\bar{\xi} \sin^{-1} \frac{\bar{\eta}}{(\bar{\xi}-Z^2)^{\frac{1}{2}}} + \bar{\eta} \cosh^{-1} \frac{\bar{\xi}}{(\bar{\eta}^2+Z^2)^{\frac{1}{2}}} \right. \right.$$
$$\left. \left. + Z \tan^{-1} \frac{\bar{\xi}\bar{\eta}}{Z\sqrt{\bar{\xi}^2-\bar{\eta}^2-Z^2}} \right]_{\bar{\eta}_l}^{\bar{\eta}_u} \right]_{\bar{\xi}_l}^{\bar{\xi}_u}$$

(ii) *Upwash*

$$W(\bar{\xi},\bar{\eta},Z) = \frac{1}{\pi} \frac{\partial}{\partial Z} \iint \frac{d\bar{\xi} d\bar{\eta}}{(\bar{\xi}^2-\bar{\eta}^2-Z^2)^{\frac{1}{2}}}$$

$$= \frac{1}{\pi} \left[\left[\tan^{-1} \left\{ \frac{\bar{\xi}\bar{\eta}}{Z(\bar{\xi}^2-\bar{\eta}^2-Z^2)^{\frac{1}{2}}} \right\} \right]_{\bar{\eta}_l}^{\bar{\eta}_u} \right]_{\bar{\xi}_l}^{\bar{\xi}_u}$$

Cut boxes: if $Z^2(\bar{\xi}^2 - \bar{\eta}^2 - Z^2) \leq 0$ then $\tan^{-1}\{\ \} = \frac{\pi}{2}$.

(iii) *Sidewash*

$$V(\bar{\xi},\bar{\eta},Z) = \frac{1}{\pi} \frac{\partial}{\partial Y} \iint \frac{d\bar{\xi} d\bar{\eta}}{(\bar{\xi}^2-\bar{\eta}^2-Z^2)^{\frac{1}{2}}}$$

$$= \frac{1}{\pi} \left[\left[\cosh^{-1} \left\{ \frac{\bar{\xi}}{(\bar{\eta}^2+Z^2)^{\frac{1}{2}}} \right\} \right]_{\bar{\eta}_l}^{\bar{\eta}_u} \right]_{\bar{\xi}_l}^{\bar{\xi}_u}$$

Cut boxes: if $\bar{\xi}/(\bar{\eta}^2 + Z^2)^{\frac{1}{2}} \leq 1$ then $\cosh^{-1}\{\ \} = 0$.

(iv) *Longitudinal "wash"*

$$U(\bar{\xi},\bar{\eta},Z) = \frac{1}{\pi} \frac{\partial}{\partial X} \iint \frac{d\bar{\xi} d\bar{\eta}}{(\bar{\xi}^2-\bar{\eta}^2-Z^2)^{\frac{1}{2}}}$$

$$= \frac{1}{\pi}\left[\left[\sin^{-1}\left\{\frac{\bar{\eta}}{(\bar{\xi}^2-z^2)^{\frac{1}{2}}}\right\}\right]_{\bar{\eta}_l}^{\bar{\eta}_u}\right]_{\bar{\xi}_l}^{\bar{\xi}_u}$$

Cut boxes: if $\dfrac{\bar{\eta}}{(\bar{\xi}^2-z^2)^{\frac{1}{2}}} \geqslant 1$ then $\sin^{-1}\{\ \} = \dfrac{\pi}{2}$.

Fig. A2 is an attempt to give a physical feel for the magnitudes of these coefficients by plotting their values for the first few rows of boxes upstream of a receiving point. The velocity influence coefficients are plotted both for a receiving point on the source sheet, and for 0 and 2 box lengths off it.

In this work, the exact expression for the velocity potential influence coefficients was not used, as a Gaussian 5-point numerical integration of (A6) was found to give results accurate to about 1 part in 10^5.

The velocity influence coefficients were, however, calculated from the analytical expressions.

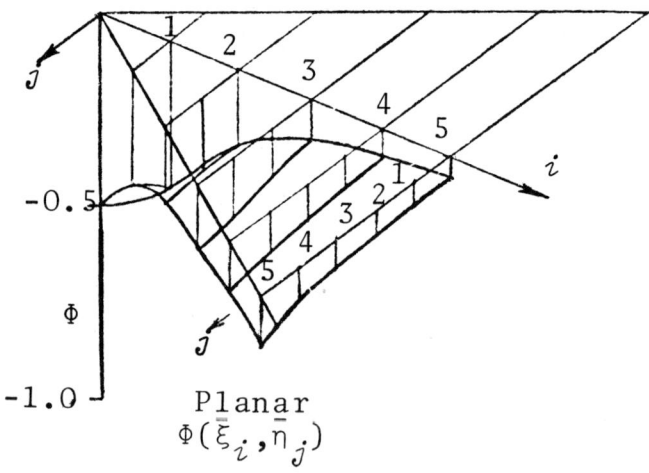

Planar $\Phi(\bar{\xi}_i, \bar{\eta}_j)$

Linearised Supersonic Wing Theory

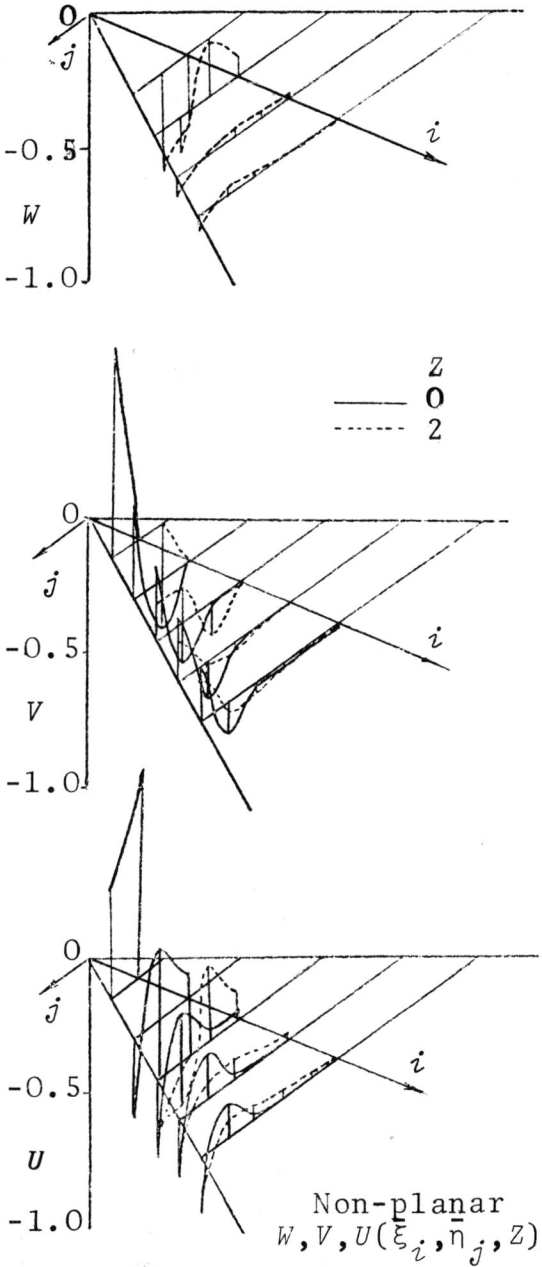

Fig. A2. Influence coefficients

APPENDIX B

LONGITUDINAL CHARACTERISTICS OF PLANAR INTERFERING WINGS

The longitudinal characteristics of two planar interfering wings can be derived with the aid of Fig. B1.

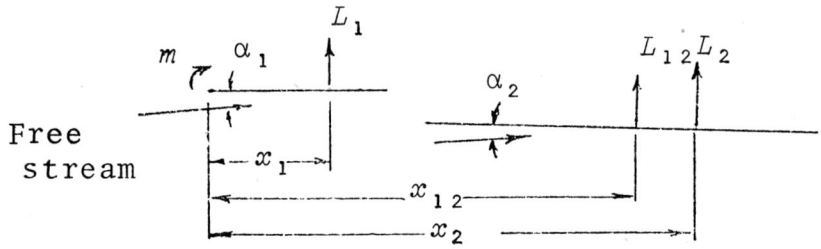

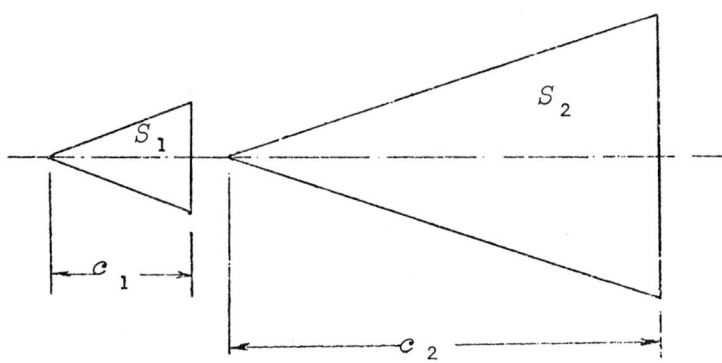

Fig. B1. Geometry - interfering wings

Let the free stream dynamic pressure be q; canard chord c_1, area S_1, main wing chord c_2 and area S_2.

Let L_1 and L_2 be the non-interference loads on the canard and the main wing at incidences α_1 and α_2, respectively. L_{12} is the interference loading induced by the canard on the wing. Let x_1, x_2 and x_{12} be the corresponding distances to the centres of the lift from the canard apex. Define the linear lift slopes as

$$a_1 = \frac{L_1}{qS_1\alpha_1}; \quad a_2 = \frac{L_2}{qS_2\alpha_2}; \quad a_{12} = \frac{L_{12}}{qS_2\alpha_1}$$

Expressions for C_L, C_{D_i} and C_m may be written all referred to canard area S_1 and chord c_1. It is assumed that vertical spacing between the two wings is small

$$C_L = a_1\alpha_1 + S_2/S_1 a_{12}\alpha_1 + S_2/S_1 a_2\alpha_2$$

$$C_{D_i} = a_1\alpha_1^2 + S_2/S_1 a_{12}\alpha_1\alpha_2 + S_2/S_1 a_2\alpha_2^2$$

$$C_m = -a_1\alpha_1\bar{x} - S_2/S_1 a_{12}\alpha\bar{x}_{12} - S_2/S_1 a_2\alpha_2\bar{x}_2$$

The position of centre of gravity is given by

$$\bar{x}_{cg} = -C_m/C_L.$$

By writing $\alpha_2 = \alpha_1 + \varepsilon$, we can differentiate with respect to α_1 and obtain

$$\frac{\partial C_L}{\partial \alpha_1} = a_1 + \frac{S_2}{S_1}a_{12} + \frac{S_2}{S_1}a_2$$

$$\frac{\partial C_m}{\partial \alpha_1} = -a_1\bar{x}_1 - a_{12}S_2/S_1\bar{x}_{12} - a_2S_2/S_1\bar{x}_2$$

From this aerodynamic centre location is

$$\bar{x}_{ac} = -\frac{\partial C_m}{\partial C_L} = \frac{a_1\bar{x}_1 + a_{12}S_2/S_1\bar{x}_{12} + a_2S_2/S_1\bar{x}_2}{a_1 + S_2/S_1 a_{12} + S_2/S_1 a_2}.$$

STEADY SUPERSONIC FLOWFIELDS WITH EMBEDDED SUBSONIC REGIONS

Robert W. MacCormack, Arthur W. Rizzi
and Mamoru Inouye

*(Ames Research Center,
National Aeronautics and Space Administration,
Moffett Field, California)*

INTRODUCTION

A supersonic flow past a blunt body contains a detached shock wave which separates the disturbed flow from the free stream (Fig. 1). The shock wave is at some point normal to the free stream direction and degenerates into a Mach wave far from this point. On the body surface there is a stagnation point from which the flow accelerates until again reaching supersonic speeds. The flow contains an embedded subsonic region which lies between the shock wave and the body surface and is also bounded by sonic lines extending from the body to the shock.

Such flows are encountered by supersonic aircraft or re-entering spacecraft in the nose, wing and tail regions, or about the entire craft if the angle of attack is large enough. The prediction of such flows for arbitrary body shapes is important to the craft designer because it provides him with information necessary to evaluate convective and radiative heat transfer rates and boundary layer effects for design configurations.

Mathematically, a steady blunt body flow is described by a set of elliptic equations in the embedded subsonic region and by a hyperbolic set elsewhere. Computationally, the flow is difficult to calculate because of the range of local Mach

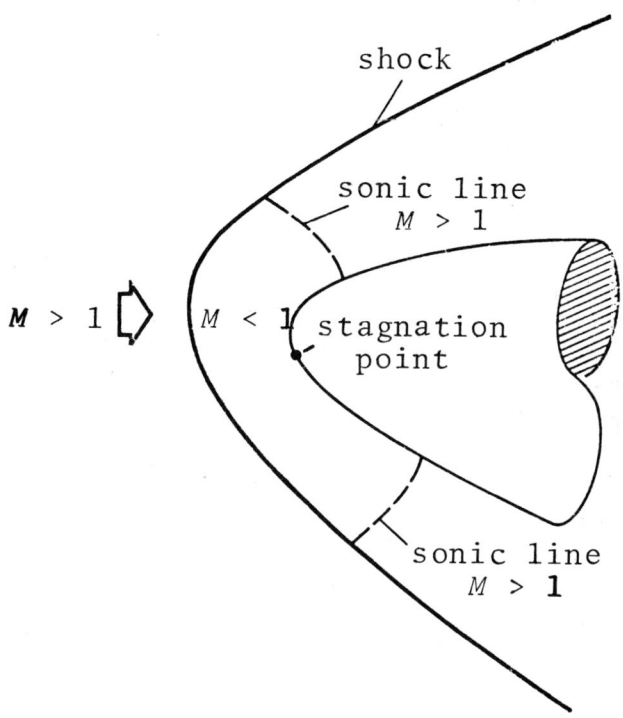

Fig. 1. Flowfield about a blunt body

numbers encountered, from zero to greater than one, and because of the shock wave discontinuity and the body geometry. This paper describes a numerical approach using the basic finite difference scheme introduced by MacCormack [1], which solves the unsteady fluid dynamic equations in integral form [2]. The integral form enables both the body surface and the shock wave discontinuity to be treated straightforwardly. The unsteady equations are everywhere hyperbolic in time so no distinction need be made between subsonic and supersonic regions. Solutions to the mixed elliptic and hyperbolic steady flow equations are approached asymptotically in time. For convenience the numerical method is described for solving two-dimensional flows. The extension to three dimensions is straightforward. The results obtained by Rizzi and Inouye [3] using this basic approach for three-dimensional flows are given.

GOVERNING EQUATIONS

The inviscid flows to be considered are governed by the Euler equations. The unsteady two-dimensional Euler equations can be written in differential vector form as

$$\frac{\partial F_t}{\partial t} + \frac{\partial F_x}{\partial x} + \frac{\partial F_y}{\partial y} = 0 \qquad (1)$$

where

$$F_t = \begin{bmatrix} \rho \\ \rho u \\ \rho v \\ e \end{bmatrix}, \quad F_x = \begin{bmatrix} \rho u \\ \rho u^2 + p \\ \rho v u \\ (e+p)u \end{bmatrix}, \quad F_y = \begin{bmatrix} \rho v \\ \rho u v \\ \rho v^2 + p \\ (e+p)v \end{bmatrix}$$

ρ is the density; u and v are the velocities in the x and y directions, respectively; e is the total energy per unit volume; and p, the pressure, is a function of ρ and the internal energy $\varepsilon = e/\rho - (u^2 + v^2)/2$.

The equations can be also written in integral vector form by first defining the double vector

$$\underset{\sim}{F} = F_t \underset{\sim}{i}_t + F_x \underset{\sim}{i}_x + F_y \underset{\sim}{i}_y$$

where $\underset{\sim}{i}_t$, $\underset{\sim}{i}_x$, and $\underset{\sim}{i}_y$ are unit vectors in the t, x, and y coordinate directions, respectively. Equation (1) can then be written in divergence law form as

$$\underset{\sim}{\nabla} \, \underset{\sim}{F} = 0$$

where

$$\underset{\sim}{\nabla} = \frac{\partial}{\partial t} \underset{\sim}{i}_t + \frac{\partial}{\partial x} \underset{\sim}{i}_x + \frac{\partial}{\partial y} \underset{\sim}{i}_y$$

Then by integrating the divergence law form equation over a volume V in x, y, t space and using the divergence theorem to reduce the volume integration to surface integrations, we obtain

$$\int_S \underset{\sim}{F} \, \underset{\sim}{n} \, ds = 0 \qquad (2)$$

where S encloses the volume V, and $\underset{\sim}{n}$ is the local unit vector normal to S.

FINITE DIFFERENCE EQUATIONS

In the next subsections, difference equations are presented for numerically approximating both the differential and integral forms of the Euler equations (1) and (2). Because of the advantage of time splitting, principally computational efficiency, only split difference equations are considered. The difference equations for the differential form are introduced to serve only as a reference point for the less familiar difference equations for the integral form used to obtain the computational results of this paper.

Differential form

To approximate numerically (1), the x, y, t space is discretised as shown in Fig. 2.

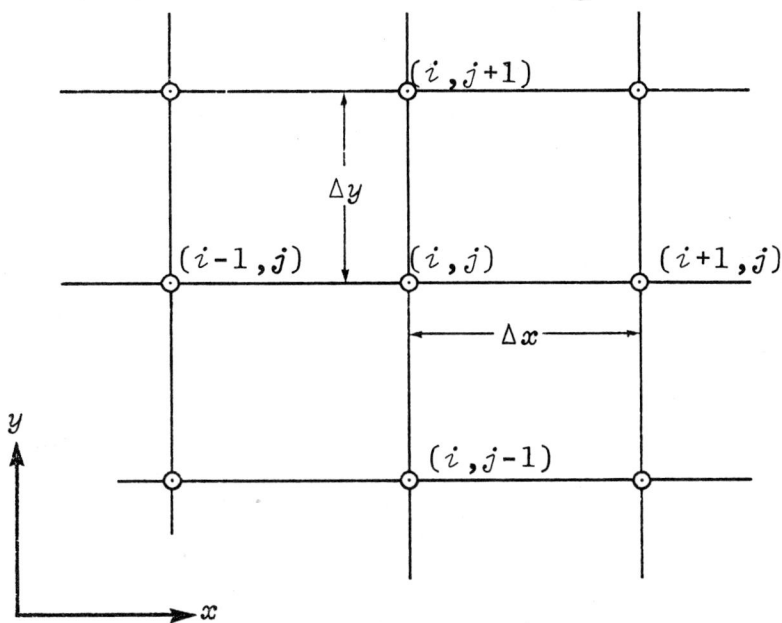

Fig. 2. Computational mesh for differential form equations

If we know the solution $F_{i,j}^n$ at time t_n at each mesh point, we can determine the solution at $t_{n+1} = t_n + \Delta t$ from the following time split difference equations [2]:

$$F^{n+\frac{1}{2}}_{t_{i,j}} = L_{\Delta x}(\Delta t) F^{n}_{t_{i,j}}$$

$$F^{n+1}_{t_{i,j}} = L_{\Delta y}(\Delta t) F^{n+\frac{1}{2}}_{t_{i,j}}$$

or

$$F^{n+1}_{t_{i,j}} = L_{\Delta y}(\Delta t) L_{\Delta x}(\Delta t) F^{n}_{t_{i,j}}$$

The operator $L_{\Delta x}(\Delta t)$ is defined by

$$F^{n+\frac{1}{2}}_{t_{i,j}} = F^{n}_{t_{i,j}} - \frac{\Delta t}{\Delta x}\left(F^{n}_{x_{i,j}} - F^{n}_{x_{i-1,j}}\right)$$

$$F^{n+\frac{1}{2}}_{t_{i,j}} = \frac{1}{2}\left[F^{n}_{t_{i,j}} + F^{n+\frac{1}{2}}_{t_{i,j}} - \frac{\Delta t}{\Delta x}\left(F^{n+\frac{1}{2}}_{x_{i+1,j}} - F^{n+\frac{1}{2}}_{x_{i,j}}\right)\right]$$

The operator $L_{\Delta y}(\Delta t)$ is defined by

$$F^{n+1}_{t_{i,j}} = F^{n+\frac{1}{2}}_{t_{i,j}} - \frac{\Delta t}{\Delta y}\left(F^{n+\frac{1}{2}}_{y_{i,j}} - F^{n+\frac{1}{2}}_{y_{i,j-1}}\right)$$

$$F^{n+1}_{t_{i,j}} = \frac{1}{2}\left[F^{n+\frac{1}{2}}_{t_{i,j}} + F^{n+1}_{t_{i,j}} - \frac{\Delta t}{\Delta y}\left(F^{n+1}_{y_{i,j+1}} - F^{n+1}_{y_{i,j}}\right)\right]$$

Because the operators $L_{\Delta x}(\Delta t)$ and $L_{\Delta y}(\Delta t)$ do not in general commute, the sequence $L_{\Delta y}(\Delta t)L_{\Delta x}(\Delta t)$ can be shown to be only first order accurate; but if the sequence is reversed at each step, forming a symmetric sequence, the calculation becomes second order accurate. Thus,

$$F^{n+2}_{t_{i,j}} = L_{\Delta x}(\Delta t) L_{\Delta y}(\Delta t) L_{\Delta y}(\Delta t) L_{\Delta x}(\Delta t) F^{n}_{t_{i,j}}$$

determines $F^{n+2}_{t_{i,j}}$ from $F^{n}_{t_{i,j}}$ to second order accuracy. Other second order accurate sequences are

$$F^{n+2}_{t_{i,j}} = L_{\Delta y}(\Delta t) L_{\Delta x}(\Delta t) L_{\Delta x}(\Delta t) L_{\Delta y}(\Delta t) F^{n}_{t_{i,j}}$$

$$F_{t_{i,j}}^{n+2} = L_{\Delta y}(\Delta t) L_{\Delta x}(2\Delta t) L_{\Delta y}(\Delta t) F_{t_{i,j}}^{n}$$

etc. Rules for a stable second order accurate operator sequence for advancing a solution by ΔT in time are as follows.

1. For stability, for each operator of the sequence, $L_{\Delta *}(\Delta t_*)$,

$$\Delta t_* \leq \frac{\Delta^*}{\max|\text{eigenvalue of } A_*|},$$

where A_* is the Jacobian of F_* with respect to F_t.

2. For consistency, $\Sigma \Delta t_* = \Delta T$.

3. For second order accuracy, the sequence should be symmetric.

For the unsteady Euler equations:

(i) $L_{\Delta x}(\Delta t)$ is stable if $\Delta t \leq \min_{i,j} \dfrac{\Delta x}{|u_{i,j}| + c_{i,j}}$

(ii) $L_{\Delta y}(\Delta t)$ is stable if $\Delta t \leq \min_{i,j} \dfrac{\Delta y}{|v_{i,j}| + c_{i,j}}$

where c is the speed of sound.

Integral form

To approximate numerically (2), the x, y, t space can be arbitrarily discretised as shown in Fig. 3. To begin we assume that the mesh is fixed in time. For the quadrilateral (i,j) bounded by the line segments ℓ_1, ℓ_2, ℓ_3, and ℓ_4 and of area $A_{i,j}$, the incremental volume $V_{i,j}$ and the six surfaces $\underline{S}_{i,j}^n$, $\underline{S}_{i,j}^{n+1}$ and $\underline{S}k_{i,j}$ for $k = 1, 2, 3$ and 4 enclosing the volume in x, y, t space are

$$V_{i,j} = A_{i,j} \Delta t$$

$$\underline{S}_{i,j}^{n+1} = -\underline{S}_{i,j}^n = A_{i,j} \underline{i}_t$$

and for $k = 1, 2, 3$ and 4

$$\underset{\sim}{S}_{k_{i,j}} = \Delta t \left[\int_{\ell_k} \underset{\sim}{i}_x \, n_\ell \, d\ell \, \underset{\sim}{i}_x + \int_{\ell_k} \underset{\sim}{i}_y \, n_\ell \, d\ell \, \underset{\sim}{i}_y \right]$$

where n_ℓ is the unit vector normal to ℓ_k.

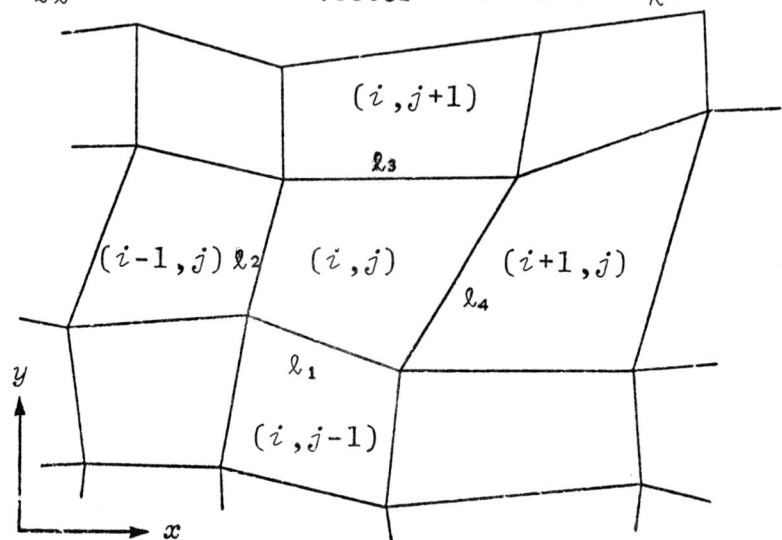

Fig. 3. Computational mesh for integral form equations

For example, from Fig. 4

$$\underset{\sim}{S}_4 = \Delta t \left[|\ell_4| \cos \theta \, \underset{\sim}{i}_x + |\ell_4| \cos (\theta + \pi/2) \underset{\sim}{i}_y \right]$$

$$= \Delta t \left[(y_{i+1,j+1} - y_{i+1,j}) \underset{\sim}{i}_x - (x_{i+1,j+1} - x_{i+1,j}) \underset{\sim}{i}_y \right]$$

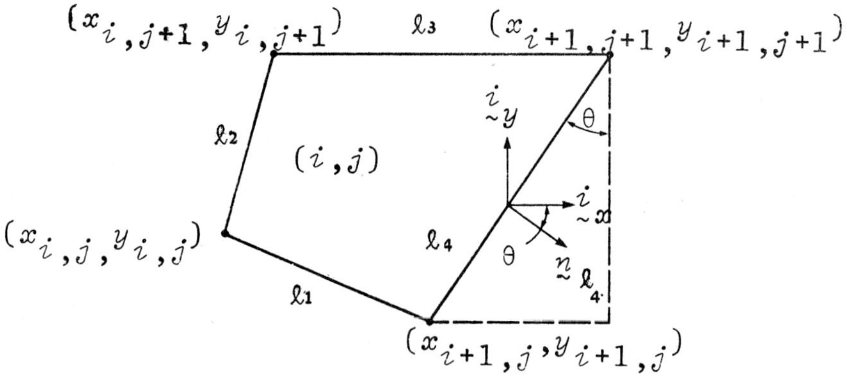

Fig. 4. Mesh quadrilateral (i,j)

If we know the solution at time t_n at each mesh quadrilateral, we can then determine the solution at time $t_{n+1} = t_n + \Delta t$ from the following time split difference equations [1], [2].

$$F^{n+1}_{t_{i,j}} = L_{\Delta J}(\Delta t) L_{\Delta I}(\Delta t) F^n_{t_{i,j}}$$

The operator $L_{\Delta I}(\Delta t)$ is defined by

$$F^{n+\frac{1}{2}}_{t_{i,j}} = F^n_{t_{i,j}} - \frac{1}{A_{i,j}}\left(F^n_{\underset{\sim}{i,j}} S_{4\ i,j} + F^n_{\underset{\sim}{i-1,j}} S_{2\ i,j}\right)$$

$$F^{n+\frac{1}{2}}_{t_{i,j}} = \frac{1}{2}\left[F^n_{t_{i,j}} + F^{n+\frac{1}{2}}_{t_{i,j}} - \frac{1}{A_{i,j}}\left(F^{n+\frac{1}{2}}_{\underset{\sim}{i+1,j}} S_{4\ i,j} + F^{n+\frac{1}{2}}_{\underset{\sim}{i,j}} S_{2\ i,j}\right)\right]$$

The operator $L_{\Delta J}(\Delta t)$ is defined by

$$F^{n+1}_{t_{i,j}} = F^{n+\frac{1}{2}}_{t_{i,j}} - \frac{1}{A_{i,j}}\left(F^{n+\frac{1}{2}}_{\underset{\sim}{i,j}} S_{3\ i,j} + F^{n+\frac{1}{2}}_{\underset{\sim}{i,j-1}} S_{1\ i,j}\right)$$

$$F^{n+1}_{t_{i,j}} = \frac{1}{2}\left[F^{n+\frac{1}{2}}_{t_{i,j}} + F^{n+1}_{t_{i,j}} - \frac{1}{A_{i,j}}\left(F^{n+1}_{\underset{\sim}{i,j+1}} S_{3\ i,j} + F^{n+1}_{\underset{\sim}{i,j}} S_{1\ i,j}\right)\right]$$

The above difference equations are similar to those of the previous subsection and even for arbitrary meshes are not much more difficult to solve. For example, a typical term of the differential form difference equation is

$$\frac{\Delta t}{\Delta x} F^n_{x_{i,j}} = \begin{Bmatrix} \rho^n_{i,j} u^n_{i,j} \Delta t/\Delta x \\ \left[\rho^n_{i,j} u^{n^2}_{i,j} + p^n_{i,j}\right] \Delta t/\Delta x \\ \rho^n_{i,j} v^n_{i,j} u^n_{i,j} \Delta t\ \Delta x \\ \left[e^n_{i,j} + p^n_{i,j}\right] u^n_{i,j} \Delta t/\Delta x \end{Bmatrix}$$

On the other hand, a typical term of the integral form difference equation is

$$\frac{1}{A_{i,j}} \underset{\sim}{F}^n_{i,j} \underset{\sim}{S}_4{}_{i,j} = \begin{Bmatrix} \rho^n_{i,j} qs^n_{i,j} \Delta t/A_{i,j} \\ \left[\rho^n_{i,j} u^n_{i,j} qs^n_{i,j} \right. \\ \left. + p^n_{i,j} \left(y_{i+1,j+1} - y_{i+1,j}\right)\right] \Delta t/A_{i,j} \\ \left[\rho^n_{i,j} v^n_{i,j} qs^n_{i,j} \right. \\ \left. - p^n_{i,j} \left(x_{i+1,j+1} - x_{i+1,j}\right)\right] \Delta t/A_{i,j} \\ \left(e^n_{i,j} + p^n_{i,j}\right) qs^n_{i,j} \Delta t/A_{i,j} \end{Bmatrix}$$

where

$$qs^n_{i,j} = u^n_{i,j}\left(y_{i+1,j+1} - y_{i+1,j}\right)$$
$$- v^n_{i,j}\left(x_{i+1,j+1} - x_{i+1,j}\right).$$

The principal advantage of the integral form is that arbitrarily shaped meshes may be used for fitting irregular boundaries without the need for determining a suitable coordinate transformation. The same rules as before apply for constructing stable second order accurate operator sequences for the integral form operators and:

(i) $L_{\Delta I}(\Delta t)$ is stable if

$$\Delta t \leq \min_{i,j} \frac{V_{i,j}}{\left|\underset{\sim}{q}_{i,j} \cdot \underset{\sim}{S}_2{}_{i,j}\right| + c_{i,j} \left|\underset{\sim}{S}_2{}_{i,j}\right|}$$

(ii) $L_{\Delta J}(\Delta t)$ is stable if

$$\Delta t \leq \min_{i,j} \frac{V_{i,j}}{\left|\underset{\sim}{q}_{i,j} \cdot \underset{\sim}{S}_1{}_{i,j}\right| + c_{i,j} \left|\underset{\sim}{S}_1{}_{i,j}\right|}$$

where $\underset{\sim}{q}_{i,j} = u_{i,j}\underset{\sim}{i}_x + v_{i,j}\underset{\sim}{i}_y$.

If the integral form mesh is chosen to be the same as the differential form mesh, then the two sets of difference equations are identical.

Moving mesh

In the previous subsection we assumed that the mesh was fixed in time. For some situations it is desirable that the mesh can be adjusted in time. For flows containing shock waves it is important for accurate calculations in the vicinity of the shock to keep the mesh aligned with the shock. Otherwise the numerical solution will be smeared across the shock rather than having a sharp jump, and will contain oscillations.

The difference operators of the previous subsection can be thought of as adjusting the solution to the movement of the fluid through a fixed mesh. The difference operators now to be defined adjust the solution because of the movement of the mesh through the frozen fluid. Thus the dynamics of the flow problem are split into four operators: the dynamics of the two spatial coordinate mesh directions and the dynamics induced in each direction by the moving coordinate mesh.

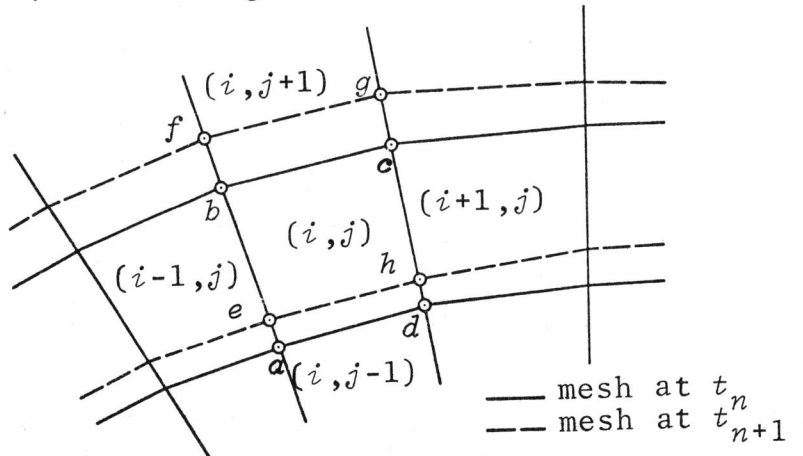

Fig. 5. Moving computational mesh

For simplicity we consider the case where only one family of mesh lines moves with time

(Fig. 5). The quadrilateral (i,j) at time t_n is defined by the points a, b, c and d, and at time $t_{n+1} = t_n + \Delta t$ by the points e, f, g and h.

$$A_{i,j}^n = \text{area of quadrilateral } abcd$$

$$A_{i,j}^{n+1} = \text{area of quadrilateral } efgh$$

$$\Delta A_{i,j} = \text{area of quadrilateral } aehd$$

$$\Delta A_{i,j+1} = \text{area of quadrilateral } bfgc$$

Thus

$$A_{i,j}^{n+1} = A_{i,j}^n + \Delta A_{i,j+1} - \Delta A_{i,j}$$

The difference equations which advance the solution because of mesh movement are

$$F_{t_{i,j}}^{n+1} = L_{\text{mesh}_J}(\Delta t) F_{t_{i,j}}^n$$

where the operator $L_{\text{mesh}_J}(\Delta t)$ is defined by

$$F_{t_{i,j}}^{n+1} = \left[F_{t_{i,j}}^n A_{i,j}^n \right.$$
$$\left. + \left(F_{t_{i,j}}^n \Delta A_{i,j+1} - F_{t_{i,j-1}}^n \Delta A_{i,j} \right) \right] / A_{i,j}^{n+1}$$

$$F_{t_{i,j}}^{n+1} = \tfrac{1}{2} \left[F_{t_{i,j}}^n A_{i,j}^n + F_{t_{i,j}}^{n+1} A_{i,j}^{n+1} \right.$$
$$\left. + \left(F_{t_{i,j+1}}^{n+1} \Delta A_{i,j+1} - F_{t_{i,j}}^{n+1} \Delta A_{i,j} \right) \right] / A_{i,j}^{n+1}$$

The operator $L_{\text{mesh}_J}(\Delta t)$, unlike $L_{\Delta I}(\Delta t)$ and $L_{\Delta J}(\Delta t)$, commutes with all operators and can be placed anywhere in an operator sequence for second order accuracy. For stability, Δt cannot be so large to allow the mesh to move so far that $\Delta A_{i,j} > A_{i,j}^n$.

Boundary conditions

There are three distinct types of boundaries for the flows under consideration: free stream entrance, supersonic exit and streamline boundaries (Fig. 6).

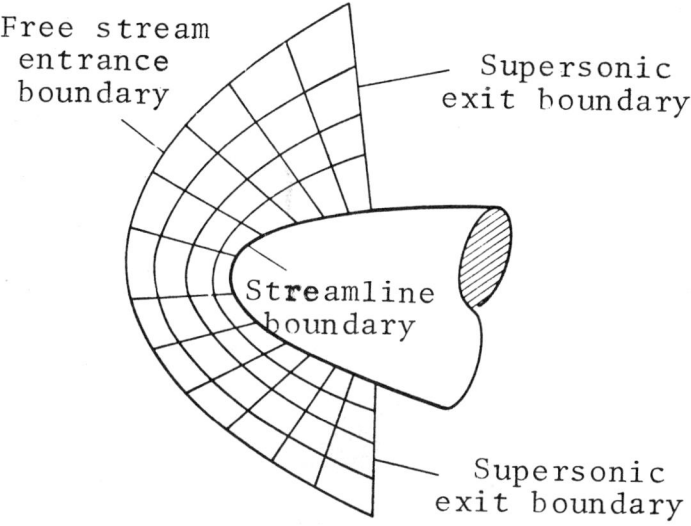

Fig. 6. Computational mesh boundaries

At mesh boundaries in the free stream across which the flow is entering, the flow variables are held fixed to their initial values during the calculation. At boundaries across which the flow is leaving supersonically, the flow variables are calculated using the difference equations given earlier except that all downstream differences are replaced by upstream differences because there are no mesh points downstream of this boundary. At streamline boundaries, for example, body surfaces, because of the flow tangency condition, $q \underset{\sim}{S}_B = 0$, the only flow variable needed is pressure (see the typical term of the integral form difference equations with $q \underset{\sim}{S}_B = 0$ implying $qs = 0$). To obtain the pressure at boundary point B (see Fig. 7), we first derive the normal momentum equation by taking the inner product between a row vector containing components of the normal to the boundary surface and the column vector consisting of the left side of the Euler equation (1)

$$[0, n_x, n_y, 0] \left(\frac{\partial F_t}{\partial t} + \frac{\partial F_x}{\partial x} + \frac{\partial F_y}{\partial y} \right)$$

$$= n_x \frac{\partial p}{\partial x} + n_y \frac{\partial p}{\partial y} - \rho \underset{\sim}{q} \left(u \frac{\partial \underset{\sim}{n}}{\partial x} + v \frac{\partial \underset{\sim}{n}}{\partial y} \right) = 0$$

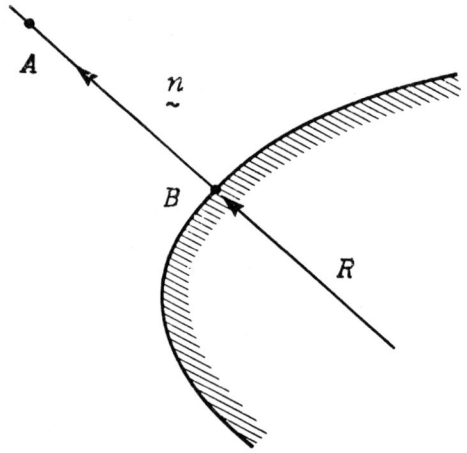

Fig. 7. Geometry for boundary condition evaluation

If the streamline at point B is a segment of a circular arc of radius R,

$$\frac{\partial p}{\partial n} = n_x \frac{\partial p}{\partial x} + n_y \frac{\partial p}{\partial y} = \frac{\rho q_T^2}{R}$$

where q_T is the velocity component tangent to the boundary at B. A finite difference approximation of first order accuracy, sufficient for our needs, for this equation is

$$p_B = p_A - \Delta n \frac{\rho_A q_{T_A}^2}{R} \qquad (3)$$

where Δn is the distance between points A and B, and the flow variables at point A are obtained by interpolation of the flow variable data at interior mesh points.

Shock wave

The shock wave is treated as an interior feature of the flowfield, and no special attention is given to it in the difference operators $L_{\Delta I}(\Delta t)$ and $L_{\Delta J}(\Delta t)$. The mesh, however, is periodically adjusted to maintain alignment with the shock (Fig. 8). With mesh shock alignment, the conservation form of the difference operators will then implicitly satisfy the Rankine-Hugoniot shock wave relations and accurately determine the solution in the vicinity of the shock.

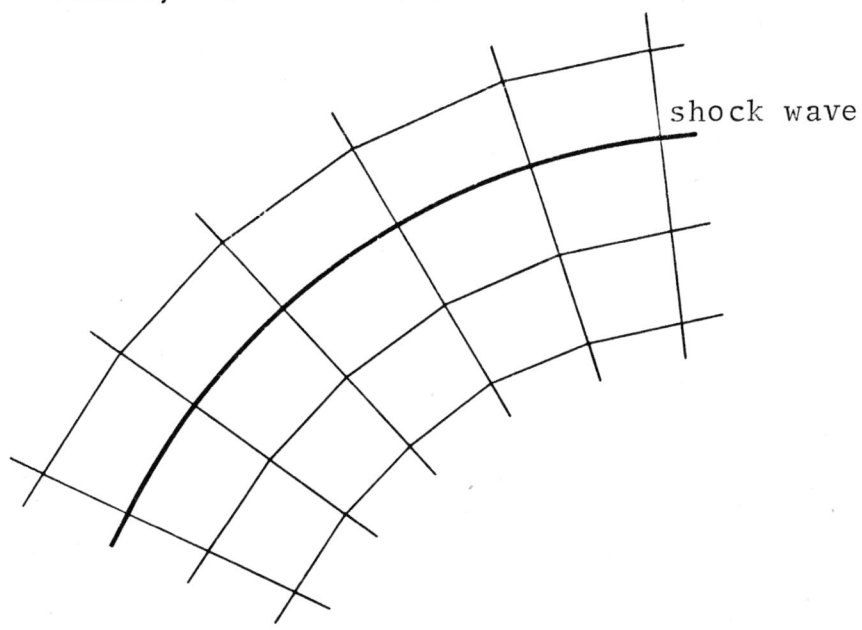

Fig. 8. Alignment of mesh with shock wave

To maintain alignment, the mesh line nearest the shock is allowed to move as if it were an unsteady shock itself. The velocity of each section of this mesh line (see Fig. 9) is determined by solving simultaneously the Rankine-Hugoniot relations and one characteristic relation. The solution is obtained by iteration for w_s the shock wave velocity, p_s the pressure at the shock, and u_s the velocity just behind and normal to the shock.

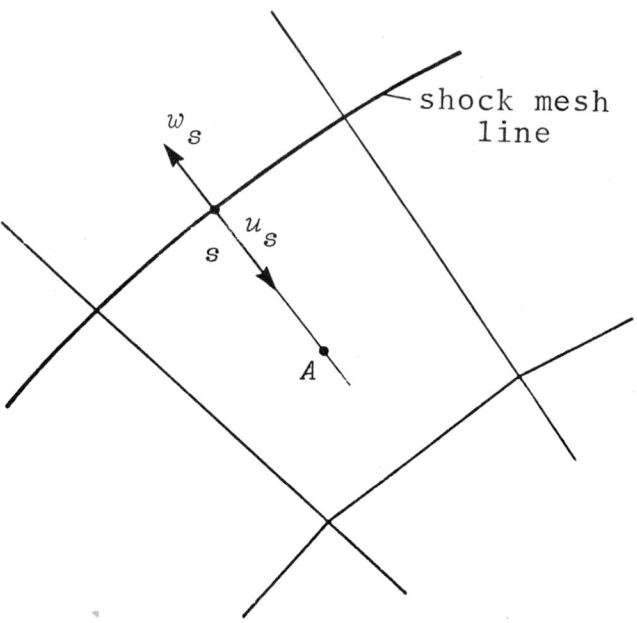

Fig. 9. Geometry for shock wave velocity evaluation

$$p_s^{(0)} = p_A$$

$$\left. \begin{array}{l} w_s^{(i)} = f\left(p_s^{(i)}, p_\infty\right) \\ u_s^{(i)} = g\left(w_s^{(i)}, u_\infty\right) \end{array} \right\} \text{Rankine-Hugoniot relations}$$

$$p_{\text{char}}^{(i)} = p_A + \rho_A c_A \left(u_s^{(i)} - u_A\right) \quad \text{Characteristic relation}$$

$$p_s^{(i+1)} = \tfrac{1}{2}\left(p_s^{(i)} + p_{\text{char}}^{(i)}\right)$$

The values of the flow variables at point A are obtained by interpolation of data at mesh points behind the shock, and those with subscript ∞ are obtained from the free stream conditions. The shock mesh line is then moved using the shock wave velocities calculated for each segment. Occasionally the entire mesh is redistributed to maintain relatively equal spacing of the mesh lines.

RESULTS

Axisymmetric flow results

The flow past a sphere is axisymmetric and can be solved completely by considering the flowfield in a plane of symmetry. At $t = 0$ the shock wave shape was approximated by a parabola about the sphere in the plane of symmetry. A computational mesh was formed so that the mesh was aligned with both the parabola and the surface of the sphere (Fig. 10(a)). The initial condition for the mesh quadrilaterals upstream of the shock was that of uniform Mach number 10, perfect gas ($\gamma = 7/5$) flow. Between the shock and the body, the initial flow values were determined by interpolation from the values obtained at the shock from the steady Rankine-Hugoniot relations and those at the body from simple Newtonian flow theory and the surface tangency condition on the flow. The solution was advanced in time using the operator sequence

$$F^{n+2}_{t_{i,j}} = L_{\mathrm{mesh}_J}(2\Delta t) L_{\Delta J}(\Delta t) L_{\Delta I}(2\Delta t) L_{\Delta J}(\Delta t) F^{n}_{t_{i,j}}$$

at interior points and the normal momentum equation (3) at body surface points. The difference operators were slightly modified to account for the axisymmetry. The converged pressure distribution through the shock wave along the dashed line indicated in Fig. 10(a) is shown in Fig. 10(b) and compared with theory [4]. Because of the shock adjusting mesh the shock wave was accurately captured by the difference equations with no smearing and virtually no oscillation. On a mesh of 12×20 quadrilaterals (12 through the shock layer and 20 about the sphere), the solution required 17.6 seconds of CDC7600 computer time for 512 time step advances, during which the free stream travelled 6.2 body radii.

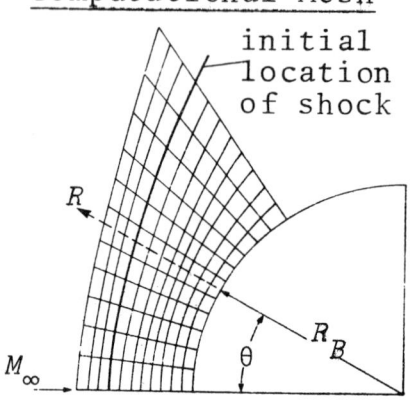

Computational Mesh
initial location of shock

Body points
  Normal momentum equation
Interior points
  $F^{n+2}_{t_{i,j}} = L_{\text{mesh}_J}(2\Delta t)L_{\Delta J} \times$
  $(\Delta t)L_{\Delta I}(2\Delta t)L_{\Delta J}(\Delta t)F^n_{t_{i,j}}$
Shock wave movement
  Rankine-Hugoniot equations
  Characteristics equations

(a) Computational mesh about sphere

Flow past a sphere

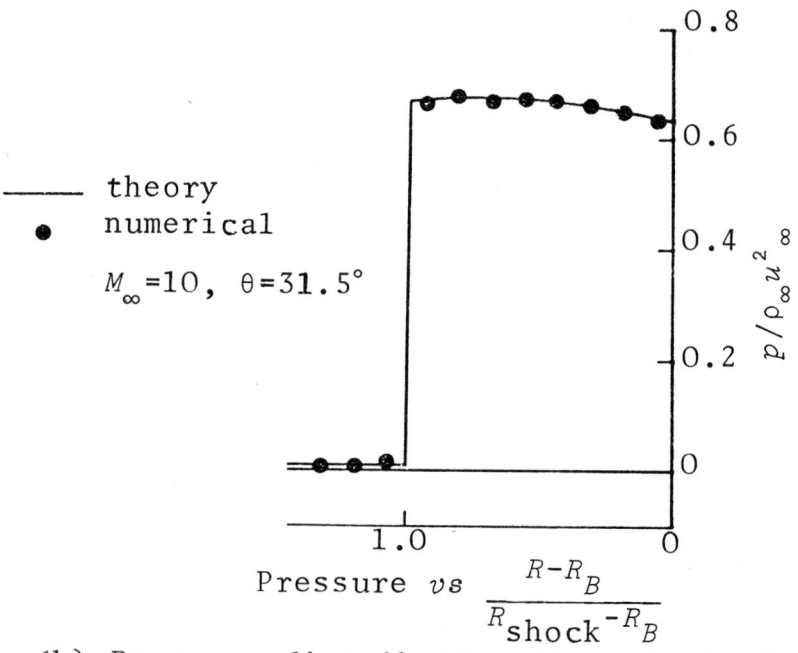

—— theory
● numerical

$M_\infty = 10$, $\theta = 31.5°$

Pressure vs $\dfrac{R - R_B}{R_{\text{shock}} - R_B}$

(b) Pressure distribution through shock

Fig. 10. Concluded

Three-dimensional flow results

The following three-dimensional results for supersonic flows past sphere cone bodies at angle of attack were recently presented by Rizzi and Inouye [3]. The flowfield was divided into 15×9×6 hexahedra (15 around the body in the latitudinal direction, 9 in the meridional direction and 6 across the shock layer). Difference operators similar to those presented earlier were devised for each of the coordinate mesh directions. Fig. 11 presents results in the pitch plane for a sphere (15° half angle) cone body at 30° angle of attack immersed in a free stream, M_∞ = 14.9, flow with specific heat ratio γ = 5/3. The calculated shape is in excellent agreement with that measured experimentally by Cleary and Duller [5].

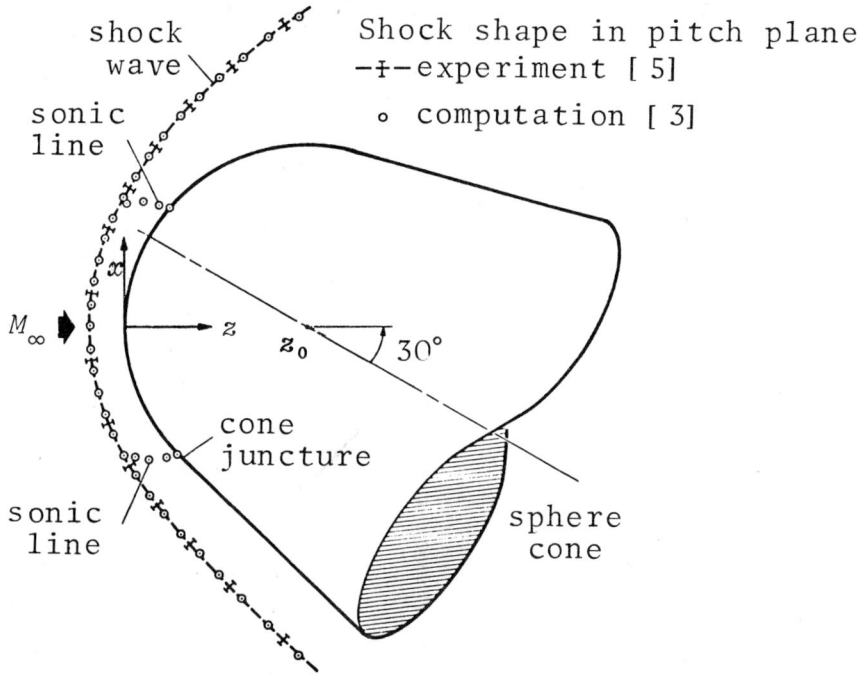

Fig. 11. Shock shape in the symmetry plane for perfect gas flow past a sphere cone body of 15° half angle at M_∞ = 14.9, γ = 5/3 and 30° angle of attack

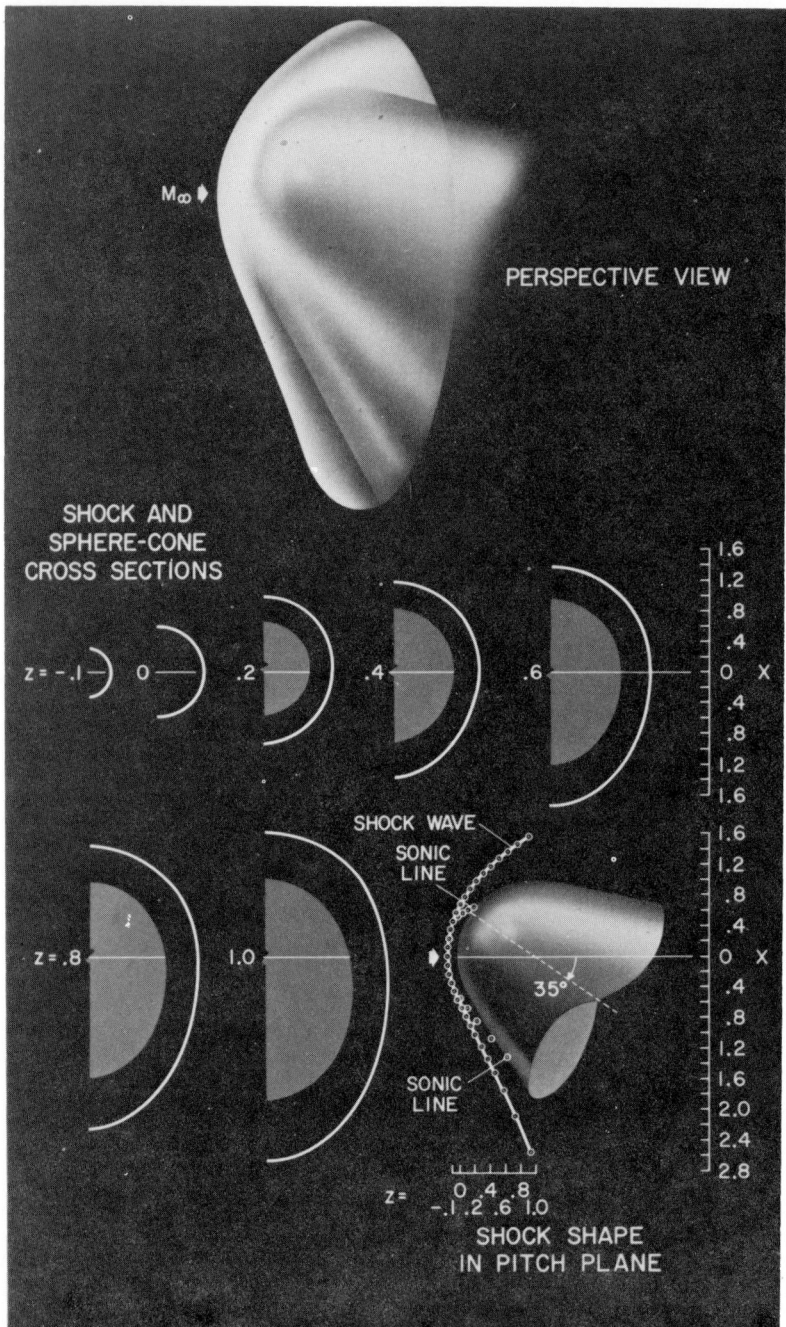

Fig. 12. Shock shape in full perspective, cross-sectional and pitch-plane views for a sphere cone body of 23° half angle in a perfect gas at $M_\infty = 10$, $\gamma = 7/5$ and 35° angle of attack

The computed shock shape and sonic line in the pitch plane as well as perspective and cross-sectional views of the full three-dimensional shock shape of a sphere cone of 23° half angle immersed in a free stream of M_∞ = 10 and specific heat ratio γ = 7/5 at 35° angle of attack are shown in Fig. 12. On the windward side in the pitch plane, the sonic line stretches far downstream on the conical portion of the body and differs totally in shape from that on the leeward part of the sphere. The flow is highly asymmetric as shown in Fig. 13 from the variation of pressure distribution on the body surface. A contour of pressure *versus* latitudinal angle θ and meridional angle ϕ, where $\phi > 90°$ indicates the windward side and $\phi < 90°$ the leeward side of the body is shown. For each value of ϕ, the pressure falls monotonically with increasing θ from its stagnation value as the flow expands around the spherical portion of the body. When the flow reaches the sphere cone juncture on the windward side, however, it encounters an abrupt change of body curvature and begins to recompress as θ increases further. For example, in the ϕ = 170° plane, the expansion begins to cease at θ = 32°, the sphere cone juncture, and then recompresses to a maximum at θ = 68°, after which the flow begins to expand again. This recompression causes a compression wave to propagate out into the flowfield from the cone sphere juncture on the body, ultimately reaching and strengthening the shock wave, and causing a change in shock curvature as shown in Fig. 12. Finally, Fig. 14 shows the calculated Mach number distribution on the body surface. The Mach number increases with θ from the stagnation point (M = 0) and leaves the computational exit mesh boundary, θ = 84°, supersonically.

CONCLUDING REMARKS

The present numerical method accurately calculates three-dimensional supersonic flows with embedded subsonic regions. We know of no method more efficient than that given here for calculating such flows. The principal reasons for this

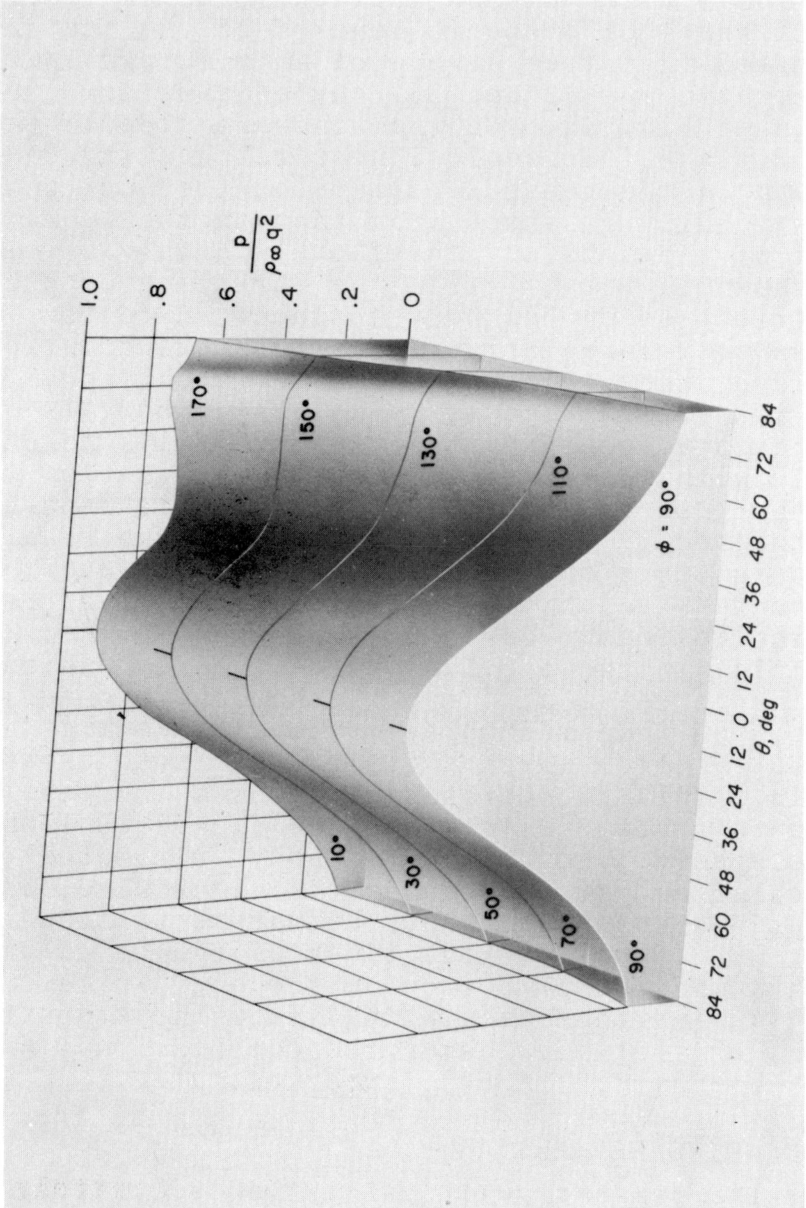

Fig. 13. Computed pressure distribution on the body surface for the flow in Fig. 12

Supersonic/Subsonic Flowfields

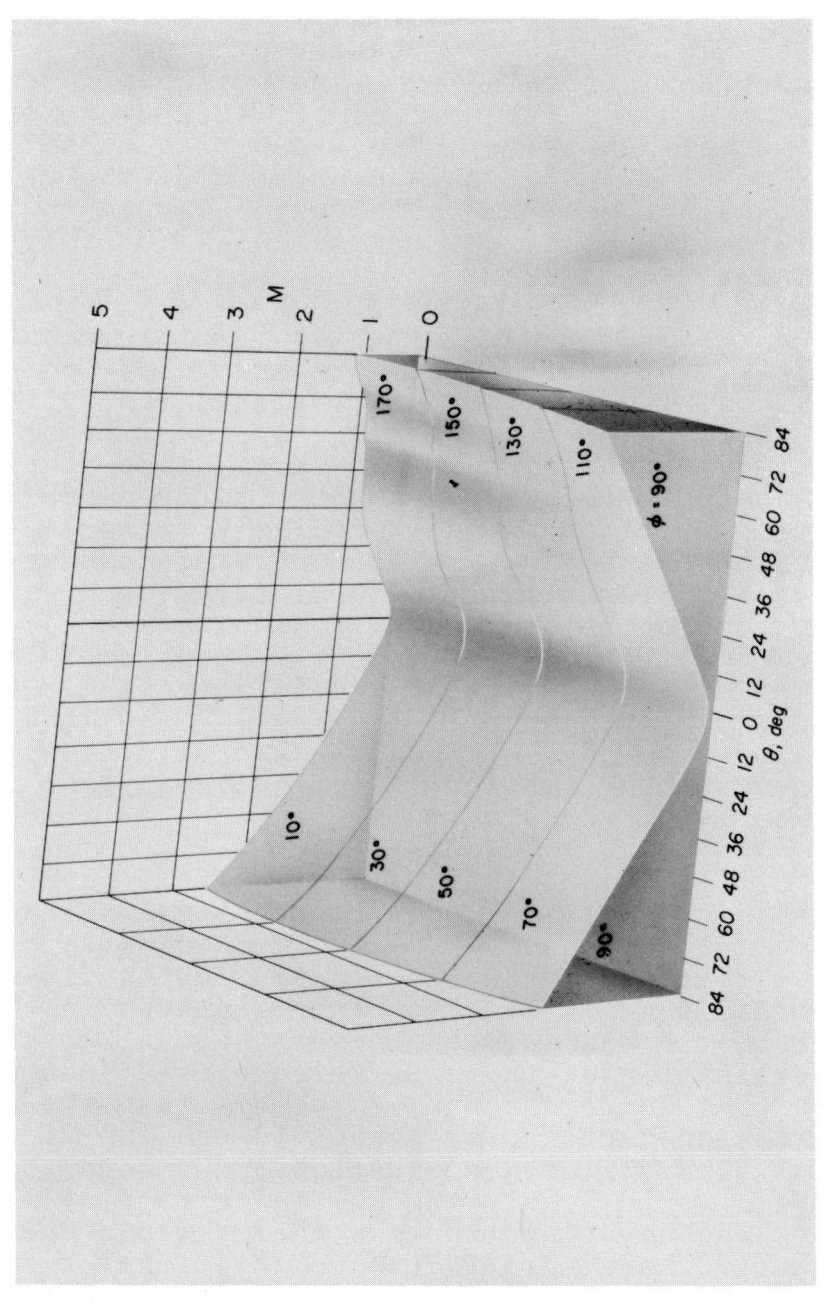

Fig. 14. Computed Mach number on the body surface for the flow in Fig. 12

efficiency are as follows.

1. *The basic finite difference scheme* – It is the simplest and most easily programmable second order accurate technique suitable for solving the nonlinear equations of fluid dynamics. Its closest competitor, "the Leap-Frog technique," loses much of its advantage because of the cumbersome dissipation term which must be added to it to counteract its inherent unstableness for solving nonlinear problems.

2. *Splitting* – For solving multidimensional problems, split techniques are more efficient than their nonsplit counterparts. First, the split techniques, being subject to only one-dimensional stability criteria, can take larger time steps than nonsplit techniques subject to the more restrictive multidimensional stability criteria. Second, the split difference operators can be arranged to form efficient operator sequences in which some operators appear less often than others and each operator is taking time steps near the maximum allowed for stability. The nonsplit schemes do not have this flexibility and in effect often advance the solution in some directions with time steps much smaller than required by the flow physics.

3. *Integral form* – The flowfield geometry and physics determine the computational mesh which, in general, is nonorthogonal. The integral form of the difference equations approximates directly the fluid dynamic differential equations in their simple Cartesian form on a nonorthogonal mesh. The evaluation of these difference equations was shown to require not much more work than those on a Cartesian rectangular mesh. To use the differential form difference equations on a nonorthogonal mesh would require a coordinate mapping which causes the equations to lose their simple Cartesian coordinate form in the computational system, and the additional terms which appear often require as much computational work for evaluation in the difference equations as the original terms.

4. *Shock wave treatment* – In the difference operators presented no special treatment is given to the shock wave. Other difference schemes, which can accurately calculate shock waves in the interior of the mesh, test for the presence of the shock and, if satisfied, treat the shock with special procedures. The shock wave treatment of the present method, mesh alignment, is done outside the difference operators and only when required by the movement of the shock wave.

REFERENCES

1. MacCormack, R.W., "The effect of viscosity in hypervelocity impact cratering," AIAA Paper No. 69-354 (1969).

2. MacCormack, R.W. and Paullay, A.J., "Computational efficiency achieved by time splitting of finite difference operators," AIAA Paper 72-154 (1972).

3. Rizzi, A.W. and Inouye, M., "Time-split finite-volume method for three-dimensional blunt body flow," *AIAA J.*, **11**, No. 11, 1478-1485 (November 1973).

4. Lomax, H. and Inouye, M., "Numerical analysis of flow properties about blunt bodies moving at supersonic speeds in an equilibrium gas," TR R-204, NASA (1964).

5. Cleary, J.W. and Duller, C.E., "Effect of angle of attack and bluntness on the hypersonic flow over a 15° semiapex cone in helium," TN D-5903, NASA (1970).

FLUID DYNAMICS APPLICATIONS OF THE ILLIAC IV COMPUTER

Robert W. MacCormack and
Kenneth G. Stevens, Jr.

*(Ames Research Center,
National Aeronautics and Space Administration,
Moffett Field, California)*

INTRODUCTION

Nearly three decades have passed since the first electronic computer, ENIAC (Electronic Numerical Integrator And Calculator), was built in 1946 at the University of Pennsylvania. It was programmed by hard wiring to integrate ordinary differential equations describing ballistic trajectories for the US Army. Even before ENIAC became operational, John von Neumann of the Institute of Advanced Study at Princeton conceived the logical design modifications required for general applications, in particular, a fluid dynamics application, the computation of shock waves in fluids. Guided by the reports von Neumann and his colleagues published in 1945 and 1946, the University of Manchester soon made important contributions toward the development of today's computers. Unlike ENIAC which required wiring changes for program modifications, their computers were the first machines whose programs could be stored and changed like data within computer memory. In addition to their relatively fast access Williams cathode ray tube memory, they developed a larger, but slower, drum memory and thus became the first to contend with the management of memory hierarchies, a problem still of much concern to us today.*

---

*For a more complete historical discussion see [1].

Fluid Dynamics of Illiac IV Computer

During the 1950's and early 1960's many advances in computer technology were made which basically improved components within the design originated by von Neumann. When computer circuitry began to reach its limits, some of which were imposed by the finite speed of light, computer engineers began to design machines with unconventional architecture. One of these is the Illiac IV at Ames Research Center. Its design development was managed by the University of Illinois during the late 1960's and its hardware was built by the Burroughs Corporation in the early 1970's. It is a new computer with over an order of magnitude more computing power than conventional computers. Its great speed is achieved by using a parallel computer architecture which essentially consists of an array of 64 arithmetic processors, each with the approximate speed of a CDC 6600 computer, under the control of a central unit. It will be used to solve problems that are inherently parallel in nature and are impractical or impossible to solve on conventional computers.

Most fluid dynamics problems have a parallel structure; the dynamics at one location in the flow is described by the same equations that apply to neighbouring locations and can be solved simultaneously using the same algorithm. The three principal fluid dynamics applications of the Illiac IV computer are:

1. to assume experimental tasks which can be computationally simulated more economically,
2. to simulate flows which cannot be simulated by experiment, and
3. to combine computer with experimental simulations to study fundamental fluid dynamics processes.

Computer flow simulations are becoming less expensive each year and many are now more economical than experimental simulations. Computer technology has increased computing speeds by a factor of about ten every 3 years. This has resulted in a reduction of the computation cost of a given problem by a factor of ten approximately every 5 years (Fig. 1).

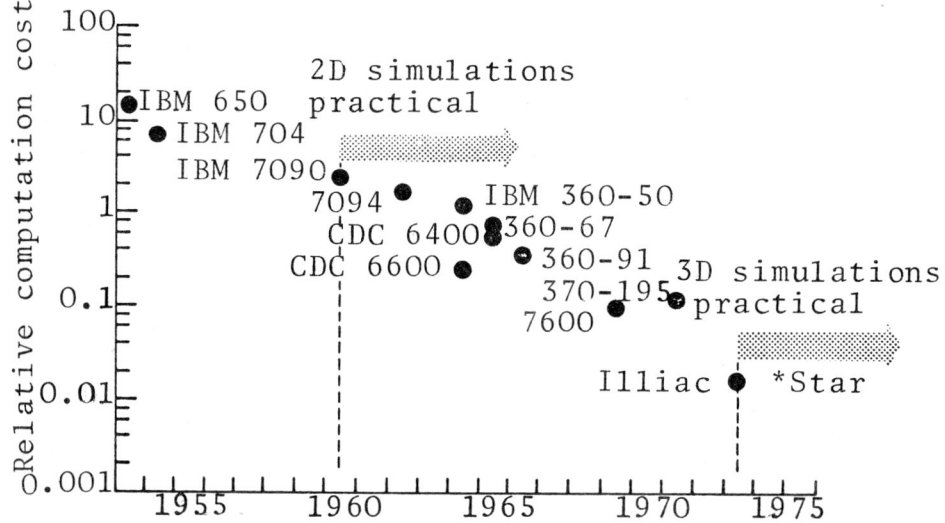

Fig. 1. Trend of computation cost for computer simulation of a given flow [2]

At the same time the number of hours and cost of wind tunnel tests required for aircraft development has been increasing each year (Fig. 2). Because of these trends the computer will be playing an ever increasing rôle in assuming wind tunnel tasks. It is already being used to reduce wind tunnel testing by the early elimination of unpromising designs tested by numerical simulation.

Computer flow simulations can provide information not obtainable by other means. These include flows outside the testing range of experimental facilities or because of their complexity beyond analytical solution. Examples of such flows are those at flight Reynolds numbers and with flight scale chemical reactions. Also, it is possible to simulate flows without wind tunnel

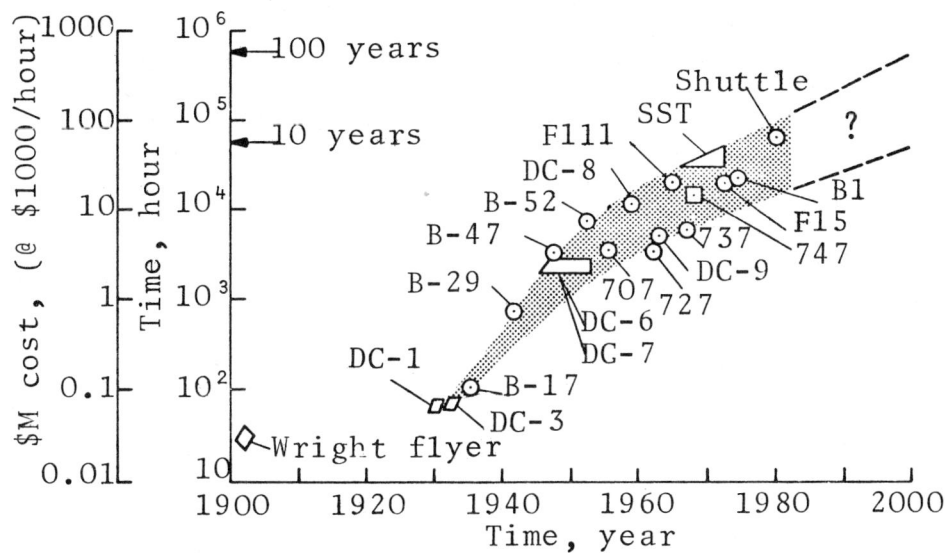

Fig. 2. Total wind tunnel test hours for development of US various aircraft

wall and model support interference effects, which are present in experiments. The application of the Illiac IV computer to obtain flowfield solutions about the space shuttle with flight air chemistry is described in detail later. In addition, although not described in detail here, considerable effort is being given to the application of the Illiac IV to problems in atmospheric research. A result promised by this research will be to increase the accuracy of 7 day weather forecasts to that of current 2 day forecasts.

Numerical simulation can provide insight for understanding the dynamics of complex flows. For example, at Ames in a combined computer and wind tunnel investigation studying turbulence, several analytical turbulence models are being numerically simulated and compared with experiments to understand further this phenomenon. Recent results obtained on the Illiac IV computer for turbulent transonic separated flow past a thick aerofoil are also described later. In a related study, Stanford University and Ames are cooperating in an effort to compute a turbulent flow in a three-dimensional box by integrating the unsteady Navier-Stokes equations. The Reynolds number for this study will be chosen sufficiently small so that there will be no appreciable turbulence of a scale smaller than the computational mesh can support. This ambitious study would be impractical on computers developed prior to Illiac IV.

THE ARCHITECTURE OF ILLIAC IV

From the user's point of view, the Illiac IV computer system (Fig. 3) consists of a central control unit, 64 processing elements each containing a processing unit (arithmetic unit) and a small random access memory containing 2048 64-bit words of storage, and a main rotating memory containing 16 million 64-bit words of storage.

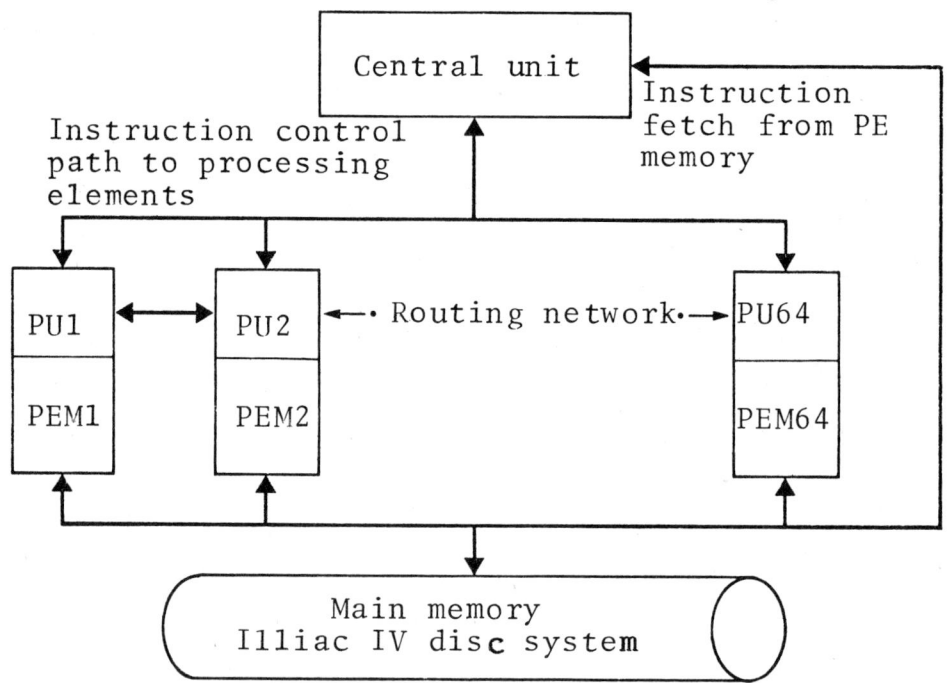

Fig. 3. Illiac IV computer system

Also connected to this system is an archival memory with an on-line storage capacity of 10^{10} words and a network for remote access from stations across the USA, Norway and the UK.

The central unit fetches instructions stored in the processing element memories, decodes them and broadcasts the decoded instructions to the processing units for execution. While the processing elements are executing a broadcast instruction, the central unit is decoding the next instruction, and thus its operations overlap those of the processing elements. The central unit is itself a small unsophisticated computer and a 64 64-bit word memory which in addition to decoding instructions can perform simple arithmetic calculations for such uses as program loop control.

Each processing unit simultaneously receives and executes the same arithmetic or logical instruction broadcast from the central unit, on data fetched from its own processing element memory or routed to it from a neighbouring processing unit. Each processing unit has two modes: enabled or disabled. If enabled it can store or modify data within its own memory; if disabled its memory is protected and cannot be changed. While disabled, however, the processing unit can fetch data from its memory and route it to its neighbours. The execution speed of the processing element is roughly equivalent to a CDC 6600 computer. Its random access memory is, however, limited and for most applications will have to be continually fed from the main rotating memory. Logically, the main memory may be thought of as a drum although, in fact, it is a set of 13 synchronised discs.

The main memory stores program instructions and data to be sent for processing to the array of processing element memories. It also receives and stores processed data to be used again later. For most large problems, such as three-dimensional flowfield calculations, only a small part of the data-base can reside in the processing element memories and efficient data management techniques must be developed so the array processing units are not kept idle for long periods waiting for data.

To illustrate the parallel operation of Illiac IV, consider the solution of a simple hyperbolic equation. We choose the wave equation in one spatial dimension,

$$\frac{\partial U}{\partial t} + C\frac{\partial U}{\partial x} = 0 \quad \text{with } C > 0.$$

We seek a solution $U(x,t)$ which satisfies this equation for all $0 < t < T$ and $0 < x < L$ with a given initial condition

$$U(x,0) = f(x)$$

and a prescribed upstream boundary condition

$$U(0,t) = g(t)$$

Computationally, the problem becomes, using simple upwind differencing,

$$U_i^{n+1} = U_i^n - \frac{C\Delta t}{\Delta x}(U_i^n - U_{i-1}^n)$$

with

U_i^1 given by $f[(i-1)\Delta x]$

and

U_1^n prescribed by $g[(n-1)\Delta t]$

for

$$1 < n \leq N, \quad (N-1)\Delta t = T$$
$$1 < i \leq I, \quad (I-1)\Delta x = L$$
$$0 < \frac{C\Delta t}{\Delta x} \leq 1.0$$

For efficient use of the Illiac IV, we should choose I to be equal to a multiple of 64. For simplicity in this example, we choose $I = 64$. The "i"th column of the array U_i^n, for $n = 1, 2, \ldots, N$ is stored in processing element memory "i," etc. To protect U_1^n, which is prescribed by the upstream boundary condition for all n, we disable the first processing element. The central unit sets up a loop on the index n from 1 to N. The integer n and the scalers C, Δt and Δx reside in central unit storage. For each n, processing element "i" fetches U_i^n from its own memory and routes it to its immediate right. Simultaneously, processing element "i" receives U_{i-1}^n. All of the processing elements then execute the above algorithm in lock-step fashion with each processing element using data from its own memory and the memory of the central unit. Then, if enabled, it stores the new result, U_i^{n+1}, in its memory.

CFD CODE - A PROGRAMMING LANGUAGE FOR ILLIAC IV

Along with the impressive technological advances in computer hardware development during the last three decades, equally important contributions have been made in programming language development which has enabled people to communicate with and control computer machinery. Today the most widespread and convenient programming language in the US is Fortran, a scaler operation type language developed for serial computers. The Illiac IV computer is an array or vector processor. It adds, subtracts, multiplies and divides on a row or vector of data simultaneously. Two languages originally developed for Illiac IV were Ask, an assembly level language, and Glypnir, a higher level Algol based language. Early experiments by the Computational Fluid Dynamics Branch at Ames found both unsuitable for solving fluid dynamics problems. Ask was too low a level of language and, therefore, inconvenient to use. Glypnir was unfamiliar to most Fortran programmers and because of its generality produced inefficient machine code. A compromise was sought between the two. Lomax, Rogallo and Stevens of Ames devised and developed CFD Code, a Fortran-like vector programming language for Illiac IV [3]. Because their primary interest was computational fluid dynamics (CFD), they designed and tailored the language to fit specific requirements of that discipline. No attempt was made to hide the architecture of Illiac IV in the language. On the contrary, because the user is the one most able to structure his program to fit the Illiac IV architecture, every attempt was made to give him access and control of the Illiac IV hardware. As a result, the CFD language is much easier to program than Ask, almost as easy as Fortran and translates into machine code four times more efficiently than Glypnir.

As a sample comparison of CFD code with Fortran, consider the programming of the algorithm of the last section in both languages.

Fortran

```
    CDTDDX = C*DT/DX
    DO 1 N = 1, NEND
    DO 1 I = 2, 64
    U(I,N+1) = U(I,N) - CDTDDX*(U(I,N) - U(I-1,N))
1   CONTINUE
```

CFD Code

```
    MODE = ON. TURN OFF. 1
    CDTDDX(*) = C*DT/DX
   *DO 1 N = 1, NEND
    U(*,N+1)=U(*,N)-CDTDDX(*)*(U(*,N)-U(*-1,N))
1  *CONTINUE
```

For the CFD coded program, the asterisk * appearing in the index of the variable is a dummy index for the processing element number, and *-1 indicates a route of the indexed variable from its immediate left neighbour processing element. The program first activates all the processing units except the first, which protects the data prescribed by the boundary condition. Then the central unit sends the data C, DT, and DX with the instructions for the parallel computation of the variable CDTDDX in the processing units. A loop is then set up in the central unit for the parallel execution of the algorithm.

The CFD language has been proved to be highly suitable for coding the majority of finite difference calculations and other calculations of a similar nature. Once the researcher has formulated his problem in terms of parallel computation, it is then simple and straightforward to code a program for its solution in the CFD language. Software has been developed for translating CFD programs into Ask and relocatable binary machine code for efficient execution on the Illiac IV. Additional software has been developed to translate CFD coded programs into serial Fortran. This enables programs to be debugged by execution on conventional serial machines prior to production job runs on the Illiac IV.

FLUID DYNAMICS APPLICATIONS

Two of the fluid dynamics applications mentioned in the introduction are described in detail here. The first is the determination of the flowfield about the space shuttle: this information cannot be simulated for flight conditions by ground based experiments. The second is the computation of transonic turbulent separated flow past a thick biconvex aerofoil. The Reynolds numbers of this study are so high that the calculation is practical only on computers with the power of an Illiac IV. These two applications are the most developed of those programmed for Illiac IV and are now beginning to yield useful results.

The other applications are also exciting undertakings and will be described in detail elsewhere. In particular, the weather prediction research effort is developing techniques which use half word operations on the Illiac IV. The purpose is to maximise both storage and execution by using one half of each Illiac IV 64-bit word for computing the weather in one hemisphere simultaneously with the computation for the other hemisphere using the other half words. Although half word arithmetic requires a longer execution time than full word arithmetic, the over-all calculation requires substantially less time because there will, in effect, be twice the number of processing units available. Another application meriting additional comment is the work at

Stanford University, for computing three-dimensional turbulent viscous flow in a box. Because of the large data-base requirements of this problem (computational mesh of size 64 × 64 × 64) disc memory mapping procedures are being developed to facilitate the timely and orderly flow of data between the main disc memory and the processing element memories. The development of these procedures is crucial for almost all three-dimensional flow applications.

Space shuttle flowfields

Fig. 4 shows an artist's conception of the shuttle orbiter spacecraft and the shock wave structure of the surrounding flowfield on re-entry. At the high velocities, altitudes and angles of attack to be encountered along the flight path, the multishocked flow will contain chemically reacting gases and cannot be simulated in the laboratory. Reinhardt and Davy of Ames are simulating this flow numerically on the Illiac IV computer [4]. They are developing numerical techniques to calculate steady, inviscid three-dimensional nonequilibrium supersonic flow about aircraft-like configurations. The equations they must solve consist of the steady Euler equations coupled to a set of partial differential equations governing species concentrations. Because of the large variation in chemical reaction rates for the five species considered, N_2, O_2, NO, N and O, Reinhardt and Davy are combining implicit and explicit finite difference techniques to solve these equations.

The flowfield solution in the nose region is obtained from the three-dimensional blunt body program of Rizzi and Inouye [5] which solves the unsteady Euler and species concentration equations. The nose region contains a subsonic part surrounded by supersonic flow. Downstream of the nose region, the flow is purely supersonic, and the governing steady Euler and species equations are hyperbolic in the flow direction. Therefore, if

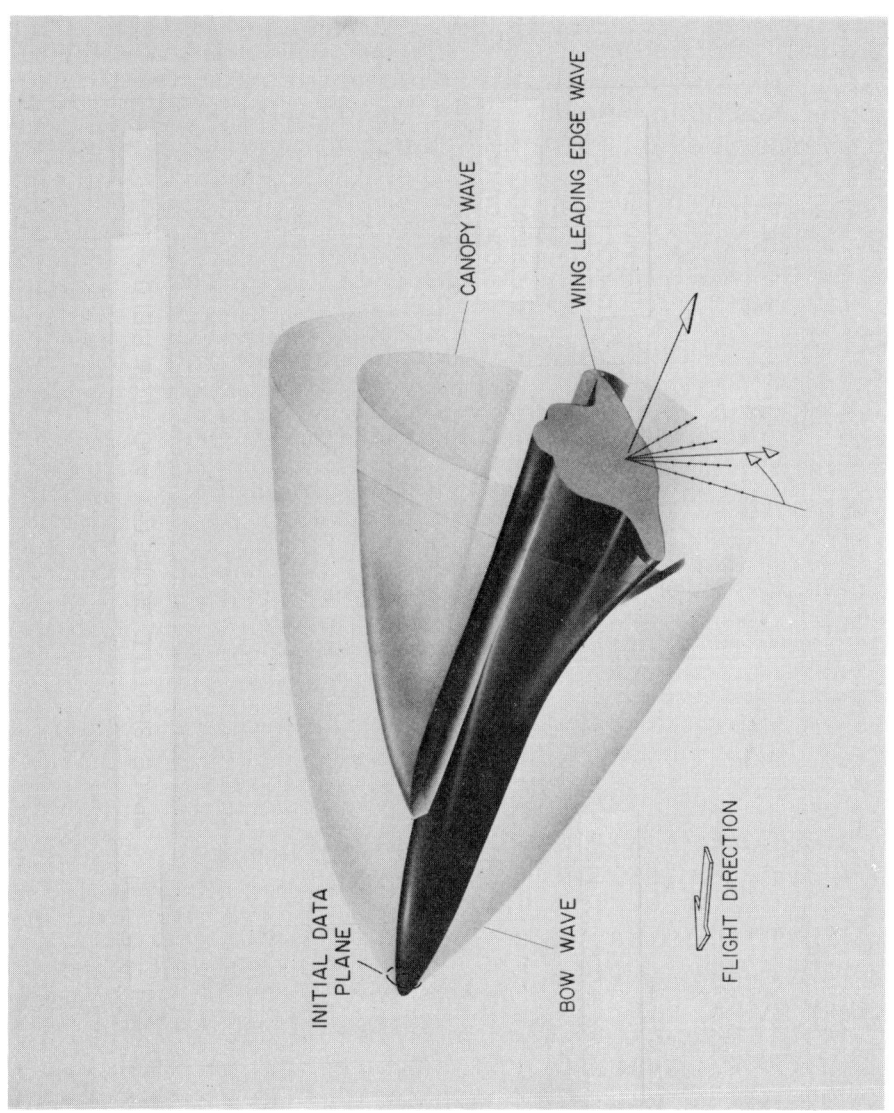

Fig. 4. Space shuttle geometry and flowfield

the solution is known at a surface spanning the flow at one upstream location, the solution can then be determined everywhere downstream by matching the solution surface in the flow direction. At each step, the numerical method solves for the dependent flow variables in terms of the variables at the previously solved for surface, by integrating numerically the steady Euler and species equations. For a three-dimensional flow the calculation is actually only two-dimensional and is updated in space. The solution surface of Reinhardt and Davy is a plane normal to the shuttle orbiter body axis shown in Fig. 5.

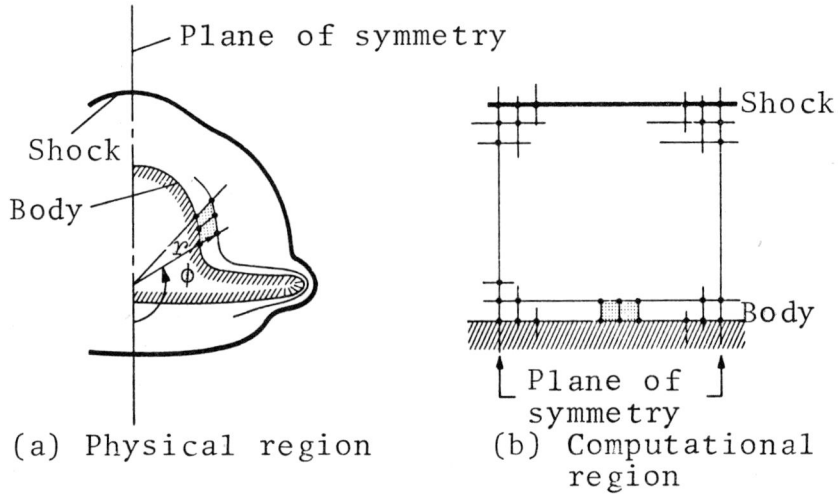

Fig. 5. Coordinate system and transformation procedure: axis normal plane

Their initial data plane was obtained by interpolation of data from the nose region solution. Also shown in Fig. 5 is the rectangular region in the computational plane into which the physical flowfield is mapped. The computational mesh contains 24 × 21 points, 24 points between the body and shock surface on each of the 21 body meridional plane intersections with the data plane. Because their computational problem is essentially two-dimensional, it fits easily within the processing element core memory of the Illiac IV. In fact for calculations requiring only 21 meridional planes for spatial resolution, three problems can be fitted into core simultaneously. Fig. 6 shows Reinhardt and Davy's storage layout in the Illiac IV array memory, which enables them to solve three problems with perhaps different angles of attack, flight speed or altitude simultaneously. Initial results indicate that the entire flow about the shuttle orbiter can be calculated in 10 to 15 minutes of Illiac IV machine time.

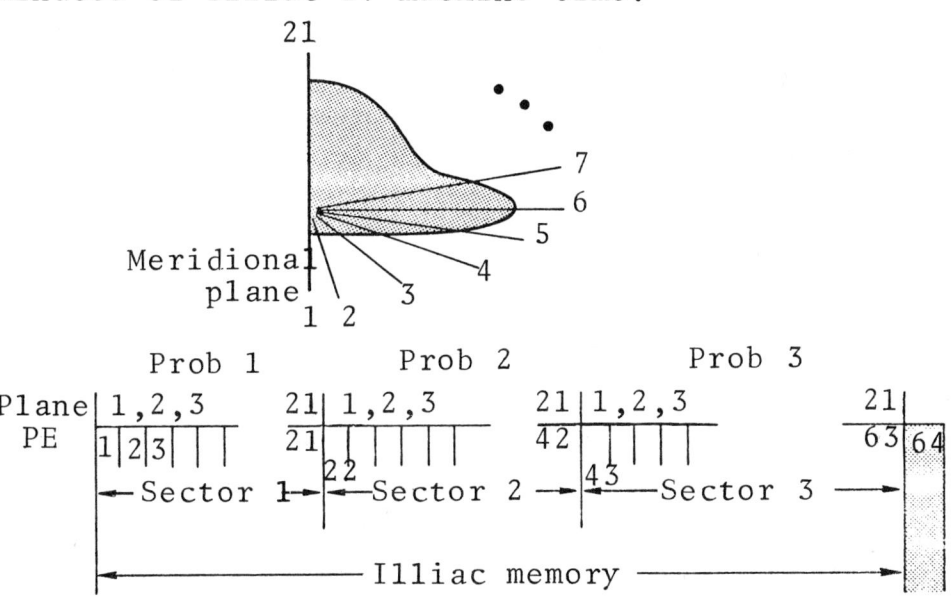

Fig. 6. Illiac IV processing element assignment

Turbulent separated flow

The increase in computing speed, resulting from the technological advances in computer hardware design during the last few years, has enabled the calculation of viscous flows at high Reynolds numbers to be made. Only a few years ago, we were beginning to solve for two-dimensional shock-induced separated laminar boundary layer flows at Reynolds numbers of 10^5, [6], [7]. Today we are on the threshold of extending our calculations to treat turbulent boundary layer separation at Reynolds numbers of 10^7. These calculations solve the compressible form of the unsteady Navier-

Stokes equations with time averaged Reynolds stress terms, and the required time for their solution increases roughly with the square root of Reynolds number. Because of the broad range of time and length scales existing in turbulent eddying flows, and because finite difference calculations are limited in both spatial and temporal resolution by the mesh spacing and time step increments, it will take at least the next generation of computers after Illiac IV to solve the Navier-Stokes equations with small scale time and space varying turbulence for flows of interest to aircraft designers. In the meantime strong efforts will be made toward the development of turbulence models for closure of the time averaged equations which accurately account for subgrid size eddies and high frequency fluctuations.

Baldwin of Ames has been testing both simple mixing length and the more sophisticated two-equation transport turbulence models for calculating the interaction of a strong shock wave with a hypersonic turbulent boundary layer at high Reynolds numbers [8]. His results show that both types of models adequately account for the effects of turbulence ahead of the interaction and within the reversed flow region. Because of the failure of the mixing length model to account for nonequilibrium levels of turbulence generated downstream of re-attachment, the agreement of the calculation using this model with experimental results is poor. The calculations using the two-equation turbulence models tested so far show improved but not total agreement with experiment.

In a related computational study at Ames for which there is a parallel experimental study, Deiwert is numerically simulating on the Illiac IV the shock induced and trailing edge separation of turbulent boundary layers for transonic flows past thick aerofoils at Reynolds numbers as high as ten million [9]. The flowfield and boundary condition treatment is shown in Fig. 7. He solves the compressible integral form of the unsteady Navier-

Stokes equations with time averaged Reynolds stresses and is currently using a simple mixing length model for closure.

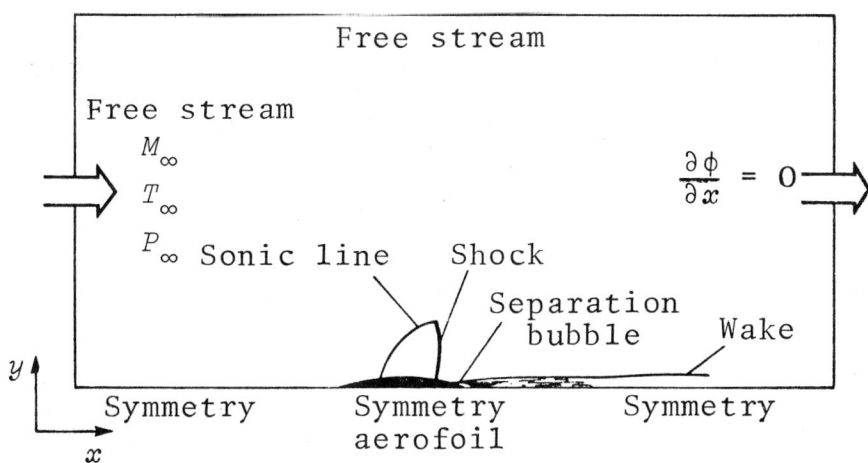

Fig. 7. Sketch of turbulent transonic flowfield about an aerofoil

His computational mesh contains 64 × 38 cells. In the stream direction the mesh is uniformly spaced over the surface of the aerofoil (20 points) and is exponentially stretched ahead of (10 points covering 6 chord lengths) and behind the aerofoil (34 points covering 30 chord lengths). Normal to the aerofoil the mesh is exponentially stretched using 38 points to cover 6 chord lengths with 14 points in the boundary layer ($\Delta y_{min} = \frac{2}{3} c/\sqrt{Re_c}$, c = chord length). To optimise the use of the Illiac IV parallel architecture, Deiwert stores his upstream boundary data which remain unchanged during the calculation in processing element "1" which is disabled, the interior data which are calculated at each step in processing elements "2" to "63," and the exit boundary data in processing element "64" which is obtained by routing the calculated data from processing element "63" after each step. Fig. 8 shows the separated velocity profile at the trailing edge of the aerofoil at a Reynolds number based on chord length of 2×10^6. These results required 2 hours of computation time using the present configuration of Illiac IV.

Optimum configuration using all of the overlap features of the Illiac IV hardware should reduce this time by a factor of three which would be approximately twenty times less than the time required for execution on the CDC 7600.

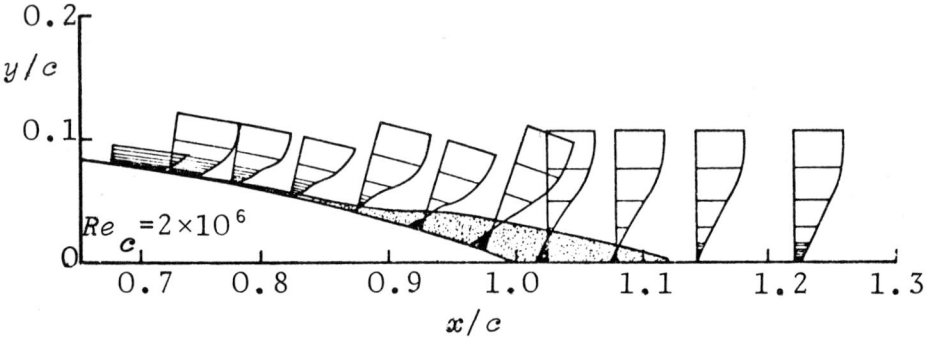

Fig. 8. Separation velocity profiles on 18% circular arc, $M_\infty = 0.775$

CONCLUDING REMARKS

The Illiac IV computer is a member of the latest generation of super computers. Its parallel architecture is unconventional, but fits well the structure of many fluid dynamics applications. A programming language CFD code has been developed to enable the programmer to use this architecture easily and efficiently. Although all of the hardware is not fully operational, Illiac IV is beginning to yield useful results at computing speeds an order of magnitude faster than the most advanced conventional computers. Even this, however, is still not fast enough to solve all fluid dynamics problems of interest.

REFERENCES

1. Rosen, Saul, "Electronic Computers: A Historical Survey," *Computing Surveys*, **1**, No. 1, 7-36 (March 1969).

2. Rakich, J.V., "Introduction to the proceedings of the symposium on Computational Fluid Dynamics," Proceedings AIAA Computational Fluid Dynamics Conference, Palm Springs, California (July 1973).

3. "CFD - A FORTRAN-based language for Illiac IV," Computational Fluid Dynamics Branch, NASA Ames, Version 2 (1974).

4. Reinhardt, W.A. and Davy, W.D., "Numerical simulation of hypersonic, three-dimensional, chemical nonequilibrium flow about shuttle configurations," paper presented at symposium on Parallelism in Computation, NASA Ames (February 1974).

5. Rizzi, A.W. and Inouye, M., "Time-split finite-volume method for three-dimensional blunt-body flows," *AIAA J.*, **11**, No. 11, 1478-1485 (November 1973).

6. MacCormack, R.W., "Numerical solution of the interaction of a shock wave with a laminar boundary layer," *in* "Lecture Notes in Physics," Volume 8, Springer-Verlag, 151 (1971).

7. Carter, J.E., "Numerical solutions of the supersonic, laminar flow over a two-dimensional compression corner," *in* "Lecture Notes in Physics," Volume 19, Springer-Verlag, 69 (1971).

8. Baldwin, B.S. and MacCormack, R.W., "Interaction of strong shock wave with turbulent boundary layer," paper presented at the 4th Int. Conf. on Numerical Methods in Fluid Dynamics, Boulder, Colorado (June 1974).

9. Deiwert, G.S., "Numerical simulation of high Reynolds number transonic flows," paper presented at the 4th Int. Conf. on Numerical Methods in Fluid Dynamics, Boulder, Colorado (June 1974).

NUMERICAL COMPUTATION OF STEADY BOUNDARY LAYERS – A SURVEY

D.B. Spalding

(Imperial College, University of London)

INTRODUCTION

The subject of boundary layer flows is so large that comprehensiveness could not be provided by even a dozen survey papers. One aspect only can, therefore, be surveyed, but which? Mathematical methods? Physical inputs? Experimental data? Unsolved problems?

The decision has been made to survey *types* of boundary layers, classified from the point of view of numerical computation. Only steady boundary layers are considered: these are classified as two, two-and-a-half and three-dimensional parabolic; and two and three-dimensional partially parabolic. Practically interesting examples are indicated, the computational problems are disclosed and methods of solution are described. Points emphasised are the following.

(*i*) For two-dimensional problems the most convenient coordinate system uses non-dimensional stream function in the cross-stream direction; but this cannot be used for three-dimensional problems.

(*ii*) For two-dimensional supersonic parabolic problems, a method has been developed for solving both momentum equations simultaneously.

(*iii*) Almost all "three-dimensional boundary layer" methods in the literature are of the restricted kind in which only two momentum equations are solved (the label "two-and-a-half" dimensional is therefore applied).

(*iv*)   Methods now exist for solving three-dimensional parabolic problems in which all three momentum equations are solved.

(*v*)   For subsonic flows, "uncoupling" of pressures is needed, but not for supersonic.

(*vi*)   For partially parabolic ("semi-elliptic") flows, new methods allow satisfaction of all momentum equations with economy of computer storage: only the pressure occupies storage of dimensionality equal to that of the flow.

(*vii*)   The latter development is likely to be especially beneficial in aeronautics, power-plant aerodynamics and ship hydrodynamics.

The main message of the paper is as follows:

(*a*)   the indicated classification scheme is a practically useful scheme because it enables the numerical analyst to choose the most economical method;

(*b*)   numerical methods do actually exist for problems of all the indicated classes; and

(*c*)   there are still several problems outstanding.

THE RELEVANT PHYSICAL LAWS AND PROCESSES

Whichever the class of boundary layer in question, the relevant basic laws are the same; they are those of momentum, chemical species, energy and turbulence-property balances; and they involve *transport* by convection and diffusion, and *generation* (or destruction) by pressure gradient, chemical reaction, heat absorption and turbulent shear, respectively. These laws are expressed by partial differential equations.

Typically, the analyst of a practically interesting problem must solve for the three velocity components, the pressure, the enthalpy, the species concentrations (perhaps numerous) and, say, two or three quantities characterising the turbulence (e.g., fluctuation energy and dissipation rate). These are the dependent variables of the differential equations.

Points deserving emphasis are:

(a) that the various differential equations, usually 8 to 12 in number, must be solved *simultaneously*, because of multiple interlinkages;

(b) that the interlinkages are of many kinds and mainly *nonlinear*;

(c) that the subject of "turbulence models" deserves several papers to itself, but must here be dismissed by way of a few references [1] to [6]

FEATURES CHARACTERISING THE NUMERICAL COMPUTATION OF STEADY BOUNDARY LAYERS

To select the optimum methods of computation for a steady boundary layer, one must consider five main features of the problem. These are as follows.

(a) *The dimensionality of the flow*. Flows are generally three-dimensional. However, there are special subgroups of two- and two-and-a-half-dimensional flows to be discussed. Plane and axisymmetric flows are two-dimensional, even although three velocity components may require computation.

(b) *The dimensionality of the associated computer storage*. This is usually one less than that of the flow. The reason is that effects from "downstream" do not propagate "upstream" (these words have meaning only when recirculation is absent); so "marching integration" is permissible. This is true of "parabolic" flows, but not entirely of the "partially parabolic" ones.

(c) *The way in which the pressure is arrived at*.

There are some boundary layers for which the whole pressure field is specified before analysis starts; but in others it must be computed.

(d) *The nature of the boundaries of the domain of integration.* The domain may be wholly bounded by solid walls or symmetry surfaces; it may be wholly surrounded by a stream of irrotational fluid of great extent; or both types of boundary may be present.

(e) *The pressure-mass-velocity relation.* At low Mach numbers a rise in pressure normally diminishes the mass velocity (*i.e.*, density velocity product) through a point; the reason is that the velocity diminishes while the density is scarcely affected. At high Mach numbers, a pressure rise *increases* the mass velocity; the increased density outweighs the diminished velocity.

SOME FEATURES CHARACTERISING SOLUTION PROCEDURES

There are also five main characteristics of the solution methods.

(a) *What variables characterise the finite difference grid.* Are grid nodes joined by stream lines, by lines of the distance coordinate mesh, or by some combination? Are the grid coordinates "natural" or "normalised"?

(b) *Grid-node arrangement.* Are the grid nodes arrayed in a regular or irregular pattern? Is this pattern topologically rectangular or triangular? Are velocities stored at the same points as pressures? or are the velocity grids "staggered"?

(c) *How many momentum equations are solved?* Although *three* momenta-per-unit-mass (*i.e.*, velocities) have to be computed, sometimes only *two* momentum equations are solved; and in plane and axisymmetrical problems, *two* momenta can be deduced, sometimes, from *one* momentum equation. Three-dimensional and two-and-a-half-dimensional methods are distinguished in this regard, as also may be "parabolic" and "partially-" parabolic flows.

(d) *Manner of integration.* The finite difference equations may be solved by a single "marching" sweep, from one end of the integration domain to the other; or iteration may be employed. Often a combination of the two techniques is used.

(e) *Matrix inversion.* It is possible to use direct methods of solving the linear finite difference equations for the region of the grid under consideration (e.g., a plane normal to the main flow direction). At the other extreme is point-by-point iterative adjustment of the Gauss-Seidel type. Intermediate methods involve use of the tri-diagonal matrix algorithm, as in the alternating-direction-implicit method [7], [8].

Evidently, choices must be made; and it is the thesis of this paper that the choices can and should be made rationally, *i.e.*, by realistic assessment of the balance of advantage in particular cases.

SOME COMPUTATIONALLY IMPORTANT FEATURES OF BOUNDARY LAYERS

Question	Parabolic	Elliptic	Partially parabolic
Do influences from downstream penetrate upstream?	No	Yes	Yes, but through p only.
Is marching integration permitted?	Yes	No	Yes, but with iteration
Is store dimensionality less than that of flow?	Yes	No	Yes, except for p.

The terms "parabolic" and "elliptic" derive from the mathematician's description of the type of differential equation. As ordinarily understood, boundary layers are parabolic, while flows exhibiting recirculation (e.g., "separated flows") are elliptic. This paper draws attention to an interesting class of flow which is parabolic in respect of all equations but that for pressure; this is the class here termed "partially parabolic".*

Recognition of this third class is important because it is often found in practice and because a partially-parabolic problem requires much less storage for its solution than an all-elliptic one.

TWO-DIMENSIONAL PARABOLIC FLOWS

PRACTICALLY INTERESTING EXAMPLES

We now begin a series of discussions on the various classes of flow beginning with two-dimensional parabolic flows.

Practically important flows in this class are conveniently grouped by reference to the type of boundary condition, as follows.

Flows with free boundaries include jets emerging into still surroundings, or into a stream in parallel motion, as in the aircraft-engine exhaust, the ventilation of rooms, and the injection of a fuel-steam-air mixture into a furnace. The turbulent "premixed" flame which spreads behind a flame-holder in an aircraft "reheat" duct is of this kind; so also is the plume of smoke rising above a bonfire on a windless day.

Fully confined two-dimensional boundary layers are found in:pipes, annuli and diffusers of circular, but not necessarily uniform, cross-section; ven-

---

* I have also used the term "semi-elliptic for the same class.

turi-meters for flow measurement, and the convergent-divergent nozzles of rocket motors; "ejectors" and "injectors," provided that the operating conditions do not give rise to recirculation; and many devices for conducting mixing with chemical reaction, whether fuel is being burnt or, for example, titanium oxide or carbon black is being produced.

Two-dimensional parabolic flows with mixed boundary conditions are exemplified by boundary layers on turbine and compressor blades; the flow around the front hemi-spherical surface of a fuel droplet moving through an oxidising atmosphere; and the region of intense gradients of velocity, gas composition and temperature near a steel block that is being cut by a jet of oxygen.

Evidently many practically interesting phenomena fall into the two-dimensional parabolic category; and turbulence, heat and mass transfer, and chemical reaction, often play significant rôles.

RECOMMENDED GRID

Many methods have been developed for solving problems of this type. I still recommend the one Patankar and I developed many years ago [9], which has been embodied in a textbook [10]; and the latest version of the relevant computer program is also generally available [11].

One of the most important features of this method is that the cross-stream independent variable is the *non-dimensional* stream function, $(\psi - \psi_I)/(\psi_E - \psi_I)$ where I and E represent the internal and external boundaries of the integration domain, and ϕ_I and ϕ_E are caused to vary with longitudinal distance x so the integration domain just encloses the region of practical interest.

The advantage of the latter feature is obvious: it is computational economy. Another less obvious advantage of using dimensionless stream function is that cross-stream convective fluxes in the differential (and finite difference) equations are known explicitly at each forward step; with other grid systems they must be guessed, and corrected by iteration.

Of course, the total width of the grid "expands" and "contracts" with longitudinal distance as the integration proceeds. Its width is calculated, like all cross-stream distances, from the continuity equation; this is done at the *end* of a forward step.

SOLUTION PROCEDURE FOR SUBSONIC FLOW

Continuing to characterise the Patankar-Spalding procedure in the manner indicated (pp. 469 - 470), we must record the following.

(*a*) The grid nodes are arranged in topologically rectangular arrays; and the longitudinal velocities are stored at the same nodes as pressure and all other variables. The cross-stream velocity is not important for subsonic flows.

(*b*) Only one momentum equation is solved, that for the longitudinal direction. The other is replaced by the presumption of pressure equilibrium across the layer; usually this means $p = p(x)$, but not always. If the cross-stream velocity is needed, it is deduced from the continuity equation.

(*c*) Integration is by a single "march" from upstream to downstream, the variables being held in one-dimensional arrays, and overwritten as integration proceeds.

(*d*) The finite difference equations are linear (or rather linearised) and fully implicit; upstream values of properties are used for calculating convection and other coefficients. Iteration is *not* employed. The finite difference equations are solved by direct use of the tri-diagonal matrix algorithm.

SOLUTION PROCEDURE FOR SUPERSONIC FLOW

Recently, I have extended the recommended procedure ("The combination of the SIMPLE algorithm with the GENMIX computer program," unpublished work at Imperial College, London, 1973) to permit supersonic flows to be handled. The special feature of these flows is that although the density and the longitudinal velocity at a point are altered by a rise in pressure at a grid point in opposite directions (ρ increases, while u decreases), the former effect dominates; so a rise in pressure increases the mass velocity (ρu) through the downstream surface of a grid cell. Now the pressure increase affects the velocities in the cross-stream direction in a similar sense, *i.e.*, so as to increase the mass flow *from* the cell; it is therefore possible to employ the SIMPLE algorithm of Patankar and Spalding [12] to compute the pressure while simultaneously satisfying *both* momentum equations and continuity.

A full paper is needed for the proper explanation of this version of SIMPLE, which has been used successfully for mixed subsonic-supersonic flows as well as for wholly supersonic ones; however, the essence of the idea may be conveyed by the following description of the execution of a single forward step

(*i*) The downstream pressure is guessed, and the resultant values of u, ρ and v (the cross-stream velocity) are computed. These values do not satisfy continuity in general; in particular there are inconsistencies in the streamline inclinations calculable from u and ρ and those calculable from u and v.

(*ii*) Truncated forms of the two momentum equations, inserted into the continuity equation, allow a set of Poisson-type finite difference equations to be solved for a *correction* to the pressures which will set to rights the above mentioned inconsistencies.

(*iii*) These equations are then solved by the tri-diagonal matrix algorithm; the resulting pressure corrections are applied and the corresponding changes to u, ρ and v also. A further forward step can then be taken.

A TYPICAL EXAMPLE

Of the many examples which could be chosen of practical use of the recommended method, one is mentioned: this is the computation of the complex flow, mixing and chemical-reaction phenomena which occur in the exhaust plume of a rocket nozzle. In this work [13] several unusual features were included, namely, computation of the concentration of a large number of chemically reacting species; the prediction of root-mean squares of concentration *fluctuations* as well as time-mean values; an account for energy transfer by emission, scattering and absorption of radiation; the computation of the amounts of solid-phase material in various size ranges, with account for condensation and sublimation; and mixing by way of a two-equation turbulence model. In all, up to 50 simultaneous partial differential equations were solved.

Despite the complexity of the situation, comparisons with experiment made by Jensen and Wilson [14] revealed satisfactory agreement.

Recently a version of the method which handles the lateral momentum equation in supersonic flow has also been adapted to the rocket exhaust plume (R.I. Issa, D.B. Spalding and D.G. Tatchell, unpublished work at Combustion Heat and Mass Transfer Ltd., 1974).

TWO-DIMENSIONAL PARTIALLY-PARABOLIC FLOWS

PRACTICALLY INTERESTING EXAMPLES

It has been mentioned that the Patankar-Spalding marching-integration procedure has been applied to mixed subsonic-supersonic situations. This is true, but the cross-stream momentum equation in the subsonic region has to be replaced by the pressure-equilibrium assumption of parabolic flow. However, there are situations, including some arising in wholly subsonic flow, for which to make this assumption misses the essence of the problem.

Examples with free boundaries include the mixing of two fluids which enter so abruptly into a high intensity exothermic chemical reaction that the streamlines must change direction rapidly.

Confined-flow examples include the flow in subsonic compressor and turbine cascades; the flow in a wide-angle diffuser, especially when swirl is present; and the movement of oil in high speed bearings. The first two are extremely important in practice.

Mixed boundary condition examples include certain events that occur when a shock impinges on a wall (but without causing recirculation); the nose-cone problem in space-vehicle re-entry; and film cooling from angled slots. Here too are phenomena which practising engineers urgently need to compute.

RECOMMENDED METHOD OF SOLUTION

Problems of this kind have been soluble by elliptic-flow methods for some time. That reported by Gosman *et al.* [15] is an example of such methods which is valid for subsonic flows; and the SIMPLE algorithm of Patankar and Spalding [12] has been adapted by Spalding and Tatchell [16] and by Issa [17] to the solution of mixed subsonic-supersonic flows. However, such methods involve the use of *two-dimensional storage for all* variables. Methods which proceed by way of the unsteady state equation, such as those described in [18] to [21] and others reviewed by Harlow [22], inevitably possess this expensive feature, which often precludes, because computer storage is limited, a sufficiently fine-scale calculation.

When recirculation is absent, however, I have shown ("A note on calculating 'semi-elliptic' flows," unpublished work at Imperial College, London, 1972) that there is no need for two-dimensional computer storage of any variable except pressure; when this is provided, the Patankar-Spalding procedure with non-dimensional stream function as coordinate can be advantageously used, but now in an iterative manner. The main elements of the method are as follows.

(*i*) The pressure field is first guessed.

(*ii*) Marching integration steps are then made, with solution of both momentum equations, in a manner similar to that described for supersonic flow (p. 474).

(*iii*) However, the pressure corrections are applied not at the station that has just been reached by the integration, but *one grid line upstream*; for now the effect of a pressure change on velocity is greater (in respect of ρu product) than that on density; so the upstream location is necessary to ensure that the procedure converges.

DISCUSSION

This short account of the new method of predicting partially-parabolic flows omits many features of interest and legitimate concern. These must be dealt with at length in a separate paper. Here only a few points need be made.

The advantages of the method are that it uses much less computer storage than do conventional methods; moreover, if the Mach number is high or the curvature of the flow small, very few iterations are needed, so computer time is also small.

Computer costs bulk so large nowadays in the total development costs of engineering equipment that every economy must be grasped; the new procedure offers large economies for partially-parabolic flows.

The convergence properties of the method depend upon the positiveness of all coefficients in the Poisson equation for the pressure correction. It is not vital (but advantageous) that these coefficients should be quantitatively precise (since it is only a *correction* that must be calculated; and this must ultimately be negligible in any case); but the coefficients must all have the right sign. Those familiar with transonic-flow computations will not be surprised that the quantity $\{1 - (\text{Mach number})^2\}$ enters significantly into the coefficients, and determines where the pressure correction is to be placed.

The status of the new procedure is that it is still experimental in several respects; but the promised advantages are certainly forthcoming. The only question is how to arrange the details of the numerical procedure to maximise them.

"TWO-AND-A-HALF-DIMENSIONAL" PARABOLIC FLOWS

PRACTICALLY INTERESTING EXAMPLES

The next class of flows contains all that many authors think of as three-dimensional boundary layers; however, the semi-humorous name "two-and-a-half-dimensional parabolic" has been chosen here to draw attention to the fact that three-dimensional parabolic phenomena are altogether more interesting.

It is true that the two-and-a-half-dimensional phenomena, like true three-dimensional ones, require three independent space coordinates for their description; and also that one of these coordinates is different from the other two, in that influences flow along it in only one direction, namely downstream. However the two-and-a-half-dimensional phenomena are characterised additionally as being "sheet-like" in shape, so that pressure equilibrium exists in the "thin" direction.

Most practical examples of this kind have mixed boundary conditions, with a solid wall on one side of the sheet and a shear-free stream on the other. Many practical boundary layers on ships' hulls and aircraft wings and fuselages are of this character; so also are some atmospheric and natural water flows. "Three-dimensional" boundary layers on compressor and turbine blades may be of the "two-and-a-half-dimensional" variety; but usually the strong effects emanating from the hub and shroud of the machine make the boundary layers three-dimensional (without quotation marks).

RECOMMENDED GRID

Let the distance x be measured along the surface in the main flow direction, let z be measured along the surface in the direction at right angles; and let the distance from the wall be measured by $\tilde{y}$, a normalised distance varying from 0 to 1. Then the normalising factor is varied with x so that the grid just encloses the physically important region of shear stress, and heat and mass transfer.

The recommended grid is then the same "staggered" one as has been employed for three-dimensional boundary layers by Gosman and Spalding [23], Caretto, Curr and Spalding [24] and Patankar and Spalding [12]. Pressures, and all fluid variables except the y-and z-direction velocities v and w, are then stored at the points of a single rectangular grid; the v and w values are stored at points *between* those grid nodes in such a manner that v-points lie on y-direction links, and w-points on z-direction links.

It is worth noting that $\tilde{y}$ is inferior as a cross-stream variable to the dimensionless stream function recommended for two-dimensional flows (p. 472); knowledge of $\tilde{y}$ does not permit the convective fluxes in the y direction to be known in advance, or even guessed with certainty. Unfortunately, there is no scalar stream function in three-dimensional flows; and no one has yet found a way of reproducing the merits of the stream-function coordinates in such flows (this is not to say that the effort is vain however; I can see *some* ways that seem fruitful, albeit clumsy ones!).

SOLUTION PROCEDURE

Because the flow is sheet-like, and adjacent to a shear-free stream, the pressure can be taken as a prescribed function of position $(x, \tilde{y}, z)$. It is this feature that gives "two-and-a-half-dimensional parabolic" phenomena their characteristic properties, and greatly simplifies the numerical computation. The recommended procedure is that of S. V. Patankar, D. Rafinejad and D. B. Spalding ("Calculation of the three-dimensional boundary layer with solution of all three momentum equations," unpublished work, Imperial College Report, 1973), which is itself a derivative of the three-dimensional procedure of Patankar and Spalding [12].

Integration is again of the marching variety; but now, of course, it is necessary to solve two-dimensional problems at each forward step. Thus, if u, v, w, p and other variables are known at all points at one x station, it is necessary to establish their values for all values of $\tilde{y}$ and z at the next downstream x.

In the recommended procedure, this is done by use of the tri-diagonal matrix algorithm along constant-y and constant-z lines, taken in alternating sequences, in a manner similar to that used in the alternating-direction-implicit method [7,8]. This is done for the u and w equations, but not for v; this third velocity field is deduced from those of u and w, and that of, ρ by way of the continuity equation.

There is thus a similarity with two-dimensional parabolic flows, in which again the cross-stream velocity is deduced without the aid of its proper momentum equation (except in supersonic flows). When we consider true three-dimensional parabolic flows, this fact may justify our use of the "two-and-a-half-dimensional" nickname for sheet-like three-dimensional boundary layers.

THREE-DIMENSIONAL PARABOLIC FLOWS

PRACTICALLY INTERESTING EXAMPLES

We turn now to flows in which, although there is one predominant direction of flow in which effects flow only from upstream to downstream, the other two directions have to be treated alike, in respect of their momentum equations.

Examples with free boundaries include the mixing of a jet with a surrounding stream to which it is slightly inclined; the mixing and eventual coalescence of an array of parallel discrete jets injected into an atmosphere at rest; and the decay of an aircraft wake or of the plume from a chimney stack in a cross-wind.

Confined-flow examples include the flow in a duct of non-circular cross-section, or in a circular-sectioned duct under the influence of laterally-directed body forces brought about, for example, by gravity, rotation or curvature of the duct.

Three-dimensional parabolic flows with mixed boundary conditions include film cooling from a row of discrete holes or slots set in the surface to be cooled; flow in the corner where an aircraft wing joins the fuselage; and the decay of the wake from a turbulence-promoting excrescence on an aircraft wing.

All these examples are characterised by the necessity to solve all three momentum equations. The pressure differences over any plane of constant x may be small; but they are vital in determining how the fluid distributes itself.

RECOMMENDED SOLUTION PROCEDURE FOR SUBSONIC FLOW

There are very few methods available for solving problems of this class; indeed, few authors have recognised the importance of the class, having perhaps had their attention diverted to, and their inquisitiveness satisfied by, the "two-and-a-half-dimensional" parabolic flows. The methods known to me are those described in [12], [23], [24], [25] and [26]: [12] is recommended for general use.

The grid is the same as for the "two-and-a-half-dimensional" parabolic method (p. 478); but all three momentum equations are solved; and, because they all contain the (unspecified) pressure, they must be solved simultaneously by way of the SIMPLE algorithm. This algorithm (semi-implicit method for pressure-linked equations) has already been described in essence (p. 474). Two special features must be mentioned here however.

(*i*) The Poisson equation for the pressure correction takes account of adjustments of only the v and w velocities which result from pressure changes; and it thus allows the continuity errors to be eliminated by adjustments to these velocities alone.

(ii) For the main-flow-direction velocity u, the pressure is either (for a flow with a free boundary) presumed known; or else (for a confined flow) a uniform pressure increment is added to a guessed value to satisfy over-all continuity.

Thus the pressure distribution for the x direction is different from that for the y and z directions; and the u equation is "uncoupled" from the v and w equations. These features cause newcomers to the subject to experience philosophic doubts; but "pressure uncoupling" is realistic, necessary and well justified.

SOLUTION PROCEDURE FOR SUPERSONIC FLOW

When the flow is supersonic (or *where* it is, because one may have supersonic regions embedded in a subsonic flow and *vice versa*), the pressure uncoupling is not necessary. What rendered it necessary for subsonic flow? It was the requirement that all the coefficients in the pressure-correction (Poisson) equation should be positive; and, when the flow is supersonic, the strong effect of pressure on density ensures this for marching integration.

Thus the supersonic case is easier to handle and freer from error than the subsonic one; but in all its main features it is the same.

So far this procedure has not been published; but my colleagues and I are currently working on two problems with its aid. One concerns the mixing and burning of hydrogen injected into a supersonic stream, where the temperature rise can bring the Mach number below unity; and the other concerns the three-dimensional mixing of supersonic and subsonic streams in an engine exhaust.

DISCUSSION

Because three-dimensional parabolic flows have been neglected in the literature, it is as well to emphasise the successes of the new procedure. One of the most striking is the successful prediction of developing laminar flow and heat transfer in a helically coiled pipe of circular cross-section [27]; this has been followed by a study, nearly as successful, of turbulent flow in the same geometry [28].

The prediction of turbulent flows is of course not just a matter of having an adequate means of solving the differential equations; these must themselves adequately express the significant features of the physics of turbulence. However Sharma [29] has shown, by combining the recommended three-dimensional parabolic procedure with a mixing-length model of turbulence, that satisfactory agreement with experiment can be achieved for three-dimensional diffusers of not too wide an angle.

This work is being pursued, under my direction, by V. S. Pratap, who has studied the *curved* diffuser. Here it appears that the pressure-uncoupling assumption is too drastic (Pratap, unpublished work at Imperial College, London, 1974); a partially-parabolic treatment of the phenomenon is much more successful in predicting what is found experimentally. The unpublished work of J. J. McGuirk (Imperial College, London, 1974), reported in a preliminary fashion [30], suggests however that the three-dimensional parabolic method suffices for the rise or fall of a turbulent jet of fluid injected into a coflowing stream of different velocity.

THREE-DIMENSIONAL PARTIALLY-PARABOLIC FLOWS

PRACTICALLY INTERESTING EXAMPLES

We now turn to the last class of flow phenomena: three-dimensional partially-parabolic flows. From what has already been said in a two-dimensional context, they will be recognised as flows in which effects can be transmitted in an upstream direction, but only by way of pressure;

recirculations and significant effects of conduction and diffusion must be absent.

Examples with free boundaries include three-dimensional mixing regions of supersonic and subsonic streams, especially when these impinge at appreciable angles, or exhibit intense exothermic chemical reaction.

Confined-flow examples of practical importance include: strongly curved ducts; turbine and compressor cascades; the centrifugal-compressor impeller; and various kinds of pumps.

Examples with mixed boundary conditions include: stratified flows in the atmosphere over hilly terrain; film cooling from angled jets, for example near the stagnation line on a gas-turbine blade; and the almost recirculating region near the stern of a ship, where fluid retarded by friction flows towards the propellers and rudder.

METHOD OF SOLUTION

It will be readily appreciated that the grid can be essentially the same as that which is used for "two-and-a-half-" and three-dimensional parabolic problems. Now, instead of a two-dimensional store in the computer for each variable, with overwriting by the next set of numbers as soon as a forward step is made, a three-dimensional store is needed for the pressure: for the pressure *only*. The other variables can be calculated by marching integration sweeps as before; but *repeated* sweeps are needed, each leading to an improved pressure distribution.

That the pressure distribution *is* improved, so that ultimately the velocity fields it generates satisfy the continuity equation, is of course the feature the procedure is designed to procure. What is wanted is *rapid* convergence, so that the fewest possible sweeps are needed.

DISCUSSION

The advantages of the partially-parabolic over the alternative fully-elliptic procedures are those of economy, and are even more desirable for three-dimensional flows than for two-dimensional ones, because the computer expenses are so much greater.

Consider for example the ship's stern problem mentioned previously (p.485). The geometry is three-dimensional, and by no means simple; so many grid points are needed to describe the flow domain. Let us suppose that 1000 points are needed in an average cross-stream plane, and that 20 such planes have to be considered. Now probably six dependent variables must be computed, namely u, v, w, p and two turbulence quantities. If all six have to be stored three-dimensionally, as in the fully-elliptic treatment, 120 000 storage locations will be required. In the partially-parabolic treatment, however, the computer needs to place only pressure in the three-dimensional store, while the other five variables are stored two-dimensionally; so only 25 000 storage locations are required, a saving of nearly 80 per cent.

Comparison with three-dimensional parabolic procedures shows that the latter are still cheaper both in storage (6000 in the above example) and computer time (only one sweep needed); so of course one will use the partially-parabolic procedure only when it is needed. However, the experience with the curved diffuser (p. 484) has shown that the partially-parabolic procedure sometimes *must* be used; it is good that it exists and is being developed further.

SUMMARY OF ACHIEVEMENTS AND OUTSTANDING TASKS

What should have emerged clearly from the foregoing survey is that methods now exist for solving problems of all the classes discussed;

and that many complex physical phenomena (turbulence, chemical reaction, radiation, etc.) can be handled at the same time.

The following relative novelties were mentioned.

(*i*) For two-dimensional supersonic flows, the lateral momentum equation has now been included in the marching-integration procedure.

(*ii*) For two-dimensional subsonic flows, the lateral momentum equation can be solved by repeated marching integration with only pressure in a two-dimensional computer store.

(*iii*) For three-dimensional parabolic flows, all the momentum equations are now being solved as a matter of routine; "uncoupling" of pressures is needed for subsonic flows.

(*iv*) The three-dimensional partially-parabolic procedure now allows many practically important three-dimensional flows to be predicted with the aid of computers of modest size.

Concentration in this paper on achievements may have given the impression that little remains to be done except to apply the existing methods. This is definitely *not* my view; it is, rather, that there is need for active and inventive research at all levels.

The aims of this research should include improvements in numerical accuracy for a fixed computer storage, in speed of convergence, in flexibility of adaptation to new geometries and phenomena, and generally in the reduction of computer costs. Non-orthogonal grids need to be looked at; general theorems of convergence are required; and a systematic comparison is needed of the many alternative formulations of the difference equations. Such problems can still occupy a generation of mathematicians.

REFERENCES

1. Harlow, F.H., "Turbulence transport modelling" Vol. XIV, AIAA, New York (1973).
2. Launder, B.E. and Spalding, D.B., "Turbulence models and their application to the prediction of internal flows," *Heat and Fluid Flow*, **2**, No. 1, 43-54 (1972).
3. Launder, B.E. and Spalding, D.B., "Mathematical models of turbulence," Academic Press, London and New York (1972).
4. Launder, B.E. and Spalding, D.B., "The numerical computation of turbulent flows," *Computer Methods in Appl. Mech. and Eng.*, **3**, 269-289 (1974).
5. Launder, B.E., Spalding, D.B. and Whitelaw, J.H., "Turbulence models and their experimental verification," Recorded in Heat Transfer Section Reports Nos. HTS/73/16, 17, 18, 19, 20, 21, 22, 23, 24, 25, 26, 27, 28 (1973).
6. Spalding, D.B., "Mathematical models of free turbulent flows," Instituto Nazionale di Alta Matematica, Symposia Matematica, **IX**, 391-416 (1972).
7. Peaceman, D.W. and Rachford, H.H., "The numerical solution of parabolic and elliptic differential equations," *J. Siam*, **3**, 28 (1955).
8. Douglas, J. and Gunn, J., "A general formulation of alternating direction methods, I," *Numer. Math.*, **6**, 428 (1964).
9. Patankar, S.V. and Spalding, D.B., "A finite-difference procedure for solving the equations of the two-dimensional boundary layer," *Int. J. Heat & Mass Transfer*, **10**, 1389-1411 (1967).
10. Patankar, S.V. and Spalding, D.B., "Heat and mass transfer in boundary layers," Second Edition, Intertext Books, London (1970).

11. Spalding, D.B., "A general computer program for two-dimensional boundary-layer problems", Imperial College, London, Mechanical Engineering Department Report No. HTS/74/32 (1974).

12. Patankar, S.V. and Spalding, D.B., "A calculation procedure for heat, mass and momentum transfer in three-dimensional parabolic flows," *Int. J. Heat & Mass Transfer*, **15**, 1787-1806 (1972).

13. Gibson, M.M., Patankar, S.V. and Spalding, D.B., "The rocket exhaust plume program REP1. Final report on project CHAM 540," Combustion Heat & Mass Transfer Ltd. Report 540 (1972).

14. Jensen, D.E. and Wilson, A.S., "Rapid computation of physical and chemical structures of rocket exhaust flames," *in* Weinberg, F., *Editor*, "Proceedings of Combustion, Institute European Symposium," Academic Press, London (1973).

15. Gosman, A.D., Pun, W.M., Runchal, A.K., Spalding, D.B. and Wolfshtein, M., "Heat and mass transfer in recirculating flows," Academic Press, London (1969).

16. Spalding, D.B. and Tatchell, D.G., "A prediction procedure for flow, combustion and heat transfer close to the base of a rocket", Imperial College, London, Mechanical Engineering Department Report No. HTS/73/42 (1973).

17. Issa, R.I., "The prediction of supersonic boundary layers with embedded pressure waves," PhD Thesis, Faculty of Engineering, University of London (1974).

18. McDonald, P.W., "The computation of transonic flow through two-dimensional gas-turbine cascades," ASME Paper 71-GT-89 (1971).

19. Harlow, F.H. and Amsden, A.A., "Numerical calculation of almost incompressible flow," *J. Computational Phys.*, **3**, No. 1 (1968).

20. Amsden, A.A. and Harlow, F.H., "A simplified MAC technique for incompressible fluid flow calculations," Los Alamos Report LA-DC-11272 (1970).

21. Gopalokrishnan, S. and Bozzola, R., "A numerical technique for the calculation of transonic flows in turbomachinery," ASME Paper 71-GT-42 (1971).

22. Harlow, F.H., "Numerical methods for fluid dynamics; an annotated bibliography," Los Alamos Laboratory Report LA - 4281 (1969).

23. Gosman, A.D. and Spalding, D.B., "The prediction of confined three-dimensional boundary layers," Symposium on Internal Flows, University of Salford. Published by Inst. Mechanical Engineers, London.

24. Caretto, L.S., Curr, R.M. and Spalding, D.B., "Two numerical methods for three-dimensional boundary layers," *Computer Methods in Appl. Mech. & Engng.*, **1**, 39-57 (1972).

25. Briley, W.R., "A numerical method for predicting three-dimensional viscous flows in ducts," United Aircraft Research Laboratories, Report No. L 110888 (1972).

26. Baker, A.J., "Finite-element solution theory for three-dimensional boundary-layer flows," *Computer Methods in Appl. Mech. & Engng.*, **4**, No. 3, 367-386 (1974).

27. Patankar, S.V., Pratap, V.S. and Spalding, D.B., "Prediction of laminar flow and heat transfer in helically-coiled pipes," *J. Fluid Mech.*, **62**, part 3, 539-551 (1974).

28. Patankar, S.V., Pratap, V.S. and Spalding, D.B., "Prediction of turbulent flow in curved pipes," Imperial College, London, Mechanical Engineering Department Report No. HTS/74/1 (1974).

29. Sharma, D., "Turbulent convective phenomena in straight, rectangular-sectioned diffusers," Imperial College, London, Mechanical Engineering Department Report No. HTS/74/26 (1974).

30. Spalding, D.B., "The mathematical modelling of rivers," Imperial College, London, Mechanical Engineering Department Report No. HTS/74/4 (1974).

NUMERICAL SOLUTION OF TURBULENT SWIRLING FLOWS

David G. Lilley

(Cranfield Institute of Technology)

INTRODUCTION

The phenomenon

As an aid to economical design and operation of combustion systems, the numerical solution of inert and reacting turbulent swirling flows is one of the foremost engineering tasks today, as the improvement and use of mathematical models will significantly reduce the cost of experimental development programmes.

Experimental studies show that swirl has large scale effects on flowfields: jet growth, entrainment and decay (for inert flows) and flame size, shape, stability and combustion intensity (for reacting flows) being affected by the degree of swirl imparted to the flow. This is characterised by the swirl number S, which is a nondimensional number representing the axial flux of angular momentum divided by the axial flux of axial momentum times nozzle radius, and the effect of swirl on the flow is related to this number. Often the local swirl number S_z (which uses the local mixing layer width r_{edge}) is used, for at any z it measures the effect of rotation on the flow. Strongly swirling flows (approximately $S > 0.6$) possess sufficient radial and axial pressure gradients to cause a central toroidal recirculation zone, which is not observed at weaker degrees of swirl [1], [2].

The problem

The swirl strength determines the degree of

upstream influence and the swirl number S measures this degree of swirl. The flow classification of parabolic (boundary layer type with a single predominant direction - weak swirl $S < 0.4$) or elliptic (recirculating type with upstream influence - strong swirl $S > 0.6$) governs the type of boundary conditions required and the solution method. Marching methods are appropriate for the former; relaxation methods for the latter.

A mathematical solution of these flows should provide results, if possible, more cheaply, quickly and correctly than by other means (for example, experimenting on real-life systems or models). In order to achieve this the model should simulate the flow in all its important respects (geometry, boundary conditions, physical properties of gases, turbulence, combustion, etc.) and quantitatively solve the governing equations, principal elements of the technique are shown in Fig. 1. Mathematical models of steadily increasing realism and refinement are being developed, both in the dimensionality of the model (together with the computational procedures) and in problems associated with the simulation of the physical processes occurring. Clearly there are two areas of difficulty: the simulation and the solution [3] to [6].

Basic Equations

The turbulent flux (Reynolds) equations of conservation of mass, momentum, stagnation enthalpy and chemical species, which govern the flow of turbulent chemically reacting multicomponent mixtures, are taken as (considering only the turbulent contribution to the fluxes, the molecular contributions being negligibly small in fully turbulent flow except near walls):

$$\begin{aligned} D\rho/Dt + \rho(\nabla \cdot \underline{v}) &= 0 \\ \rho D\underline{v}/Dt &= -\nabla p + \nabla \cdot \tau \\ \rho Dh/Dt &= -\nabla \cdot J_h - p(\nabla \cdot \underline{v}) + \nabla \underline{v} : \tau \\ \rho Dm_j/Dt &= -\nabla \cdot J_j + R_j \end{aligned} \quad (1)$$

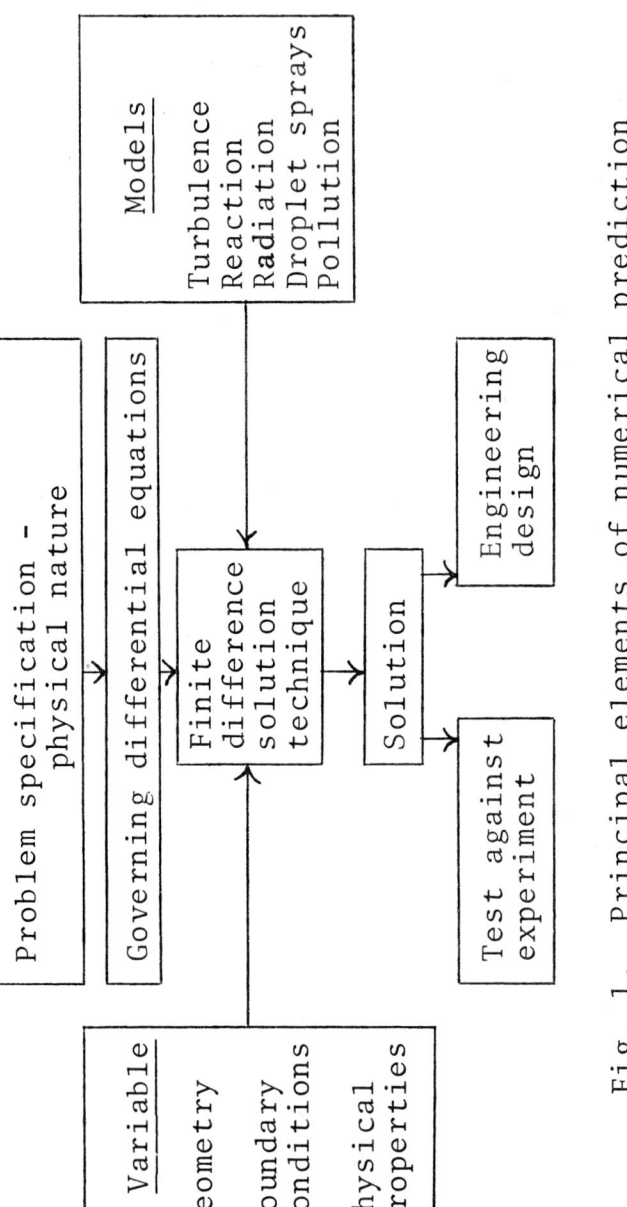

Fig. 1. Principal elements of numerical prediction

If p, v, T and m_j are considered as unknowns the equation system is not closed. Further unknowns are ρ, h and R_j (thermodynamic) and τ, J_h and J_j (turbulent fluxes), which must be specified prior to solution of the equations. Thermodynamic considerations provide some of the necessary extra equations to close the system; these are taken as

$$p = \rho RT/M$$
$$h = c_p T + \sum_j (H_j m_j) + v^2/2$$
$$R_j = \text{provided by reaction model} \tag{2}$$
$$\sum_j m_j = 1$$

Further equations are required to simulate two-phase effects, turbulence, pollution, etc., but a central point of the theory is that all these equations are like laminar flow equations, but the variables are time-mean quantities, and the turbulent fluxes τ, J_h and J_j for momentum, stagnation enthalpy and chemical species are related to correlations of turbulent fluctuations. These must be specified via a turbulence model prior to solution of the equations. By analogy with laminar flows, extensions of Newton's constitutive stress-strain relation, Fourier's law of heat conduction and Fick's law of diffusion have been postulated and used in the past with variable turbulent exchange coefficients μ, Γ_h and Γ_j; the equations are

$$\tau = -\rho \,\overline{v'v'} = 2\mu \Delta$$
$$J_h = \rho \,\overline{v'h'} = -\Gamma_h \nabla h \tag{3}$$
$$J_j = \rho \,\overline{v'm'_j} = -\Gamma_j \nabla m_j$$

Prandtl, Schmidt and $r\theta$ (and other) viscosity numbers relate other exchange coefficients to the primary component of turbulent viscosity μ_{rz}; these are defined by

$$\sigma_h = \mu_{rz}/(\Gamma_h)_r \qquad \sigma_j = \mu_{rz}/(\Gamma_j)_r \qquad \sigma_{r\theta} = \mu_{rz}/\mu_{r\theta} \tag{4}$$

Often consideration is given to a simplified main exothermic reaction between just two species, fuel and oxidant. This and other assumptions lead to the simplifying equations:

$$1 \text{ kg of fuel} + i \text{ kg of oxidant} \rightarrow 1+i \text{ kg of product} + H_{fu}$$

$$h = c_p T + H_{fu} m_{fu} + v^2/2$$
$$m_{fu} + m_{ox} + m_{pr} = 1 \tag{5}$$
$$m_{ox} - i m_{fu} \text{ is a conserved property}$$

Outline of the Paper

Consideration is given to the major features of the numerical solution of two-dimensional swirl flows. Since equation solution yields predictions which are realistic only if the physical processes are correctly expressed in mathematical form, suitable turbulence (nonisotropic) and reaction (eddy-break-up) models to effect closure are discussed under Physical Simulation Problems. Various forms of the governing equations and computational methods themselves are discussed in the section on Numerical Solution Procedures, the message being that they are well advanced. For strongly swirling recirculating flows both the stream function vorticity and primitive pressure velocity approaches are considered, the merits and disadvantages of each being discussed. It is possible to predict major features of swirling flows; some numerical solutions are exhibited in Computations and Discussion, where the theme is that both weak and strong swirling flows are now predictable.

PHYSICAL SIMULATION PROBLEMS

Turbulence

Closure of the time-mean equation system is effected by means of a turbulence model and models are classified according to the turbulent flux hypothesis (whether or not turbulent exchange coefficients are introduced) and the number of extra differential equations to be solved [7],[8]. If introduced, exchange coefficients have generally been assumed isotropic until recently, even in flows with swirl, but recent experimental, inverse and prediction works have disputed this for swirling flows. Briefly the choice available is:

1. Prandtl mixing length $\mu_{rz} = \rho l^2$ $(2\Delta:\Delta)$
2. Energy length $\mu_{rz} = \rho k^{\frac{1}{2}} l$ (6)
3. Stress modelling $D\tau_{rz}/Dt = P_{rz} + D_{rz} + R_{rz} + \varepsilon_{rz}$
4. Algebraic stress modelling $\tau_{rz} = f$ (other τ's, k, l, Δ)

and currently energy length models are to be recommended, in particular the k-ε model [7]. The first two are examples of theories of the exchange coefficient type; the second two are of direct stress specification type. For nonswirling flow the familiar Prandtl model is

$$\mu_{rz} = \rho l^2 \left|\frac{\partial u}{\partial r}\right|, \quad l = \lambda r_{edge}, \quad \lambda = \text{constant}$$

Various extensions of Prandtl's mixing length model to swirling flows have been proposed. Generally μ_{rz} is taken to be proportional to the second invariant of the mean flow rate of deformation tensor and nonisotropy obtained by use of a variable $r\theta$-viscosity number, etc. [9]. The first of the equations of (6), which in axisymmetric cylindrical polar form, is

$$\mu_{rz} = \rho l^2 (2(\partial u/\partial z)^2 + 2(\partial v/\partial r)^2 + 2(v/r)^2 + (\partial u/\partial r + \partial v/\partial z)^2 + (r\partial(w/r)/\partial r)^2 + (\partial w/\partial z)^2)^{\frac{1}{2}} \quad (7)$$

and which under boundary layer assumptions reduces to

$$\mu_{rz} = \rho l^2 ((\frac{\partial u}{\partial r})^2 + (r\frac{\partial}{\partial r}(w/r))^2)^{\frac{1}{2}} \quad (8)$$

is used together with

$l = \lambda r_{edge}$, $\lambda = 0.08 (1 + \lambda_s S_z)$, $\lambda_s = $ constant

$\sigma_{r\theta} = $ constant dependent on S or variable dependent on S_z

More recent work on calculating turbulent flows has postulated that the turbulence may be described adequately by two quantities: the kinetic energy and the length scale. The second of the equations of (6) is used and two extra differential equations are required, one for k itself and the other for any variable $Z = k^m \ell^n$. The equation for Z may be taken for Z equal to $k\ell$, $W = k/\ell^2$ or $\varepsilon = k^{1.5}/\ell$ and the indications are that for general flows the ε-equation is preferable [7]. For boundary layer flows any of these is suitable. These equations are of the standard type with Schmidt numbers near unity. With an energy length model the effect of swirl may be represented via an additional source term $C_R \, \rho R i k^{1.5}$ in the Z-equation [9]. A characterisation of the form C_R = constant and

$\sigma_{r\theta}$ = constant dependent on S or variable dependent on S_Z

is used.

Strongly swirling enclosed flow, for example reacting flow in an assumed axisymmetric cylindrical combustion chamber (where in any case laterally induced secondary air creates extra turbulence) may be well predicted on the basis of a simple algebraic turbulence model which uses initial velocities and mass flow rates to give approximate turbulent jet values of μ [10]:

$$\mu = K(D^2 \rho^2 \{\dot{m}_P u_P^2 + \dot{m}_S (u_S^2 + w_S^2)\}/L)^{\frac{1}{3}} \qquad (9)$$

where K is a constant = 0.012.

More advanced alternatives never introduce exchange coefficients and specification of the stresses is direct from solution of the stress transport equations, see part 3 of (6). The stresses are arguments of differential equations which contain other second and third order correlations. Rather than solve differential equations for high order correlations, modelling of the correlations is often preferred, this modelling involving mean velocities, k and ℓ. On local equilibrium assumptions the stress transport differential equations reduce to algebraic ones, which combine with differential equations for k and ℓ, and form a model which has some merits (see the last of (6)) and its application to swirling flows indicates a promising improvement in universality of predictive power over these exchange coefficient models [11]. However indications are that research is now being concentrated on differential stress models.

Inverse calculations

As an intermediate contribution to the full prediction of turbulent flows, a general inverse solution procedure has been developed for non-recirculating swirling jets and flames [12]. The method allows certain components of the turbulent fluxes and associated exchange coefficients to be calculated directly from limited experimental time-mean data without the need for a complete solution to the problem. It is an intermediate step which has been used on both swirling isothermal and combustion systems. The method allows distributions of τ_{rz}, $\tau_{r\theta}$, $(J_h)_r$ and $(J_j)_r$ and associated exchange coefficients (isotropy is not assumed) to be determined from the experimental mean distributions of u, w, T and m_j. It thus provides a link between mean measurements and certain correlations of turbulent fluctuation components and throws light on the appropriateness or otherwise of any given turbulence model for the flow under consideration. From the calculated values of turbulence quantities in swirling systems, modifications to turbulence models can be deduced.

Inverse calculations [12] for weakly swirling jets and flames show generally that μ_{rz} and $\sigma_{r\theta}$ increase with the local swirl number S_z and that the turbulent stress distribution is nonisotropic. Typical inverse calculations are shown in Fig. 2 for isothermal swirling jets at $z/d = 6$ and similar results have been obtained in swirling flames.

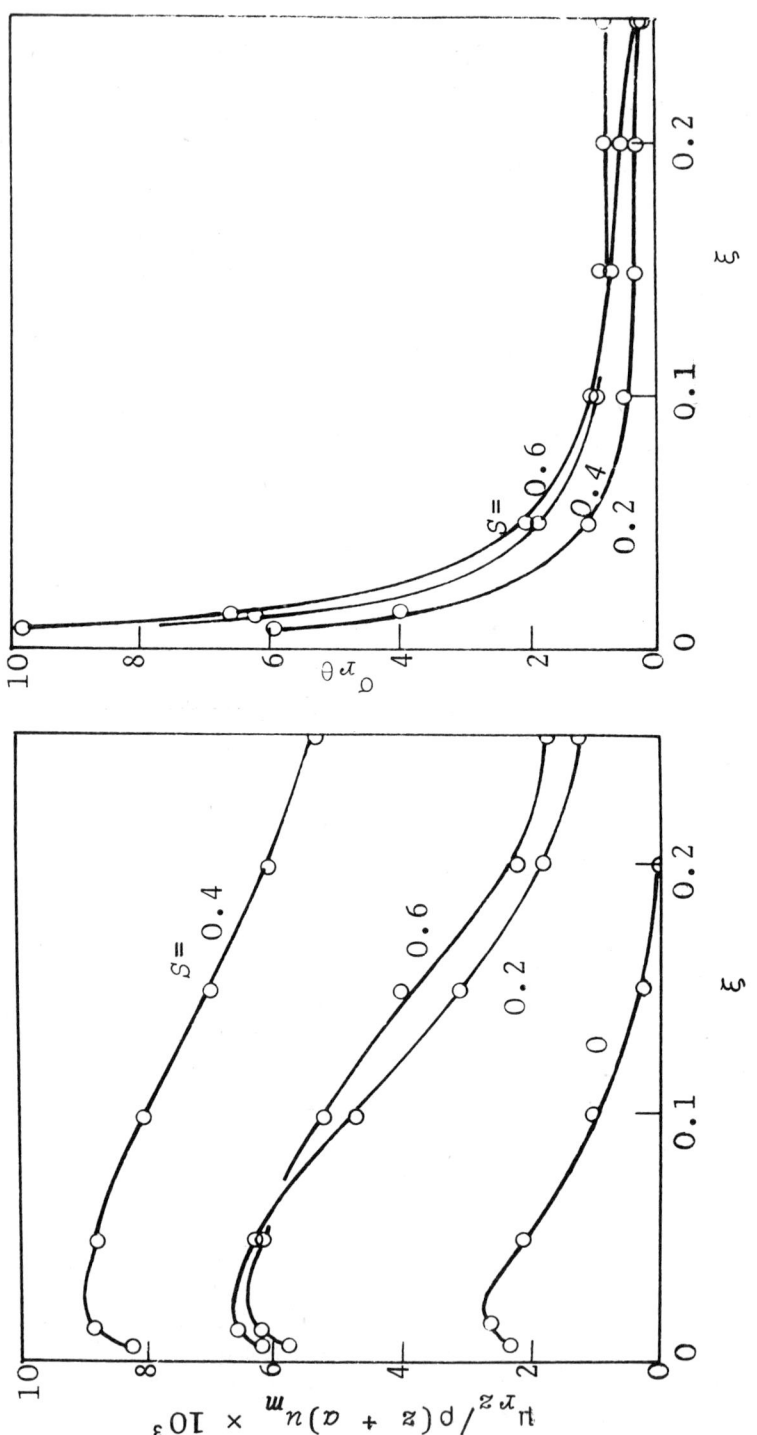

Fig. 2. Inverse calculation of μ_{rz} and $\sigma_{r\theta}$ for isothermal swirling jet at $z/d = 6$ (o predicted)

There are variations of μ_{rz} with downstream distance but at this axial station μ_{rz} increases with swirl up to $S = 0.4$ and then decreases as S further increases. This is because in the initial region of the jet the viscosity for weakly swirling flows is small under the influence of the central core and there is a progressive increase as S increases, whereas further downstream in the fully developed region the reverse occurs, as a result of the more rapid decay of velocities associated with more strongly swirling flows. The computations shown are made near the change-over point. The variation with axial position of the $\sigma_{r\theta}$ values shown is far less striking and shows the remarkable degree of nonisotropy that occurs in some regions of swirling flows.

Chemical reaction

When chemical reaction takes place it is often necessary to solve extra differential equations for h, m_{fu}, $m_{ox} - im_{fu}$ and possibly g, the last of these being required in the eddy-break-up premixed and thick flame front diffusion flame models [13]. However, if the chemical reaction is sufficiently rapid for combustion to be controlled by mixing, as in a diffusion flame, solution for h and the time-mean mixture fraction f which is related to $m_{ox} - im_{fu}$ is sufficient. If this reaction, with negligible kinetic heating, takes place in a chamber with walls impervious to both heat and matter, then h and f are linearly related and one need only solve for one of them, f [10]. But for premixed flames, where the equation for m_{fu} is solved, some model for the reaction rate R_{fu} must be incorporated.

The Arrhenius model. Reaction rates are specified via simple time averaging of the instantaneous Arrhenius expression

$$R_{fu} = - Pp^2 \, m_{fu} m_{ox} \exp(- E/RT) \qquad (10)$$

where time-mean values are used on the right. The equation for g is not required. The model is naive, because fluctuations in temperature and concentrations can be extremely large and the instantaneous reaction rate function is highly nonlinear. Moreover, the heating of fuel from hot combustion product prior to burning is strongly related to the turbulence structure. A model which promotes the effect of turbulence on the reaction rate and reduces the chemical kinetic influence would be more realistic.

The eddy-break-up model. The model involves the calculation of g. The fluctuations of T and m_{ox} are supposed to be perfectly correlated with those of m_{fu}. Hence a chemical kinetic limit is set to the reaction rate but a central point of the model is that there is a second limit, often much lower, set by the rate of dissipation of the fluctuations. This is a process which involves break-up of large eddies into small ones and although molecular processes are involved in the final stages of dissipation they are not the controlling ones. It is the fuel concentration dissipation rate which controls the reaction process and the eddy-break-up reaction rate expression

$$R_{fu} = -C_{EBU} \, \rho g^{\frac{1}{2}} k^{\frac{1}{2}} / \ell \tag{11}$$

has been proposed for a turbulent flame of high temperature [13]. When both the chemical kinetic and eddy-break-up limits are of comparable magnitude the lower value is used.

When the reaction is diffusion controlled, one can still solve for g and its value is that it gives fluctuations in m_{fu}, m_{ox} and T, so allowing nonzero time-average concentrations and a thick flame region.

Simulation of other Processes

Other processes in combustors include two-phase phenomena, particle size calculation for fuel droplet sprays, radiative transfer and pollutant formation. Some details of their simulation have already been reviewed [4].

NUMERICAL SOLUTION PROCEDURES

Assumptions of two-dimensional parabolic or elliptic flow simplify the basic equations and determine the type of boundary conditions and solution method, as discussed in the Introduction.

Marching methods for two-dimensional axisymmetric parabolic boundary layer flow: weak swirl (S < 0.4)

Weakly swirling flows, with a single predominant direction and no recirculation regions, are examples of two-dimensional axisymmetric boundary layer flow. They may be simulated by a simplified form of the governing equations (1) without much loss in accuracy. Application of the well known boundary layer approximations results in truncation of the full elliptic equations to parabolic form; fewer terms and unknown fluxes are left in the equations; and a simpler, quicker forward-marching solution procedure can be applied. With these approximations the basic equation system (in the cylindrical polar coordinate system, assuming the flow is quasi-steady ($\partial/\partial t = 0$) and axisymmetric ($\partial/\partial \theta = 0$) and neglecting compression and kinetic heating) reduces to

$$r\partial(\rho u)/\partial z + \partial(r\rho v)/\partial r = 0$$
$$\rho(u\partial u/\partial z + v\partial u/\partial r) = 1/r\ \partial(r\tau_{rz})/\partial r - \partial p/\partial z$$
$$\rho w^2/r = \partial p/\partial r$$
$$\rho(u\partial w/\partial z + v\partial w/\partial r) = 1/r^2 \partial(r^2 \tau_{r\theta})/\partial r - \rho vw/r \quad (12)$$
$$\rho(u\partial h/\partial z + v\partial h/\partial r) = -1/r\ \partial\{r(J_h)_r\}/\partial r$$
$$\rho(u\partial m_j/\partial z + v\partial m_j/\partial r) = -1/r\partial\{r(J_j)_r\}/\partial r + R_j$$

in which there are now only four significant flux components, simplifying the simulation requirements.

The nonisotropic boundary layer equivalents of the constitutive assumptions (3) for these fluxes are

$$\tau_{rz} = \mu_{rz} \, \partial u/\partial r \qquad \tau_{r\theta} = \mu_{r\theta} \, r\partial(w/r)/\partial r$$
$$(J_h)_r = -(\Gamma_h)_r \, \partial h/\partial r \qquad (J_j)_r = -(\Gamma_j)_r \, \partial m_j/\partial r \tag{13}$$

A standard computer program is available for solving this type of problem; details are shown in Fig. 3 [14].

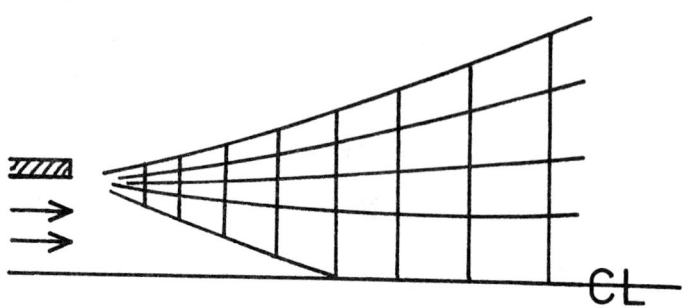

Marching integration
Automatic expanding grid
Implicit
One-dimensional storage

Fig. 3. Computational details for two-dimensional parabolic flows

The marching integration proceeds on an automatic expanding grid and the finite difference formulation is fully implicit in character; this combination ensures economy and stability, accuracy is achieved by using a fine grid. Of all the techniques available for this type of flow, this is noteworthy in that the program is generally formulated and includes the novelty of a nondimensional stream function which replaces the radial coordinate. It is defined by

$$\eta = (\psi - \psi_I)/(\psi_E - \psi_I)$$
$$\psi(r) = \int_0^r \rho u r \, dr \quad \text{(for fixed } x\text{)} \tag{14}$$

and in this von Mises coordinate system the equations possess the common form

$$\partial \phi/\partial x + (a + b\eta) \, \partial \phi/\partial \eta = (\partial/\partial \eta)(c \partial \phi/\partial \eta) + d \tag{15}$$

except for the swirl equation which becomes

$$(\partial/\partial x)(rw) + (a+b\eta)(\partial/\partial \eta)(rw) = (\partial/\partial \eta)\{cr^2(\partial/\partial \eta)(w/r)\} \tag{16}$$

These equations, where ϕ is a general dependent variable, differ primarily in their final source terms d and pressure gradients, generation and consumption expressions, etc. find their place here. A fully implicit micro-integral method is used to obtain the corresponding finite difference equation

$$\phi_i = A\phi_{i+1} + B\phi_{i-1} + C \qquad (17)$$

and the equations are solved for each forward step by use of the tri-diagonal matrix algorithm.

Relaxation methods for two-dimensional axisymmetric recirculating flow: strong swirl (S > 0.6)

Most combustion systems exhibit recirculation. They are therefore not amenable to the kind of marching integration process mentioned; instead iterative procedures are essential. Strongly swirling flames with recirculation are examples of general two-dimensional axisymmetric flows. The equations (from (1)) contain six different components of τ, two of J_h and two of J_j.

They are (with the same assumptions of the previous section)

$$r\partial(\rho u)/\partial z + \partial(r\rho v)/\partial r = 0$$

$$\rho(u\partial u/\partial z + v\partial u/\partial r) = \partial \tau_{zz}/\partial z$$
$$\qquad + 1/r\,\partial(r\tau_{rz})/\partial r - \partial p/\partial z$$

$$\rho(u\partial v/\partial z + v\partial v/\partial r - w^2/r) = \partial \tau_{rz}/\partial z$$
$$\qquad + 1/r\,\partial(r\tau_{rr})/\partial r - \tau_{\theta\theta}/r - \partial p/\partial r$$

$$\rho(u\partial w/\partial z + v\partial w/\partial r + vw/r) = \partial \tau_{z\theta}/\partial z \qquad (18)$$
$$\qquad + 1/r^2\,\partial(r^2 \tau_{r\theta})/\partial r$$

$$\rho(u\partial h/\partial z + v\partial h/\partial r) = -\partial\{(J_h)_z\}/\partial z$$
$$\qquad - 1/r\,\partial\{r(J_h)_r\}/\partial r$$

$$\rho(u\partial m_j/\partial z + v\partial m_j/\partial r) = -\partial\{(J_j)_z\}/\partial z$$
$$\qquad - 1/r\,\partial\{r(J_j)_r\}/\partial r + R_j$$

The component forms of the constitutive assumptions (3), assuming isotropy, are:

$$\tau_{zz} = 2\mu\, \partial u/\partial z \qquad \tau_{rz} = \tau_{zr} = \mu(\partial u/\partial r + \partial v/\partial z)$$
$$\tau_{rr} = 2\mu\, \partial v/\partial r \qquad \tau_{r\theta} = \tau_{\theta r} = \mu r\, \partial(w/r)/\partial r$$
$$\tau_{\theta\theta} = 2\mu v/r \qquad \tau_{z\theta} = \tau_{\theta z} = \mu\, \partial w/\partial z \qquad (19)$$
$$(J_h)_z = -\Gamma_h \partial h/\partial z \qquad (J_h)_r = -\Gamma_h \partial h/\partial r$$
$$(J_j)_z = -\Gamma_j \partial m_j/\partial z \qquad (J_j)_r = -\Gamma_j \partial m_j/\partial r$$

If isotropy is not assumed, different values of μ, Γ_h and Γ_j may be appropriate to different equations of this set. This may be obtained by the use of variable Prandtl, Schmidt and $r\theta$ (and other) viscosity numbers, previously defined.

These equations, which govern general strongly swirling flows perhaps with recirculation, are elliptic in character and together with this simulation problem is the necessity to solve the equations; a lengthy numerical relaxation method is appropriate. The solution may be obtained using the stream function vorticity [10] or primitive pressure velocity [15] approach. Iteration proceeds on a predetermined variable size grid which is made fine in regions of great activity. All values of dependent variables at grid points must be simultaneously in store and memory requirements are therefore greater (two-dimensional) than those of the previous section (one-dimensional). General methods are available for these problems and details are shown in Fig. 4.

In the former technique [10] the equations are reduced to the general form, in cylindrical polar coordinates:

$$a\{\frac{\partial}{\partial z}(\phi\frac{\partial\psi}{\partial r}) - \frac{\partial}{\partial r}(\phi\frac{\partial\psi}{\partial z})\} - \frac{\partial}{\partial z}\{br\frac{d}{\partial z}(c\phi)\} - \frac{\partial}{\partial r}\{br\frac{\partial}{\partial r}(c\phi)\} + rd = 0 \qquad (20)$$

where ϕ represents a general dependent variable and, again, the equations differ primarily in their final source terms d. These equations involve ψ and ω, which are defined by

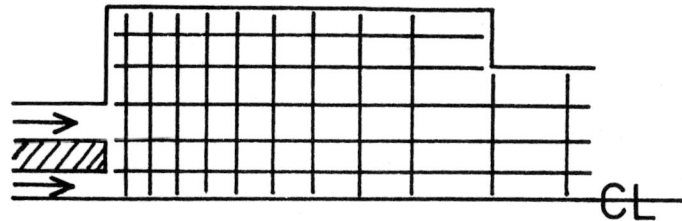

Stream function vorticity or
pressure-velocity equations

Iterative solution -
G-S or LBL method

Predetermined variable
size grid

Two-dimensional storage

Fig. 4. Computational details for two-dimensional elliptic flows

$$r\rho u = \partial \psi/\partial r \qquad r\rho v = -\partial \psi/\partial z$$
$$\omega = \partial v/\partial z - \partial u/\partial r \qquad (21)$$

A tank and tube micro-integral method is used to obtain the corresponding finite difference equation

$$\phi_P = \sum_j c_j \phi_j + D_P \qquad (22)$$

relating the value of a dependent variable at a node point P to its values at its four neighbouring points of a variable size rectangular grid, which is made fine in regions of a great activity. These equations are then solved iteratively (by Gauss-Seidel technique) using under-relaxation (to procure convergence) or over-relaxation (to speed convergence) as required.

The technique described in [15] solves directly for the primitive pressure and velocity variables, unlike the one just described which obtains these by way of stream function and vorticity. In addition the u and v velocities are positioned between the nodes where p and other variables are stored and the combination of staggered grid and a line relaxation method [16] leads to rapid solution. This method has the advantage of wider application and the capability of extension to three-dimensional flows. The equations are reduced to the form

$$\frac{\partial}{\partial z}(r\rho u\phi) + \frac{\partial}{\partial r}(r\rho v\phi) - \frac{\partial}{\partial z}(\frac{r\mu}{\sigma_\phi}\frac{\partial \phi}{\partial z}) - \frac{\partial}{\partial r}(\frac{r\mu}{\sigma_\phi}\frac{\partial \phi}{\partial r}) = rs^\phi \quad (23)$$

where ϕ is a general dependent variable and the equations differ primarily in their final source terms s^ϕ. The corresponding finite difference equations

$$a_P^\phi \phi_P = \sum_j a_j^\phi \phi_j + s_u^\phi \quad (24)$$

are solved by a semi-implicit line-by-line method [16] for values at points of a variable size rectangular grid, again under- or over-relaxing as required.

COMPUTATIONS AND DISCUSSION

Weakly Swirling Flows

Predictions are made [9] with both the Prandtl mixing length and energy length turbulence models to confirm suitable extensions of the models to swirl flows. Extensions of the form

$$\lambda_s = 0.6 \text{ for the Prandtl mixing length model}$$

and $\quad(25)$

$$C_R = 0.06 \text{ for the } k\text{-}Z \text{ model}$$

have been developed (see section on Turbulence) for isothermal swirl flows, together with

$$\sigma_{r\theta} = 1 + CS_z^{\frac{1}{3}} \quad (26)$$

where C takes a value between 2 and 5. Computations with either model produce longitudinal decays with different degrees of swirl as shown in Fig. 5. To make such isothermal computations simultaneous parabolic partial differential equations are solved for u, rw, k and $Z = k\ell$.

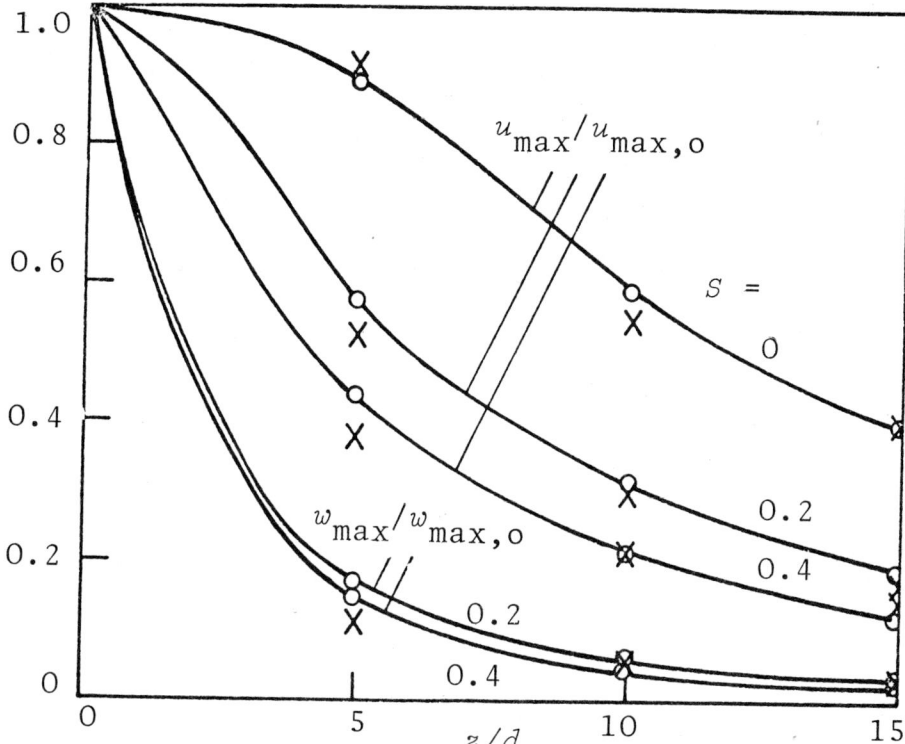

Fig. 5. Prediction of longitudinal decays with swirl for inert jets (x experimental, o predicted)

The decays of u_{max} and w_{max} compare well with experimental data [1]. The progressive increase in the u_{max} decay as S increases and the approximately non-swirl-dependent w_{max} decays are clearly evident. The primary use of swirl in a jet is to increase the angle of spread and rate of decay of axial velocity. Results agree well with the data and show that the effect of swirl on isothermal jet growth, entrainment and decay may be predicted well. For example Fig. 6 shows and compares the half angle predictions with experimental evidence.

Premixed flames (m_{fu} = 0.245) with initial fuel/air ratio well outside the flammability limits have been studied experimentally [1]. With turbulence simulated in a manner similar to the form used in the prediction of inert jets, predic-

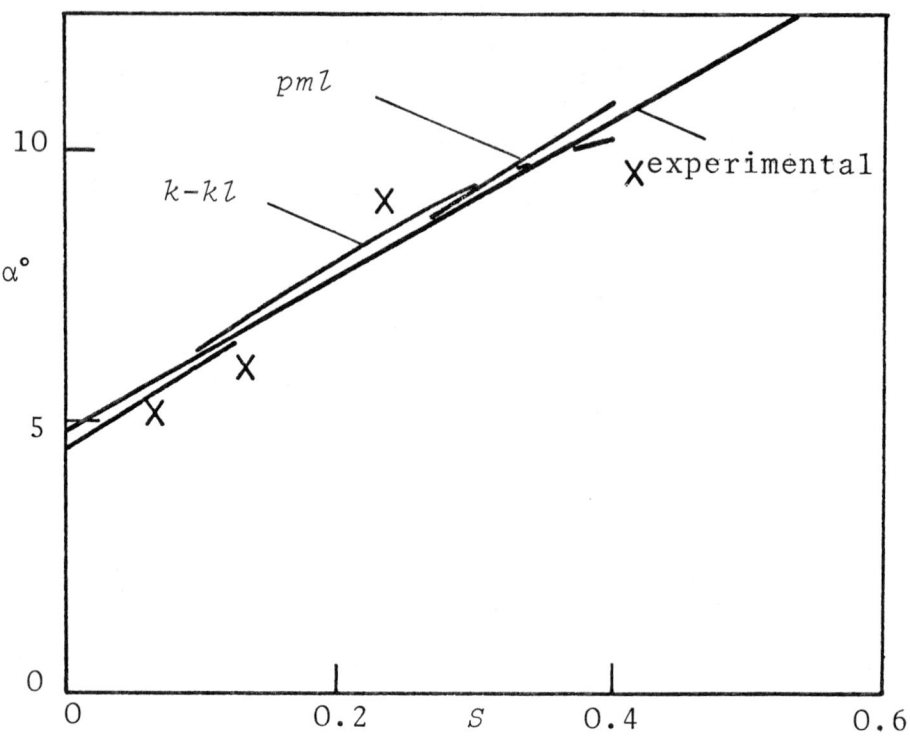

Fig. 6. Prediction of variation of jet half angle with swirl for inert jets (x experimental)

tions have been made [17] with the eddy-break-up reaction model (section on Chemical reaction). Eight parabolic partial differential equations are solved for u, rw, k, $k\ell$, h, m_{fu}, $m_{ox} - im_{fu}$ and g. Predicted longitudinal decays of u_{max} and w_{max} compare well with experimental results and there is a progressive increase in them as S increases. It is to be noted, however, that the velocity decays are slower than in cold swirling jets. This is largely a result of the temperature and density changes and a consequence of this gas expansion is increased axial and radial velocities, giving a reduced rate of decay of u_{max} and a wider jet initially.

Fig. 7 shows longitudinal variations with swirl of T_c, the centreline temperature, and $\dot{m}_{fu}$, the total mass flow rate of unburned fuel. Notice

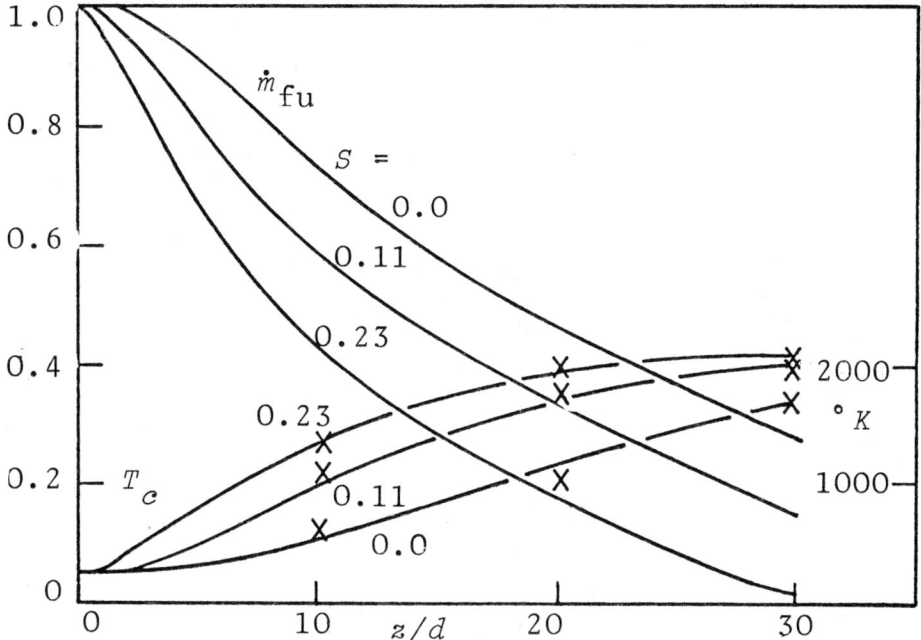

Fig. 7. Longitudinal T_c and $\dot{m}_{fu}$ variations with k-$k\ell$ turbulence model for weakly swirling premixed jet flames (m_{fu} = 0.245) (x experimental)

that the centreline temperature increases more rapidly as S increases, indicating the more rapid mixing of hot combustion products with the cooler higher velocity core region. Observe also that $\dot{m}_{fu}$ decreases more rapidly as S increases, indicating the more rapid consumption of fuel per unit length of flame. Temperature field predictions clarify the effect of swirl on flame size, shape and combustion intensity. Fig. 8 shows the predicted flame front lines (loci of temperature maxima) and it is clear that the length of the flame decreases markedly with swirl and that there is a progressive increase in the initial width (at z/d = 10) of the flame as S increases [17]. Flame lengths as determined from the figure may be compared favourably with experimental values.

Downstream development of these jets and flames is often characterised by the parameters A

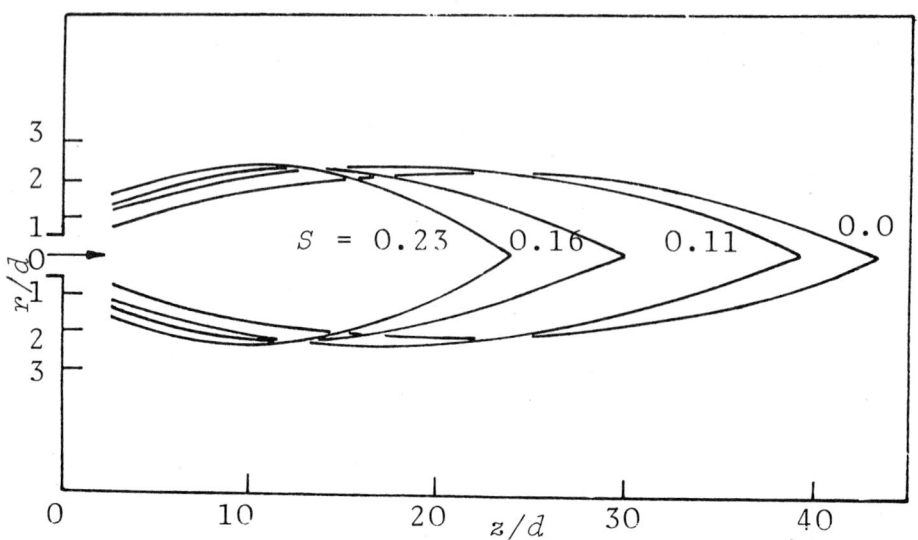

Fig. 8. Predicted flame front lines for weakly swirling premixed jet flames ($m_{fu}=0.245$)

(for axial velocity decay) and α (for jet half angle) defined by

$$u_{max}/u_{max,o} = A(\rho_\infty/\rho_{min})^{\frac{1}{2}} d/(z + a) \qquad (27)$$

$$\tan \alpha = r_{0.5}/(z + a) \qquad (28)$$

Recommendations for a, A, α and flame lengths are given in Table I.

Table I

Development parameters for jets and premixed ($m_{fu} = 0.245$) jet flames

	Jet ($S < 0.4$)	Flame ($S < 0.23$)
a/d	2.3	$35 + 100S$
A	$6.8/(1 + 6.8S^2)$	$15 + 10S$
α	$4.8 + 14S$	$2.2 + S$
Flame length $/d$	—	$43 - 100S$

Strongly swirling flows

Strongly swirling flows, which give rise to axial recirculation, are generally created by the use of a swirl generator or swirl vanes at angle ϕ to the main flow direction. In this section two such flows are considered: a free swirling jet from a swirl generator and an enclosed swirling annular jet, reacting with a central fuel gas jet, from swirl vanes. For comparison purposes it is worth noting that ϕ and S are related approximately by

$$S = \tfrac{2}{3}\left\{\frac{1 - (d_h/d)^3}{1 - (d_h/d)^2}\right\}\tan \phi \sim \tfrac{2}{3} \tan \phi \qquad (29)$$

for $d_h \ll d$ so that vane angles of 15, 30, 45, 60, 70 and 80, for example, correspond to S values of 0.179, 0.385, 0.667, 1.155, 1.832 and 3.781.

A tangential and axial entry swirl generator gives a strongly swirling flow with exit u and w velocities progressing peaked toward the outer edge of the nozzle as S increases. As S increases the w profile tends to be more peaked than solid body rotation and the u profile ensures that much of the flow leaves near the outer edge. Such a flow, expanding into free air, has been investigated experimentally [18] in the presence of a central hub of diameter $d_h = 0.24d$ in the swirl generator of exit diameter $d = 0.254$m. The mean exit velocity $u_{av,o}$ averaged 10 ms^{-1} and the effect of the degree of swirl on the size and shape of the recirculation zone and the decay of axial and swirl velocities are made with the two-dimensional elliptic stream function vorticity programme with a simple viscosity formula, similar to (9) with average exit velocities.

Predicted streamlines and recirculation zone for the $S = 1.57$ jet show striking resemblance to the experimental figure [18] but the zone is predicted slightly too wide and too short, see Fig. 9. Decreasing μ or increasing $\sigma_{r\theta}$ acts as a corrector

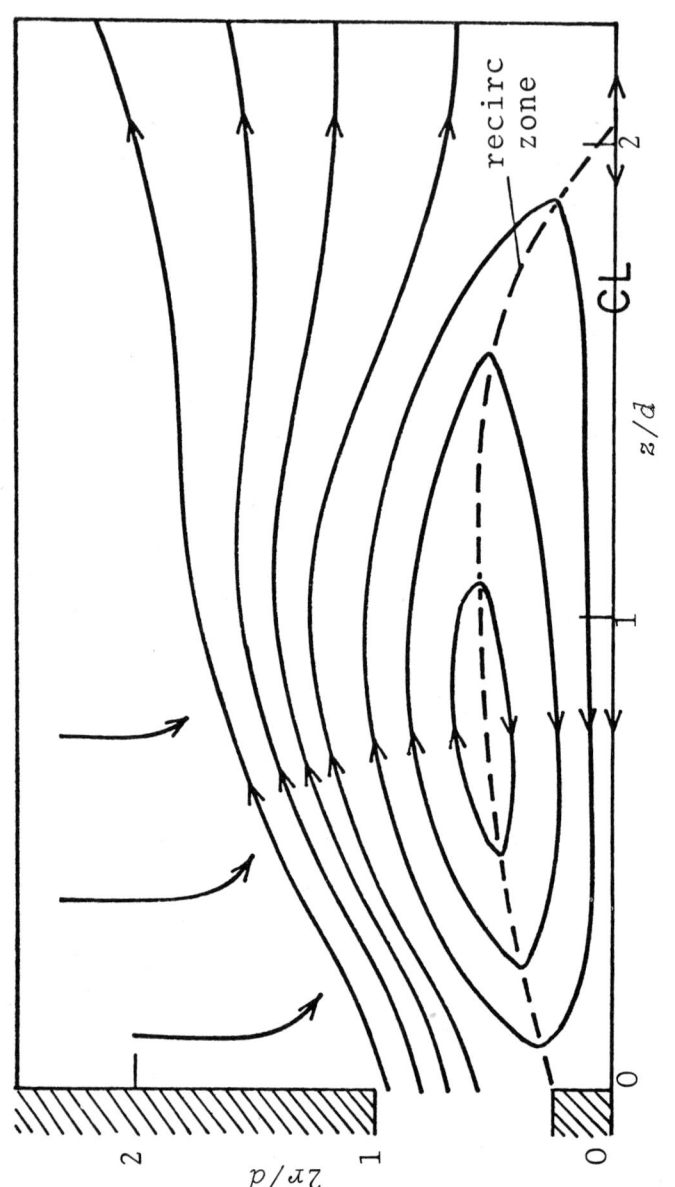

Fig. 9. Predicted streamlines (———) and recirculation zone (---) for $S = 1.57$ jet

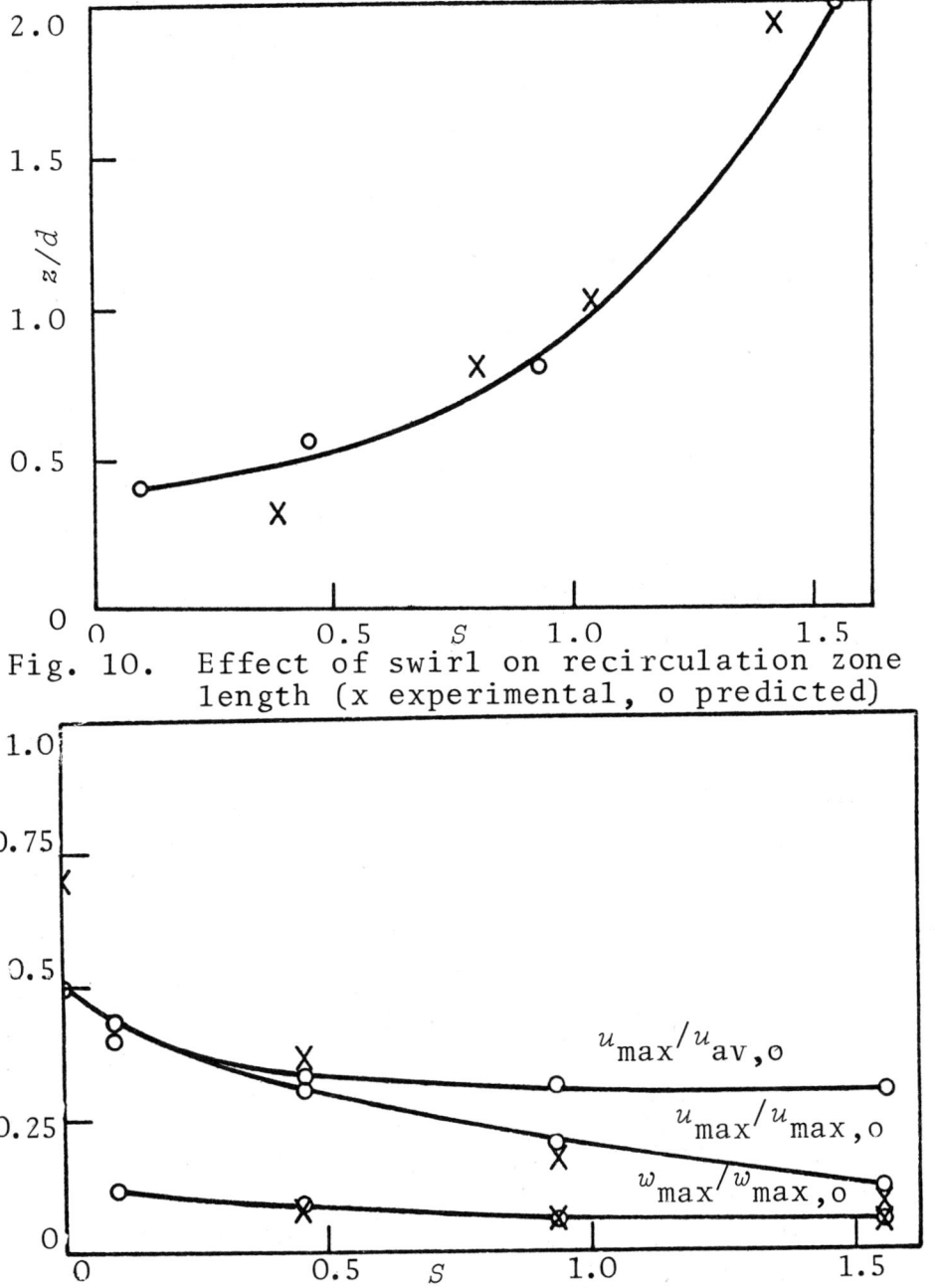

Fig. 10. Effect of swirl on recirculation zone length (x experimental, o predicted)

Fig. 11. Effect of swirl on axial and swirl velocity decays at $z/d = 6$ (x experimental, o predicted)

for this but makes the w velocity decay increasingly inaccurate. Without any corrector the recirculation zone length and velocity decays predicted and shown in Figs. 10 and 11 compare well with the experimental evidence. The u and w velocities (see Fig. 11) both decay more rapidly with higher values of S when compared with their initial peaked maximum values, but the effect is far less dramatic when, for example, u_{max} is compared with initial mass average velocity $u_{av,o}$.

Fig. 12 illustrates a computation for a cylindrical combustion chamber, which has many of the features of practical equipment: fuel gas and air input at one end, exit for combustion products at the other, arrangement for the air inlet as an annular orifice (outer diameter d = 0.0762m) surrounding the fuel inlet, swirl vanes for causing the air to enter with a swirling motion (w constant profile) and a thick annular lip between the air and fuel inlets, its thickness so chosen that equal axial velocities of the primary fuel stream u_p and the secondary air stream u_S give over-all stoicheiometric conditions (air fuel ratio = 15). For simplicity the chemical reaction rate is presumed to be rapid enough for combustion to be controlled by mixing and the turbulent viscosity formula (9) is used. Results were not found to be sensitive to any *ad hoc* nonisotropic tests. More satisfactory turbulence and reaction models now exist; the predictions shown are intended as demonstrative and as such the following conditions are taken at the inlet: p = 25 atmospheres, T = 850° K and u_S = 36 ms^{-1}. For given flow conditions (vane angle ϕ, primary fuel velocity u_p [which also governs the air fuel ratio] and allowing or disallowing chemical reaction) one wishes to compute the distributions of stream function, velocities, temperature and composition within the chamber. As a first step it is necessary only to solve differential equations for ψ, ω/r, rw and f using the two-dimensional elliptic programme.

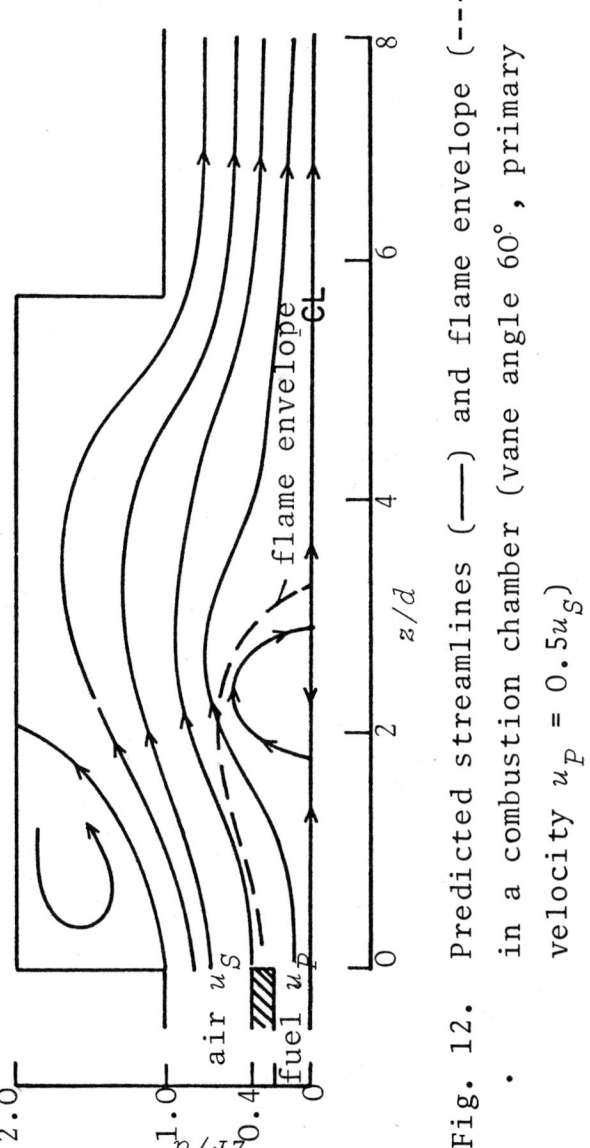

Fig. 12. Predicted streamlines (——) and flame envelope (---) in a combustion chamber (vane angle 60°, primary velocity $u_P = 0.5 u_S$)

The figure shows the predicted streamlines and flame envelope for a swirl vane angle of 60° and primary fuel velocity u_p of 18ms^{-1} (= $0.5u_S$) giving an inlet air fuel ratio of 30:1. The presence of this degree of swirl ($S \sim 1.1$) causes a toroidal vortex to form in the middle of the combustion chamber, in addition to the corner recirculation near the entrance provoked by the sudden enlargement of the cross-sectional area. It has also shortened the flame, as compared with a non-swirling case, but still the flame engulfs the central recirculation zone. A more uniform temperature across the exit from the chamber has also resulted. All these effects are well known to combustion engineers [1] who strive to utilise the recirculation of hot combustion products and the bluff body effect of this zone as an aid to the combustion process.

The vane angle has a strong effect on the existence (or otherwise) and size of the central recirculation zone. Fig. 13 shows the length of this zone as a function of vane angle for a variety of values of u_p (and hence air fuel ratio): on the left when no reaction is allowed to take place and on the right with reaction. In both cases the zone lengthens as ϕ increases, for a given u_p (air fuel ratio). Reduction of u_p promotes the existence of the zone and lengthens it, for a given vane angle. The third point is that the lines on the right are lower than those on the left: the presence of reaction, gas expansion and increased velocity tend to shorten, and in some cases destroy (for example, 60° vanes with $u_p = u_S$), the zone as compared with its nonreacting counterpart. Predicted also, and as expected experimentally, is that reduction of u_p (increase in air fuel ratio) and/or increase in vane angle progressively reduces the flame length and improves the temperature traverse quality (uniformity of temperature) across the exit from the combustor.

The presence and increase in size of a central recirculation zone is encouraged by:

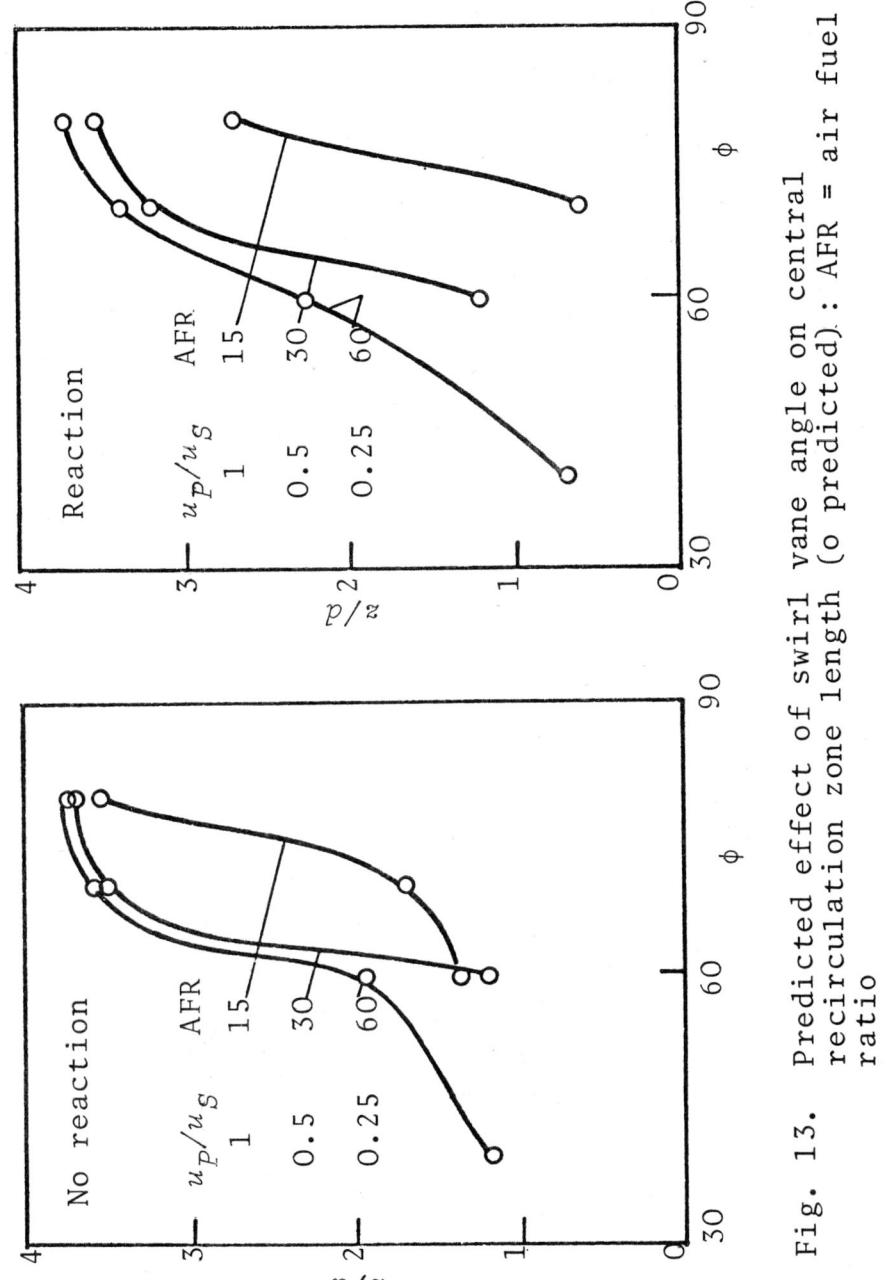

Fig. 13. Predicted effect of swirl vane angle on central recirculation zone length (o predicted).: AFR = air fuel ratio

(i)   increasing D/d
(ii) having a central hub in the inlet flow $d_h > 0$
(iii) increasing d_h/d
(iv) reducing primary velocity (also shifts zone more upstream)
(v)   increasing swirl number or swirl vane angle
(vi) suppressing chemical reaction

For combustors an optimum swirl strength for reasonable recirculation zone and pressure loss is about $S = 1.15$ or swirl vane angle $\phi = 60°$ when $D/d = 2$ although recent work [19] suggests that S based on chamber diameter D rather than nozzle diameter d gives a better guide to similarity of flow downstream of the inlet for different D/d systems. In making predictions the recirculation zone lengthens and narrows if μ is reduced or $\sigma_{r\theta}$ increased. For example, a 25 per cent. reduction in μ gives a zone 30 per cent. longer, $u_{max}/u_{max,o}$ 10 per cent. less and $w_{max}/w_{max,o}$ 60 per cent. more and vice versa. Increasing $\sigma_{r\theta}$ to, say, 30 increases $w_{max}/w_{max,o}$ by about a factor of 10 at $z/d = 6$ and doubles the zone length. There is insufficient evidence, however, to suggest a high constant value of $\sigma_{r\theta}$ although in some regions of the flow this is the case. Work is in progress at Cranfield on the development of a general computer program which solves axisymmetric swirling flows in terms of the primitive variables and includes solution of additional equations for k and ε (turbulence), g (turbulent diffusion flame) and others for eddy-break-up premixed reaction and the presence of a fuel droplet spray.

CONCLUSIONS

Finite difference predictions of inert and reacting axisymmetric turbulent flows are now possible, which show the effect of swirl on jet growth, entrainment and decay and flame size, shape and combustion intensity. Both weakly swirling boundary layer flows and strongly swirling recirculating flows are amenable to numerical solution, and indeed published computer programs are available. Adequate models to simulate turbulence and reaction in these flows are developing rapidly: nonisotropic mixing length and energy length turbulence models and eddy-break-up reaction models are becoming available and recent progress is encouraging. Soon numerical prediction will provide designers of practical equipment with results more cheaply, quickly and correctly than currently possible by experimental means. Techniques are well advanced and their use in design will provide a powerful stimulus to their further development.

REFERENCES

1. Beér, J.M. and Chigier, N.A., "Combustion Aerodynamics," Applied Science, London (1972).
2. Chigier, N.A., "Gasdynamics of swirling flow in combustion systems," *Astronautica Acta*, **17**, 387-395 (1972).
3. Roache, P.J., "Computational Fluid Dynamics," Hermosa, Albuquerque, New Mexico (1972).
4. Lilley, D.G., "Modelling of combustor swirl flows," *Acta Astronautica*, **1**, 1129-1147 (1974).
5. Patankar, S.V. and Spalding, D.B., "Simultaneous prediction of flow pattern and radiation for three-dimensional flames," invited lecture at Int. Centre for Heat and Mass Transfer 1973 Seminar, *Scripta Technica* (1974).
6. Spalding, D.B., "Numerical computation of practical combustion-chamber flows," Paper presented at AGARD Propulsion and Energetics Panel, Liege, Belgium (1974).
7. Launder, B.E. and Spalding, D.B., "Mathematical Models of Turbulence," Academic Press, London (1972).

8. Harlow, F.H., *Editor*, "Turbulence Transport Modelling," AIAA Selected Reprint Series, Vol. XIV (1973).

9. Lilley, D.G., "Prediction of inert turbulent swirl flows," *AIAA J.*, **11**, No. 7, 955-960 (July 1973).

10. Gosman, A.D., Pun, W.M., Runchal, A.K., Spalding, D.B. and Wolfshtein, M.W., "Heat and Mass Transfer in Recirculating Flows," Academic Press, London (1969).

11. Koosinlin, M.L. and Lockwood, F.C., "The prediction of axisymmetric turbulent swirling boundary layers," *AIAA J.*, **12**, No. 4, 547-554 (April 1974).

12. Lilley, D.G. and Chigier, N.A., "An inverse solution procedure for turbulent swirling boundary layer combustion flow," *J. Comp. Physics*, **9**, 237-253 (1972).

13. Mason, H.B. and Spalding, D.B., "Prediction of reaction rates in turbulent premixed boundary-layer flows," *in* Weinberg, F.J., *Editor*, Proc. of Combustion Inst., European Symp., Academic Press, London, 601-606 (1973).

14. Patankar, S.V. and Spalding, D.B., "Heat and Mass Transfer in Boundary Layers," Second Edition, Intertext, London (1970).

15. Gosman, A.D. and Pun, W.M., "Lecture notes for course entitled, 'Calculation of Recirculating Flows'," Report No. HTS/74/2, Dept. of Mech. Eng., I.C., London (1974).

16. Patankar, S.V. and Spalding, D.B., "A calculation procedure for heat, mass and momentum transfer in three-dimensional parabolic flows," *Int. J. Heat Mass Transfer*, **15**, 1787-1806 (1972).

17. Lilley, D.G., "Turbulent swirling flame prediction," *AIAA J.*, **12**, No. 2, 219-223 (1974).

18. Chigier, N.A. and Beér, J.M., "Velocity and static-pressure distributions in swirling air

jets issuing from annular and divergent nozzles," *J. Basic Eng.*, **86**, No. 4, 788-798 (1974).
19. Beltagui, S.A. and Maccallum, N.R.L., "Aerodynamics of swirling flames - vane generated type," *in* Weinberg, F.J., *Editor,* Proc. of Comb. Inst., European Symp., Academic Press, London, 559-564 (1973).

NOMENCLATURE

A	axial velocity decay parameter
a	apparent origin distance
c	specific heat
D	diffusion term of stress component equation, chamber diameter
D/Dt	substantial time derivative
d	nozzle diameter
E,P	Arrhenius constants
f	time-mean mixture fraction
G	axial flux or momentum
g	mean square fluctuating component of fuel concentration
H	heat of combustion
h	stagnation enthalpy
i	stoicheiometric ratio
J	turbulent flux vector
k	kinetic energy of turbulence
l	Prandtl mixing length
ℓ	length scale of turbulence
M	mean mixture molecular weight
m	time-mean chemical mass fraction
P	production term of stress component equation

p	time-mean pressure
R	universal gas constant, mass rate of creation per unit volume or redistributive term of stress component equation (with subscript)
Ri	Richardson number
S	swirl number $= 2G_\theta/(G_z d)$
T	time-mean temperature
$\underline{v}=(u,v,w)$	time-mean velocity (in z,r,θ directions)
W	mean square vorticity fluctuation
z,r,θ	axial, radial, polar coordinates
α	jet half angle ($u/u_{max} = 0.5$)
Γ	turbulent exchange coefficient
Δ	mean flow rate of strain tensor
$\delta r, \delta z$	small distances in r and z directions
ε	turbulence energy dissipation rate
η	nondimensional stream function
λ	Prandtl mixing length parameter
μ	turbulent viscosity
$\xi=r/(z+a)$	nondimensional radial coordinate
ρ	time-mean density
σ	Prandtl-Schmidt number
τ	turbulent stress tensor
ϕ	swirl vane angle, general dependent variable
ψ	stream function
ω	vorticity
∇	vector differential operator

<u>Superscripts</u>

$(\ )'$	turbulent fluctuating component

$(\bar{})$	time average
$(\dot{})$	flow rate

Subscripts

c	centreline value
EBU	eddy-break-up model
edge	position where $u/u_{max} = 0.01$
fu,ox,pr	fuel, oxidant, product (including inerts)
h	enthalpy equation, relating to hub
j	jth specie equation
max,min	maximum, minimum value at a particular axial station
o	value at orifice of jet
p	constant pressure
R	constant in Z equation
rz, etc.	rz-component of second order tensor, etc.
s	swirl
z	value at axial station z
z, θ	directions z, θ
∞	ambient conditions